Modeling and Simulation of Polymer Composites

Modeling and Simulation of Polymer Composites

Guest Editors

Valeriy V. Ginzburg
Alexey V. Lyulin

Basel • Beijing • Wuhan • Barcelona • Belgrade • Novi Sad • Cluj • Manchester

Guest Editors

Valeriy V. Ginzburg
Chemical Engineering and
Materials Science
Michigan State University
East Lansing
United States

Alexey V. Lyulin
Department of Applied
Physics and Science Education
Technische Universiteit Eindhoven
Eindhoven
Netherlands

Editorial Office
MDPI AG
Grosspeteranlage 5
4052 Basel, Switzerland

This is a reprint of the Special Issue, published open access by the journal *Polymers* (ISSN 2073-4360), freely accessible at: www.mdpi.com/journal/polymers/special_issues/C5CG12WK30.

For citation purposes, cite each article independently as indicated on the article page online and using the guide below:

Lastname, A.A.; Lastname, B.B. Article Title. *Journal Name* **Year**, *Volume Number*, Page Range.

ISBN 978-3-7258-2928-6 (Hbk)
ISBN 978-3-7258-2927-9 (PDF)
https://doi.org/10.3390/books978-3-7258-2927-9

© 2024 by the authors. Articles in this book are Open Access and distributed under the Creative Commons Attribution (CC BY) license. The book as a whole is distributed by MDPI under the terms and conditions of the Creative Commons Attribution-NonCommercial-NoDerivs (CC BY-NC-ND) license (https://creativecommons.org/licenses/by-nc-nd/4.0/).

Contents

About the Editors vii

Preface ix

John P. Mikhail and Gregory C. Rutledge
Mechanisms of Shock Dissipation in Semicrystalline Polyethylene
Reprinted from: *Polymers* **2023**, *15*, 4262, https://doi.org/10.3390/polym15214262 1

Kevin A. Redosado Leon, Alexey Lyulin and Bernard J. Geurts
Computational Requirements for Modeling Thermal Conduction in Polymeric Phase-Change Materials: Periodic Hard Spheres Case
Reprinted from: *Polymers* **2024**, *16*, 1015, https://doi.org/10.3390/polym16071015 25

Jesper K. Jørgensen, Vincent K. Maes, Lars P. Mikkelsen and Tom L. Andersen
A Precise Prediction of the Chemical and Thermal Shrinkage during Curing of an Epoxy Resin
Reprinted from: *Polymers* **2024**, *16*, 2435, https://doi.org/10.3390/polym16172435 46

Xingwen Qiu, Haishan Yin, Qicheng Xing and Qi Jin
Development of Fatigue Life Model for Rubber Materials Based on Fracture Mechanics
Reprinted from: *Polymers* **2023**, *15*, 2746, https://doi.org/10.3390/polym15122746 68

Angela Russo, Rossana Castaldo, Concetta Palumbo and Aniello Riccio
Influence of Delamination Size and Depth on the Compression Fatigue Behaviour of a Stiffened Aerospace Composite Panel
Reprinted from: *Polymers* **2023**, *15*, 4559, https://doi.org/10.3390/polym15234559 98

Zhining Huang, Caixia Gu, Jiahao Li, Peng Xiang, Yanda Liao and Bang-Ping Jiang et al.
Surface-Initiated Polymerization with an Initiator Gradient: A Monte Carlo Simulation
Reprinted from: *Polymers* **2024**, *16*, 1203, https://doi.org/10.3390/polym16091203 112

Demetrios A. Tzelepis, Arman Khoshnevis, Mohsen Zayernouri and Valeriy V. Ginzburg
Polyurea–Graphene Nanocomposites—The Influence of Hard-Segment Content and Nanoparticle Loading on Mechanical Properties
Reprinted from: *Polymers* **2023**, *15*, 4434, https://doi.org/10.3390/polym15224434 124

Xianfeng Pei, Xiaoyu Huang, Houmin Li, Zhou Cao, Zijiang Yang and Dingyi Hao et al.
Numerical Simulation of Fatigue Life of Rubber Concrete on the Mesoscale
Reprinted from: *Polymers* **2023**, *15*, 2048, https://doi.org/10.3390/polym15092048 142

Lu Wei, Yanan Chen, Junjie Hu, Xueao Hu, Yunlong Qiu and Kai Li
The Light-Fueled Self-Rotation of a Liquid Crystal Elastomer Fiber-Propelled Slider on a Circular Track
Reprinted from: *Polymers* **2024**, *16*, 2263, https://doi.org/10.3390/polym16162263 164

Haiyang Wu, Yuntong Dai and Kai Li
Self-Vibration of Liquid Crystal Elastomer Strings under Steady Illumination
Reprinted from: *Polymers* **2023**, *15*, 3483, https://doi.org/10.3390/polym15163483 182

Kai Li, Yufeng Liu, Yuntong Dai and Yong Yu
Self-Vibration of a Liquid Crystal Elastomer Fiber-Cantilever System under Steady Illumination
Reprinted from: *Polymers* **2023**, *15*, 3397, https://doi.org/10.3390/polym15163397 198

About the Editors

Valeriy V. Ginzburg

Valeriy Ginzburg was born in Kharkiv (USSR, now Ukraine) in 1966. He earned his B.S. (Physics) in 1989 and his Ph.D. (Polymer Physics) in 1992 at the Moscow Institute of Physics and Technology ("FizTech") in Russia. After postdoctoral fellowships at the University of Colorado (1993–1997) and the University of Pittsburgh (1998–2000), he worked at The Dow Chemical Company (2001–2020). Currently, he is a visiting professor at Michigan State University and founder of a consulting company, VVG Physics Consulting LLC. Dr. Ginzburg is a co-inventor on 15 US patents and author or co-author of about 100 journal publications, including several in *Physical Review Letters*, *Science*, and *Progress in Polymer Science*. Dr. Ginzburg is a co-editor of a Book titled *"Theory and modeling of nanocomposites"* (published by Springer Nature in 2020). He has been elected a Fellow of the American Physical Society (2014) and was awarded the Dow Core R&D Excellence in Science award (2015). His main research interests are polymer glass transition and polymer statistical physics.

Alexey V. Lyulin

Alexey Lyulin studied physics at St. Petersburg State University and obtained his Ph.D. from the Institute of Macromolecular Compounds of the Russian Academy of Sciences in 1992. He has held postdoc positions with leading European simulations groups at the Université Libre de Bruxelles, the University of Bristol, the Max-Planck Institute of Polymer Research (Mainz), and the University of Leeds. In April 2000 he was appointed as a senior fellow of the Dutch Polymer Institute (Eindhoven) and obtained a tenure track position at Eindhoven University of Technology (TU/e). In 2016, he was a visiting scientist at Stanford University with the group of Prof. C. Frank (Chemical Engineering Department). Since 2019, he has been an Associate Professor at TU/e. Since 2023, he has been a Professor of Multiscale Molecular Dynamics at the University of Twente. His research mainly includes atomic-scale modeling of macromolecules, including glass and polymer nanocomposites, dendrimers, and thin polymer films, with a particular focus on polymer dynamics and mechanics. His aim is to develop a long-standing research program in Eindhoven for multi-scale computer simulations (from macroscale to nanoscale, in time and space) of disordered materials for new energy. His group is an important part of the Center for Computational Energy Research (CCER).

Preface

This reprint contains eleven original articles from the Special Issue titled "Modeling and Simulation of Polymer Composites". The included articles address various aspects of composite theory and modeling, ranging from processing (polymerization kinetics) to mechanical and physical properties, including viscoelasticity, fatigue, and fracture.

Valeriy V. Ginzburg and Alexey V. Lyulin
Guest Editors

Article

Mechanisms of Shock Dissipation in Semicrystalline Polyethylene

John P. Mikhail [1,2] and Gregory C. Rutledge [1,2,*]

[1] Department of Chemical Engineering, Massachusetts Institute of Technology, 77 Massachusetts Avenue, Cambridge, MA 02139, USA
[2] Institute for Soldier Nanotechnologies, Massachusetts Institute of Technology, 500 Technology Square, Cambridge, MA 02139, USA
* Correspondence: rutledge@mit.edu

Abstract: Semicrystalline polymers are lightweight, multiphase materials that exhibit attractive shock dissipation characteristics and have potential applications as protective armor for people and equipment. For shocks of 10 GPa or less, we analyzed various mechanisms for the storage and dissipation of shock wave energy in a realistic, united atom (UA) model of semicrystalline polyethylene. Systems characterized by different levels of crystallinity were simulated using equilibrium molecular dynamics with a Hugoniostat to ensure that the resulting states conform to the Rankine–Hugoniot conditions. To determine the role of structural rearrangements, order parameters and configuration time series were collected during the course of the shock simulations. We conclude that the major mechanisms responsible for the storage and dissipation of shock energy in semicrystalline polyethylene are those associated with plastic deformation and melting of the crystalline domain. For this UA model, plastic deformation occurs primarily through fine crystallographic slip and the formation of kink bands, whose long period decreases with increasing shock pressure.

Keywords: molecular simulation; semicrystalline; polyethylene; shock; deformation mechanism; slip; kink band

Citation: Mikhail, J.P.; Rutledge, G.C. Mechanisms of Shock Dissipation in Semicrystalline Polyethylene. *Polymers* **2023**, *15*, 4262. https://doi.org/10.3390/polym15214262

Academic Editors: Alexey V. Lyulin and Valeriy V. Ginzburg

Received: 29 September 2023
Revised: 24 October 2023
Accepted: 26 October 2023
Published: 30 October 2023

Copyright: © 2023 by the authors. Licensee MDPI, Basel, Switzerland. This article is an open access article distributed under the terms and conditions of the Creative Commons Attribution (CC BY) license (https://creativecommons.org/licenses/by/4.0/).

1. Introduction

Shock waves are supersonic, high pressure waves that propagate through a material as a result of an extreme deformation or disturbance [1,2]. They are encountered in military settings, resulting from ballistic or explosive impact, and pose major safety hazards to people and equipment. Additionally, they are an important safety consideration when designing supersonic aircraft [3] and in controlling ignition or pressure waves from certain chemical processes [4,5]. The design of materials that are capable of withstanding and dissipating the energy from these shock waves decreases the danger to the user; however, many traditional materials are incapable of either dissipating the shock energy effectively or maintaining their structural integrity after shock for continued use. For this purpose, polymeric materials offer a promising area of design due to their wide diversity of useful material properties, a result of flexibility in both chemical composition and molecular organization.

The design of materials capable of withstanding extreme shock pressure requires knowledge of the relevant shock dissipation mechanisms, in order to anticipate the amount of energy that can be absorbed by the material. At high shock pressures, chemical dissociation is a significant mechanism for energy dissipation. In fact, there are certain chemical reaction pathways that are unique to shock events [6]. At low shock pressures below the threshold for breaking chemical bonds, other essentially thermophysical mechanisms to dissipate energy must be activated. For example, simulations of shocked diblock copolymers in a lamellar morphology revealed that the polymers can absorb the energy of a shock wave by decreasing the segregation of their initially distinct phases [7].

Shock waves induce nonlinear responses in materials due to the extreme pressure and temperature applied, complicating a mechanistic analysis. The first step in understanding shock response is the construction of an equation of state for a particular material; this equation gives the relationship between pressure and a specific volume, or between the shock velocity and the particle velocity, for a material undergoing shock deformation. The velocities can be derived from the pressure–volume description using the Rankine–Hugoniot (RH) conditions, which describe the relationship between states on either side of the shock wave [1,2,8]. The RH conditions are [8]

$$\rho_0 u_s = \rho(u_s - u_p) \tag{1}$$

$$P_{zz} - P_0 = \rho_0 u_s u_p \tag{2}$$

$$\Delta E = \frac{(P_{zz} + P_0)(v_0 - v)}{2}, \tag{3}$$

where the subscript 0 designates the unshocked, or pre-shocked, state. ρ is the density, $v = \rho^{-1}$ is the specific volume, $P_{zz} = -\sigma_{zz}$ where σ is the stress tensor and the subscript zz indicates the normal component of the stress tensor in the direction of the shock wave, in this case in the z-plane. u_p is the particle velocity, u_s is the shock velocity, and ΔE is the change in total internal energy as a result of the shock. For typical shock pressures, $P_{zz} - P_0$ is well approximated as simply P_{zz}.

Pressure–volume relationships associated with shock in a specific material can be measured experimentally, and they can be estimated theoretically or computationally. Nonequilibrium molecular dynamics (NEMD) is typically used to simulate systems under the application of a driving force such as a piston colliding with the system and forming a shock wave from the resulting impact. Equilibrium simulation methods have also been developed to study state points along the Hugoniot, a curve that satisfies the RH conditions for all points along the curve. Examples include the NP_{zz}Hug method of Ravelo et al. [8] and the Multiscale Shock Technique (MSST) of Reed et al. [9], each of which modify the equations of motion to simulate a shocked system at equilibrium that lies on the Hugoniot. The NP_{zz}Hug method of Ravelo, used in this work, employs a "uniaxial Hugoniostat," similar to a single (non-chain) Nosé–Hoover thermostat and barostat [10] in which the target pressure is specified and the target energy at each time step is computed as a function of the current configuration; the equilibrium value of the target energy is also consistent with the RH conditions [8].

Semicrystalline polyethylene (e.g., Dyneema® or Spectra®) is a material that is commonly used in soft and hard body armor because it can be spun into fibers with exceptionally high specific strength and specific modulus, resulting in lightweight fabrics that can be cut and sewn or laminated as reinforcing elements in composites. Polyethylene is widely used in engineering materials to withstand extreme impacts; gel-spun polyethylene strands have stiffnesses comparable to that of steel, while maintaining light weight and ease of manufacturing [11]. Other applications of polyethylene that take advantage of its high strength and toughness include the structural engineering of aerospace and military components [12], packaging of consumer products, films, water and gas pipelines [13], and components of artificial joints [14]. Polyethylene is also the prototype for many other semicrystalline polymers. On the length scale of micrometers, semicrystalline polymers comprise domains of both crystalline and noncrystalline materials, which differ in their mechanical compliances (ease of deformation under an applied stress) and can contribute to the dissipation of energy during shock wave progression. At the nanoscale, the representative motif of the system consists of alternating layers of crystalline and noncrystalline material. Mechanical properties vary with the thickness of the crystalline lamellae [15,16]. Importantly, covalently bonded chains weave back and forth between crystalline and noncrystalline domains, giving rise to a unique interfacial region called the "interphase," in which the constraints of connection to the crystalline domain strongly influence the topology of the chains [17], making this region distinct from an amorphous melt or glass. Chains in the noncrystalline domain consist of loops (chain segments with typically non-adjacent connections to the same crystal lamella), tails (chain segments that connect to the lamella at one end and terminate in the noncrystalline domain at the other end), and bridges (chain segments that traverse the noncrystalline domain to connect to distinct lamellae). For the purposes of modeling the coupling between crystalline and noncrystalline domains, the simplest representative volume element that includes both types of domains is the "lamellar stack" model [18].

The first simulations of semicrystalline polyethylene were those reported by in't Veld et al. [19] using the Interphase Monte Carlo (IMC) method [16,17] to sample the distributions of loops, tails, and bridges in a thermodynamically consistent manner. In that method, nonlocal reptation and end-bridging moves were introduced to sample different topologies within a single Monte Carlo simulation. The resulting configurations were then used in a series of studies of isothermal deformations [20–25]. For nonisothermal deformations like shock, chemical as well as thermophysical rearrangements can occur at sufficiently high pressure, necessitating the use of bond-breaking methods like density functional theory (DFT) or reactive force fields such as ReaxFF [26] or AIREBO-M [27]. Shock studies of PE using DFT simulations have been used to obtain chemical [28] and thermodynamic [29] information for shock pressures up to 250 GPa. Typically, simulations of shock waves in polyethylene consider crystalline and noncrystalline domains separately. Elder et al. [30] first considered semicrystalline polyethylene (SCPE) models that comprised the two types of domains together, using a method involving *deletion*, *cutting*, and *melting* (DCM) to reduce density and introduce conformational disorder to the noncrystalline domain. They then used NEMD simulations to investigate how the interfaces between crystalline and noncrystalline domains of SCPE transmit and reflect propagating shock waves, based on the impedance of each region. The DCM method is analogous in many respects to the IMC method, except that it lacks the ability to sample alternative connectivities efficiently once the initial structure is generated. As a result, the interphase topology obtained does not minimize free energy. The DCM method also retains some memory in the noncrystalline region of the crystalline region from which it was generated. It remains an open question whether the shock response of a semicrystalline polymer is sensitive to the topological nature of the interphase.

Crystalline regions in general deform through a variety of mechanisms, including defect-mediated mechanisms (slip, kinking, twinning, Martensitic transformation, etc.) and melting–recrystallization [21,31]. Deformation of noncrystalline regions are relatively simpler, only straining due to interlamellar compression and shear. Previous studies of crystalline polyethylene have found that the (100)[001] and (100)[010] fine crystallographic slip mechanisms are dominant in compression because they have the lowest activation energy barriers [31]. The notation $(hkl)[uvw]$ refers to slip in the (hkl) plane and $[uvw]$ direction, where h, k, l, u, v, and w are Miller indices. Galeski et al. showed that, for plane strain compression of high-density polyethylene (HDPE), spontaneous generation of dislocations within polyethylene lamellae sufficient to cause coarse crystallographic slip, involving the translation of blocks of material within the crystal phase, only occur for compression ratios greater than three [32], which is far greater than those considered for this work.

There is much prior research on the sub-shock compression of SCPE under various deformation modes; one common method of deformation in both experiments and simulations is isothermal uniaxial compression—also called unconfined compression—where the system is deformed along one axis, labelled z, while the x and y axes have a constant stress condition, which allows them to expand according to the Poisson's ratio of the material [21]. In contrast, shock simulations typically consider confined compression under a uniaxial Hugoniostat that keeps dimensions transverse to the compression at a fixed length. This is done to isolate the study to a 1D propagating shock wave in the z-direction, which avoids complications relating to the nonlinear superposition of shock waves [1,33]. When the transverse (x and y) lengths are kept fixed, the total system pressure naturally increases to a much greater level than when the transverse stresses are controlled at some small value, e.g., atmospheric or vacuum pressure.

Several studies involving unconfined compression have been used to identify deformation mechanisms as functions of strain rate. Kazmierczak et al. studied the mechanisms of plastic deformation of polyethylene crystals for strain rates of 5.5×10^{-5}, 1.1×10^{-3}, and 5.5×10^{-3} s^{-1}, and different crystal thicknesses [34]. For uniaxial compression of HDPE, the relationship between true stress and strain rate was shown to follow a logarithmic dependence for a wide range of strain rates between 10^{-4} and 2.6×10^{3} s^{-1} [35]. Furthermore, Brown et al. show that the relationship between true stress and temperature follows a linear trend [35]. Kim et al. simulated SCPE models under unconfined compression at two different strain rates, 5×10^{6} s^{-1} and 5×10^{7} s^{-1} [21]. They found that the crystallographic slip mechanism dominated the deformation response for the slower strain rate. For the faster strain rate, they first observed an increase in stress and then a subsequent crystallographic slip. Jordan et al. examined the behavior of the speed of sound in polyethylene, elastic moduli, unit cell parameters, and other variables, as a function of pressure, using confined compression [36].

In this work, we examine the effect of shock deformation on lamellar stacks of semicrystalline polyethylene with realistic topological distributions in the noncrystalline regions. Uniaxial Hugoniostatted (NP_{zz}Hug) equilibrium molecular dynamics simulations are used to sample state points along the Hugoniot curve for shock pressures up to 10 GPa. From these state points, measures of orientational and nematic order are obtained. The evolution of density and stress profiles during the transient equilibration period are also examined. Changes in potential energy as a result of shock are analyzed according to the contributions from the different terms of the potential. From such analyses, we propose some mechanistic interpretations for the storage and dissipation of shock wave energy in a prototypical semicrystalline polyethylene lamellar stack model.

2. Materials and Methods

2.1. Model Generation

The united atom (UA) force field used in this work was adapted from the original Transferable Potential for Phase Equilibria (TraPPE-UA) [37] by including a harmonic bond potential, as in Bolton et al. [38]. The TraPPE-UA potential was parameterized to capture realistic behavior of vapor–liquid coexistence curves as well as densities at pressures of several hundred MPa [37].

Following the work of Lee et al. [20], semicrystalline polyethylene systems were generated using the Interface Monte Carlo (IMC) method [16,19]. Building and pre-equilibration of the PE systems were conducted using the Enhanced Monte Carlo (EMC) software (version 9.3.4) [39], which has been shown to realistically simulate the crystalline and noncrystalline (i.e., amorphous plus interphase) domains of semicrystalline polyethylene [16,21]. Following the procedures of Ranganathan et al. [25] and Kumar et al. [24], all systems were generated in EMC by first creating a fully crystalline system of $4 \times 6 \times 112$ ($a \times b \times c$) orthorhombic PE unit cells. For the fully crystalline system ($\chi^c = 1.0$), henceforth referred to as crystalline polyethylene (CPE), the a, b, and c axes of the orthorhombic unit cells were aligned with the x, y, and z Cartesian axes, respectively. χ^c is the mass-weighted crystallinity fraction as defined in Section 2.5.2. The semicrystalline polyethylene systems with mean, pre-shock values of χ^c equal to approximately 0.44 and 0.81, are henceforth referred to as SCPE44 and SCPE81, respectively. The crystal unit cells were oriented such that the {201} Miller plane was perpendicular to the z-axis, where the crystal–amorphous interface is eventually formed; experimental studies by Bassett et al. determined the mean angle between crystalline chains and the normal vector to the crystalline–amorphous interface to be 35°, approximately corresponding with the {201} facet [40]. Subsequent computational studies also showed that this facet resulted in the lowest interfacial energy [16,41]. Next, central layers of amorphous-like density, 72 and 35 unit cells thick between fixed crystals, were created by cutting UA sites from each of the 16 chains, for a total of 2265 and

1046 methylene sites removed from SCPE44 and SCPE81, respectively. The 32 methylene sites at the end of each cut were replaced with methyl sites for both systems. The central layer was then amorphized using 10,000 cycles of both local and global Monte Carlo moves at 10,000 K. A set of MC moves was chosen that preserves the number of tails and the sum of loop and bridge segments while changing the overall topology of segments in the amorphous domain (the NN_eVT ensemble, where N_e is the number of methyl sites) [16]. This step was followed by a step-wise cooling sequence at temperatures of 10,000, 5000, 2000, 1000, 750, 500, 400, and 300 K, each step lasting for 20,000 Monte Carlo (MC) simulation cycles. Ten independent configurations for each SCPE system were generated in this way.

After generation, SCPE44 and SCPE81 were simulated using molecular dynamics (MD) in the canonical (NVT) ensemble for 2 ps to stabilize the temperature at approximately 300 K. For CPE, one perfectly crystalline configuration was created and 10 different trajectories were initiated by assigning velocity distributions at 300 K with different starting seeds and allowing each to equilibrate under isothermal–isobaric (NPT) conditions. All molecular dynamics (MD) simulations were conducted using the LAMMPS software package [42,43] and thermalized throughout both crystalline and noncrystalline layers by MD in either the NVT or NPT ensemble. The time step of integration was 2 fs. To control pressure and temperature, respectively, the barostat and thermostat methods implemented in LAMMPS follow the form of Shinoda et al. [44], which combines the Nosé–Hoover and Parrinello–Rahman methods; the pressure damping parameter was 2000 fs and the temperature damping parameter was 200 fs. Equilibration was confirmed by ensuring that the thermodynamic parameters (total energy, enthalpy, pressure, and density) of the system fluctuated about the mean values with negligible drift over a period of at least 10 ns. The deviation from the mean was measured by calculating the coefficient of determination, r^2, for the thermodynamic parameters vs. time; if this value is small (<0.01 for this work), then the deviation of the trend from its mean value is better explained by random fluctuations rather than any change in the mean value itself.

2.2. Shock Simulation

Following equilibration in the unshocked state (pressure $P = 0$ GPa), the systems were then re-equilibrated to a new state consistent with uniaxial shock using the NP_{zz}Hug method of Ravelo et al. [8], which is an equilibrium Hugoniostat method that drives the system to a new equilibrium state consistent with the RH conditions. Compression was limited to the lamellar stack direction for the SCPE models, and the crystallographic chain direction for the CPE model. Lateral dimensions were held at fixed length. The method approaches the Hugoniot state by adjusting the equations of motion in a manner similar to the Nosé–Hoover barostat and thermostat, such that the system pressure and energy oppose deviations from values prescribed by the RH conditions. To avoid complications due to bond breaking and the more intricate reactive force fields required to describe them, 10 GPa was chosen as the upper limit of shock pressures, at or below which chemical reactivity is insignificant in real polyethylene systems (c.f. [28]). The Hugoniot state up to 10 GPa was also validated against Hugoniot curve data from both experiments and density functional theory simulations [28,45,46]. Each Hugoniostat simulation was carried out in seven subsequent levels of 11 output steps each; at each level, k, data were output every 10^k time steps for $0 \leq k \leq 6$. This was done in order to probe long-term behavior while also focusing on trends that may occur at intermediate and short time scales. Averages reported henceforth for each system either consider the final equilibrium state of the Hugoniostat simulation or are temporal averages at intermediate points during extended Hugoniostat trajectories after equilibration. The pressure and temperature damping parameters for the Hugoniostat simulations were the same as those used for NPT simulations.

For Hugoniostat shock simulations in which the z-axis is compressed while the x- and y-axes are held at constant length, symmetry prevents any significant transverse slip mechanisms, e.g., (100)[010]. According to Bartczak and Galeski, coarse crystallographic slip is generally caused by the heterogeneous nucleation of dislocations [31]; one should note that, in this somewhat idealized computational model, there are relatively few crystal defects that would encourage such a coarse slip. The only ones that may occur are methyl groups that moved into the crystalline region near the lamellar interface.

2.3. All-Atom Models

To check the validity of simulations conducted using the TraPPE-UA force field in select situations, a few representative configurations were converted to all-atom (AA) representations and modeled using the OPLS-AA force field [47]. Because of the considerably larger computational and memory costs of the AA models compared to the UA models, only one configuration for each system type was converted and then run using the Hugoniostat.

To convert a UA model to AA, each UA site was first converted to either a methylene or methyl carbon. Next, explicit hydrogens were inserted using geometric criteria based on the local configuration of the alkane chain, in a manner similar to the reverse-mapping procedure described by Brayton et al. [48]. Newly formed angles and dihedrals were identified based on the bond connectivity of the AA representation of the chain. Then, the potential energy was minimized, followed by MD simulation using the OPLS-AA force field with a timestep of 1 fs for 1 ps in an NVT ensemble to stabilize the temperature. Finally, the output data file from the NVT run was used as input for the Hugoniostat simulations.

2.4. Order Parameters

Using position and velocity information from the MD trajectory, three order parameters are calculated on a per-UA basis. These are the nematic order parameter, p_2, the orientational order parameter, S_z, and the specific volume, v. Here, the nematic order describes the degree of coalignment of nearby bond chords with a reference bond chord within a local region of space, whereas the orientational order describes the degree of alignment of each bond chord with a reference direction, in this case the direction in which shock pressure is applied. Specifically, the p_2 order parameter for atom i was calculated using [49]

$$p_{2,i} = \frac{3}{2} \left\langle \cos^2 \theta_{ij} \right\rangle_j - \frac{1}{2}, \quad (4)$$

where i is the index of the bond chord from atom $i-1$ to atom $i+1$ within the chain under consideration, and j indexes the neighboring chords within a cutoff radius $r_{ij} < r_{p2}$, here taken to be 1 nm. p_2 takes values close to 1 for chords oriented nearly parallel (or antiparallel) to their neighbors, 0 for randomly oriented chords, and $-1/2$ for chords oriented perpendicular to their neighbors. S_z for atom i takes a similar form except that the angle, ϕ, is that between the bond chord and the Cartesian unit vector $\hat{z}$:

$$S_{z,i} = \frac{3}{2} \cos^2 \phi_i - \frac{1}{2}. \quad (5)$$

The third order parameter, v, is determined via Voronoi tessellation [50], which determines the convex polyhedron surrounding each UA containing the space closer to that UA than any other UA in the system. The specific volume defined on a per-UA basis is then the ratio of the volume occupied by that polyhedron to the mass of the UA. The periodic boundary conditions in the system are accounted for by first replicating the system across each plane of the simulation box (resulting in 3^3 identical subsystems) and then computing the Voronoi tessellation for this larger system, using the Voronoi polyhedra of the central subsystem for the calculation of specific volumes.

2.5. Clustering Analysis

2.5.1. Selection of the Clustering Method

To distinguish trends in the different regions of the system (crystalline vs. non-crystalline), a clustering algorithm was used to segregate the UAs into the two different populations. A clustering algorithm was chosen for this purpose, primarily due to its ability to classify atoms optimally into a finite set of distinct populations. This approach avoids the requirement of selecting, a priori, a threshold value for the classification of sites into one cluster or the other. For example, in prior work we have used a local nematic order parameter (p_2) to classify UAs as crystalline or noncrystalline, with a threshold based on the minimum in the distribution function of p_2 for a thermally equilibrated, partially crystallized system [49]. However, under nonequilibrium conditions such as flow, this distribution function changes dynamically, so that the threshold value should change as well [51]. The clustering algorithm avoids this difficulty by defining clusters such that a loss function $L(p_2)$, defined as the sum of squared distances in the p_2 space from each UA to the mean p_2 value for the cluster to which it is assigned, is minimal [50,52]. One needs only to specify the number of clusters a priori and provide an initial guess for the mean of each cluster, which can be handled automatically by algorithms such as kmeans++ [53].

Both fuzzy c-means (FCM) [54] and k-means [53,55] were employed for clustering, but it was found that FCM provided more consistent results among different initial configurations, resulting in lower standard errors of several variables as functions of pressure (see Supplementary Materials for extended discussion). FCM assigns to each UA a probability of membership in each cluster (see Section 2.5.2). The improved consistency of FCM makes physical sense because SCPE contains interphase regions where a transition occurs between fully crystalline and fully noncrystalline UAs; FCM can account for the partially crystalline character of sites within the interphase by assigning to each UA a finite probability of being crystalline, with the complementary probability of being noncrystalline, whereas k-means uses a strictly binary classification (i.e., a UA is either crystalline or noncrystalline). Thus, to calculate the mean values of variables not included in the clustering, FCM weights contribute to the mean via membership probability, so that outliers have less influence on the statistic. FCM requires an additional adjustable parameter, m, that is the exponent of the fuzzy partition matrix; the exact meaning of this parameter is clarified in Section 2.5.2. Different values of the exponent could be chosen for a particular problem, but in this work reasonable results were obtained using a constant value of 2. This value is consistent with the sum of the squared errors objective function [56].

In addition to the nematic order parameter, p_2, two other order parameters were considered individually or together for clustering purposes: specific volume, v, and orientational order, S_z. Using different combinations of these order parameters in the clustering leads to 2^3 possible clustering criteria. Silhouette plots [57] were used to evaluate the different combinations, leading to the conclusion that p_2 alone provides the best quality clustering. All Silhouette plots used to judge the clustering quality are shown in the Supplementary Materials. Figure 1 illustrates a typical result of this clustering method for an SCPE system. Importantly, clustering in this way has no explicit dependence on the UAs' positions in Cartesian space and in fact does not guarantee spatial contiguity. However, by definition, a UA with a p_2 value near 1 must be in a local neighborhood with consistent alignment of its bond chords, so spatial segregation typically accompanies nematic order segregation, as is depicted in Figure 1.

Before clustering with the above variables, they were converted to Z-scores on a per-variable basis. In other words, the data were centered using their mean value and then scaled using the inverse of their standard deviation. This was carried out because the clustering algorithms operate by minimizing the sum of squared distances from each observation to the centroid of its cluster (the mean position of all members within the cluster); if the data were given in different units, the weighting of that variable would be affected. Z-scores, on the other hand, are dimensionless and weight each variable roughly equally. For a different application, it may be desirable to control the weighting of each variable, in which case a different scaling may be used.

2.5.2. Statistics of Order Parameters Using Clustering

FCM assigns a probability f_i^k that the ith UA belongs to the kth cluster for all $i \in \{1, \ldots, n_i\}$ and $k \in \{1, \ldots, n_k\}$. $\sum_k f_i^k = 1$ for all i, so that the ith UA must be fully accounted for in the defined clusters. The centroid of each cluster is denoted by C^k. C^k and f_i^k are defined by the following simultaneous equations [56]:

$$f_i^k = \left[\sum_{j=1}^{n_k} \left(\frac{\|x_i - C_k\|^2}{\|x_i - C_j\|^2} \right)^{1/(m-1)} \right]^{-1} \tag{6}$$

$$C_k = \frac{\sum_{i=1}^{n_i} \left(f_i^k\right)^m x_i}{\sum_{i=1}^{n_i} \left(f_i^k\right)^m}, \tag{7}$$

where m is the exponent of the fuzzy partition matrix ($m = 2$ in this work) and x_i is the datum for the ith UA used for clustering. The dimensionalities of both C^k and x_i are equal to the number of variables used for clustering. For any order parameter q_i assignable to UA i, the mean value of that order parameter for each cluster k is defined as

$$\left\langle q^k \right\rangle = \frac{\sum_i f_i^k q_i}{\sum_i f_i^k}. \tag{8}$$

Equation (8) is used in this work to calculate averages separately for crystalline and noncrystalline populations of SCPE systems. The probability assignments to each UA can also be used to define a cluster fraction, χ^k, representing the contribution of the cluster k to the entire system. The cluster fraction is defined as

$$\chi^k = \frac{1}{n_i} \sum_i f_i^k. \tag{9}$$

Note that, in the UA representation of the PE systems, all methylene UAs have the same mass. Thus, the definition in Equation (9) is essentially the mass-weighted cluster fraction of the system.

In certain contexts, especially when comparing computational and experimental methods of partitioning systems into distinct populations, it is desirable to calculate a cluster fraction that is weighted by a specific order parameter, q. Different experimental techniques may naturally measure crystallinity in terms of volume fraction (e.g., with Raman scattering [58]) or other intensive properties [59]. The output of a clustering algorithm also provides a means to compute such parameter-weighted cluster fractions as

$$\chi_q^k = \frac{\sum_i f_i^k q_i}{\sum_i q_i} \tag{10}$$

for a specific order parameter, q, and any cluster, k. Note that the definitions of cluster fractions in Equations (9) and (10) have the property that $\sum_k \chi_q^k = 1$. Henceforth, for $n_k = 2$, the superscript "c" denotes the crystalline cluster while the superscript "nc" denotes the noncrystalline cluster.

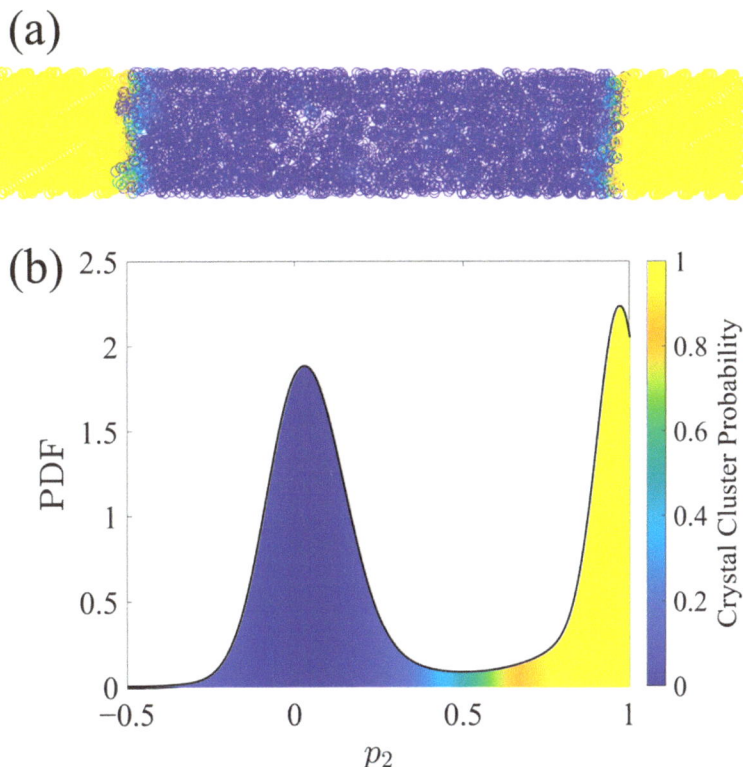

Figure 1. An example of clustering an SCPE44 configuration by the per-atom variable p_2 using the FCM algorithm. United atoms are colored by their probability of being included in the crystalline population. (**a**) shows the UAs in Cartesian space while (**b**) shows a probability density function (PDF) of p_2 with shading under the curve denoting the crystal cluster probability (f_i^c) as calculated via FCM. The PDF was smoothed using a kernel density estimate with normal distribution kernel functions [60]. Kernel density estimation is a nonparametric method to estimate a PDF using kernel functions as weights for the contributions from each of the discrete sample points.

3. Results and Discussion
3.1. Hugoniot Post-Shock States
3.1.1. Pressure versus Specific Volume

To validate the models and simulations used in this work over the pressure range of interest, Hugoniot curves were constructed using the equilibrated states of systems shocked to different pressures. Figure 2a shows a comparison of these curves in P-v space to experimental data and other simulation results previously reported in the literature for polyethylene. The CPE data in Figure 2a are essentially the same as those previously reported in Hsieh et al. [61]. The results obtained in this work are quite close to the experimental trend reported by Marsh [46]. The results of MD simulations by Agrawal et al. [48] deviate the most from experimental data, behavior which they attribute to a low initial density. However, Agrawal et al. noted that scaling all specific volumes by their respective ambient or zero-pressure values brought their data into accord with the theoretical trends of Pastine [62].

The work of Chantawansri et al. also presented results for a model of semicrystalline polyethylene [45]; they used a simplified "layered" structure that fused together purely crystalline and purely amorphous PE (APE) chains such that the crystallographic c-axis was perpendicular to the interface between the two regions. They observed larger shifts in the curves of pressure vs. volume with increased crystallinity compared to that observed here. To provide a closer look at the effect of crystallinity, Figure 2b compares the simulation data obtained in this work, converted to u_s-u_p space using Equations (1)–(3), with the simulation data from Chantawansri [45] as well as the theoretical curves for purely amorphous and purely crystalline PE from Pastine [62]. For crystalline and noncrystalline regions of SCPE simulations, the conversion is applied using the mean specific volumes of the corresponding clusters. For CPE and the crystalline regions of semicrystalline models, all three sets of data are fairly consistent for u_p approximately equal to or exceeding 1 km/s. SCPE81 shows a lower shock speed than the other data sources for lesser values of u_p. For APE and the noncrystalline populations of the current semicrystalline models, the simulation data from Chantawansri et al. and the current work are fairly consistent, while the Pastine curve shows greater values of u_s. Plots of u_s vs. u_p tend to decrease emphasis on the initial density of the system because both speeds are linearly proportional to the square root of initial density [8]; trends in u_s vs. u_p are empirically known to often follow a linear relationship [63], although the intercept of the relationship does also scale with the square root of initial density.

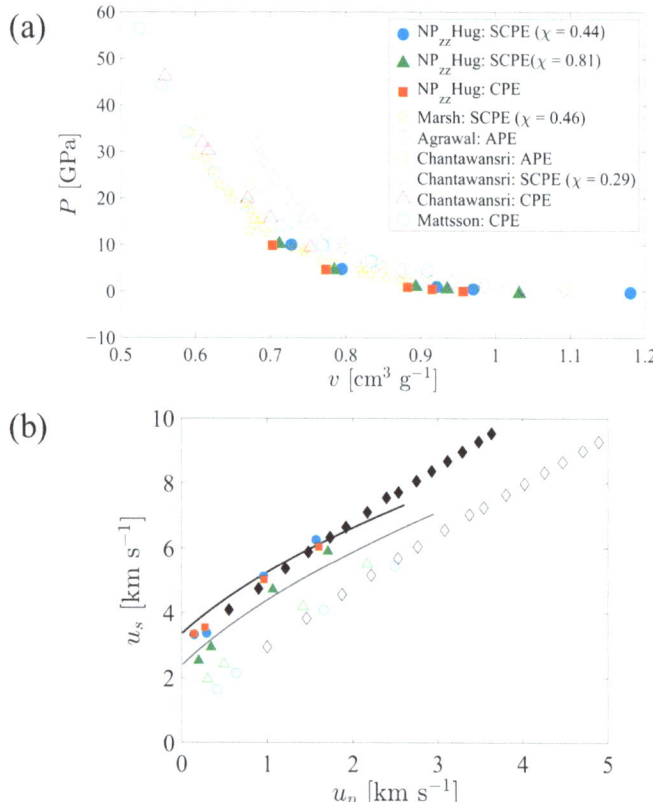

Figure 2. (a) Hugoniot P-v curves for SCPE models compared to experimental data and other simulations results reported in the literature. Values for SCPE (χ^c = 0.44 and 0.81) and CPE are from the current work (filled symbols). Experimental values (gold stars) were obtained from the LASL Shock Handbook [46], where data are reported for experiments in which an explosively driven flying plate was used to induce shock waves in bulk polyethylene (ρ_0 = 0.916 g/cm^3). MSST molecular

dynamics simulation data of an AA model of amorphous PE (magenta diamonds) come from Agrawal et al. [64]. DFT simulation data for three different crystallinities of PE (triangles) come from Chantawansri et al. [45]. DFT-AM05 temperature ramp simulation data of CPE (blue circles) come from Mattsson et al. [28]. (**b**) Hugoniot u_s vs. u_p curves for SCPE models compared with results reported in the literature. Values for SCPE44 (blue circles), SCPE81 (green triangles), and CPE (orange squares) are from the current work. Simulation data used for comparison come from Chantawansri et al. [45] (black diamonds). For all of the simulation data, filled symbols indicate CPE (or crystalline regions, in the case of semicrystalline models) while empty symbols indicate APE (or noncrystalline regions, in the case of semicrystalline models). Data are also compared with the theoretical curves of Pastine [62] for CPE (black line) and APE (grey line).

3.1.2. Temperature

The temperature increase associated with the application of shock pressure depends on the heat capacity of the system. Heat capacity tends to be strongly model-dependent. Molecular dynamics simulations of fully flexible AA models tend to overestimate heat capacity because they treat all degrees of freedom classically, including those associated with high frequency vibrations that would be more properly considered as quantum mechanical in nature [65]. On the other hand, UA models tend to underestimate heat capacity because they eliminate numerous vibrational degrees of freedom, including some that would be activated at the temperatures experienced by the system. Thus, the temperature of the UA systems under shock always increases more than that of the AA systems. To bracket the actual temperature increase, a few Hugoniostat simulations using the AA force field were performed. The variation of temperature with shock pressure is shown in Figure 3. When comparing the temperature increases in SCPE44 and SCPE81 as functions of the shock pressure, the UA systems increase by approximately twice as much as the corresponding AA systems. However, this trend was not observed for the CPE systems. The temperature increase in the AA systems is very small for all pressures up to approximately 10 GPa, where it is approximately 30 K. The corresponding UA system increased by approximately 260 K, or approximately 8.6 times the AA pressure increase. Along a Hugoniot curve,

$$\left(\frac{\partial T}{\partial P}\right)_{Hug} = \left(\frac{\partial T}{\partial v}\right)_{Hug} \left(\frac{\partial v}{\partial P}\right)_{Hug}, \qquad (11)$$

where v is the specific volume. From Figure 4, the compressibility along the Hugoniot curve $((\partial v/\partial P)_{Hug})$ is only about 10% different between the UA and AA systems, clearly not enough to compensate for the temperature difference. We hypothesize that the ratio of constant-pressure to constant-volume heat capacities is closer to unity in the AA CPE than it is for the other systems, so the temperature increase in this system along the Hugoniot curve is correspondingly reduced. The basis for this hypothesis lies in consideration of a much simpler system—an ideal gas heated adiabatically. Although this simpler system is unsuitable for quantitative comparison with CPE in the current work, the temperature increase in CPE may be strongly dependent on the heat capacity ratio, by analogy to the ideal gas.

3.1.3. Orientational Order Parameter

The orientational order parameter, S_z, is used to track the shift in orientation of the bond chords in the crystalline domain with respect to the z-axis, which is both the direction in which the shock compression is applied and the direction normal to the interfaces between the crystalline and noncrystalline regions in the lamellar stack. Changes in values of the mean orientational order parameter for the crystalline domains, $\langle S_z^c \rangle$, are potentially indicators of crystallographic slip + compression in the SCPE systems, necessitated by the geometric confinement placed on the bond chords due to the compression. It is important to note that UA models tend to underestimate the energy barrier preventing crystallographic slip; see, e.g., Olsson et al. [66]. Thus, while configurational changes to the UA PE systems due to crystallographic slip are realistic, it is possible that the role of crystallographic slip may be overemphasized in these systems relative to other deformation mechanisms that would be more prominent in a more detailed atomistic model.

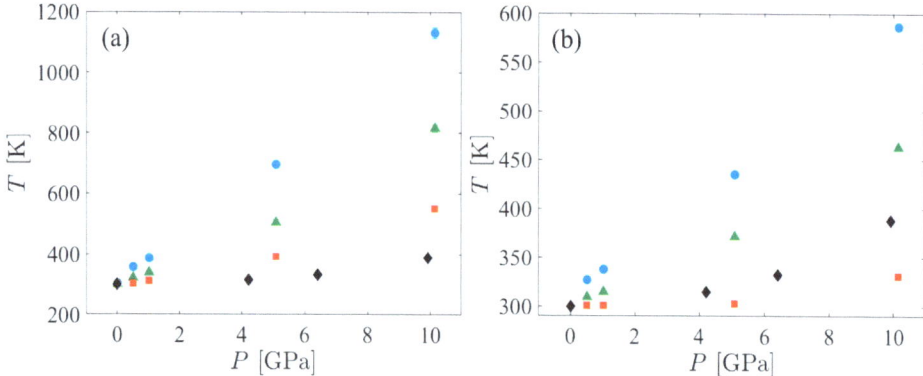

Figure 3. T vs. P for the SCPE44 (blue circles), SCPE81 (green triangles), and CPE (orange squares) systems. Also shown are the T vs. P data for polyethylene from DFT-AM05 temperature ramp simulations conducted by Mattsson et al. [28] (black diamonds). (**a**) shows the UA data while (**b**) shows the AA data.

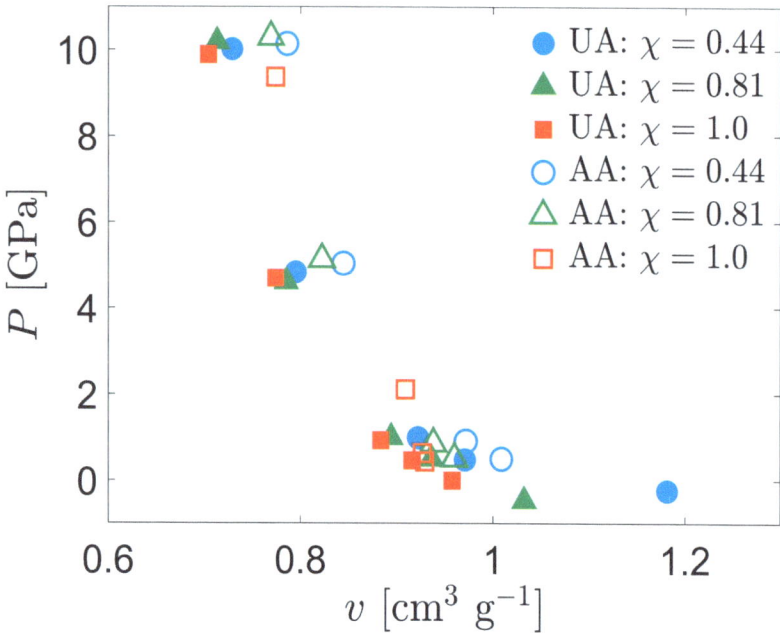

Figure 4. Comparison of Hugoniot P-v curves for UA and AA models of SCPE.

Figure 5 shows all of the methylene UAs in an SCPE44 system colored according to the S_z value of the associated bond chord. Methyl UAs are not shown because they cannot be assigned a bond chord in the same way. At low pressures, as the shock pressure increases, bond chords in the crystalline region tilt uniformly away from the z-axis, so their S_z values decrease on average. At the highest pressure (10 GPa) the crystalline clusters begin to exhibit greater variance in S_z, an indication of the decrease in crystallographic order.

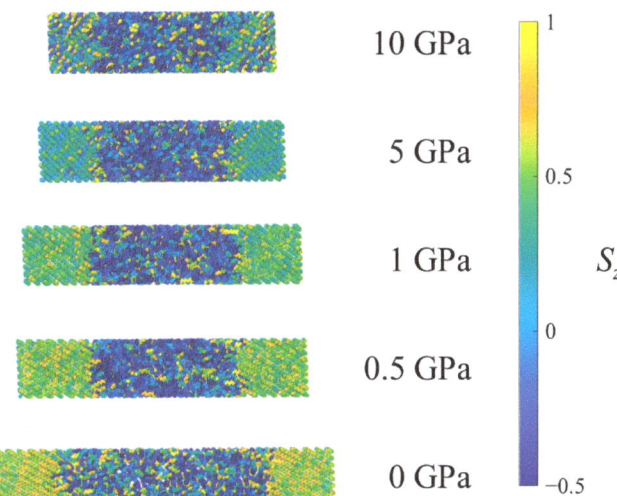

Figure 5. Example SCPE44 system under ambient conditions (0 GPa) and after equilibration at several pressures in Hugoniostatted shock simulations. Atoms are colored according to the orientational parameter S_z.

Figure 6 shows the UAs of one system of SCPE81 colored according to the S_z for different pressures. Interestingly, kink band formation is observed in SCPE81 at 5 GPa for this configurational seed. A kink band forms when crystallographic slip is inadequate to accommodate compression within a crystal uniformly, so that a section of the crystal rotates cooperatively, localizing the deformation [31]. For different configurational seeds (not shown), kink bands occur in different locations and with different numbers of bands. Kink bands with widths of roughly 10 bond lengths across were observed in 7 out of 10 configurational seeds at 5 GPa, while the remaining seeds exhibited shorter crystalline defects roughly 2 bond lengths across. This kink band formation in the crystalline structure is identified as another feature of shock energy absorption that appears under special conditions. For $P < 5$ GPa, the bond chords tilt uniformly, and thus crystallographic slip + compression is the dominant mechanism at low shock pressures. At 10 GPa, the combination of the decreased nematic order of the crystal and higher temperature apparently decrease the barrier to tilt and disrupt the organized formation of large kink bands; only 2 out of 10 configurational seeds exhibit kink bands with widths of roughly ten bond lengths across while the remaining seeds exhibit shorter crystalline defects roughly two bond lengths across. Thus, prominent kink bands are observed mainly at intermediate shock pressures and in systems with sufficiently thick crystalline lamellae.

Figure 7 shows the UAs of one system of CPE colored according to the S_z for different pressures. An important observation is that the CPE system, due to constraints on the system geometry, is not free to tilt by large angles as the crystalline regions of SCPE systems are. In CPE, UAs on each side of the periodic boundaries perpendicular to the z-direction must be bonded, so tilting in one direction must be accompanied by tilting in the opposite direction elsewhere in the crystal and the formation of kink bands, such that the long period of deformation is commensurate with the simulation cell size. Such kink band formation is a typical case of buckling in response to compression. For low pressures, the systems have only two bends separating regions of tilt by different angles but similar S_z values. At 5 GPa, the long period or wavelength of buckling is reduced, resulting in the formation of multiple kink bands and higher tilt angles. Finally, at 10 GPa most of the nematic order in the crystal has diminished and the system loses long-range spatial correlations, forming numerous small regions of different alignment. Thus, for fully crystalline systems, kink bands are observed at all shock pressures, but the long period decreases with increasing pressure and eventually breaks up into disordered domains at the highest shock pressure, analogous to the large crystalline domain in SCPE81.

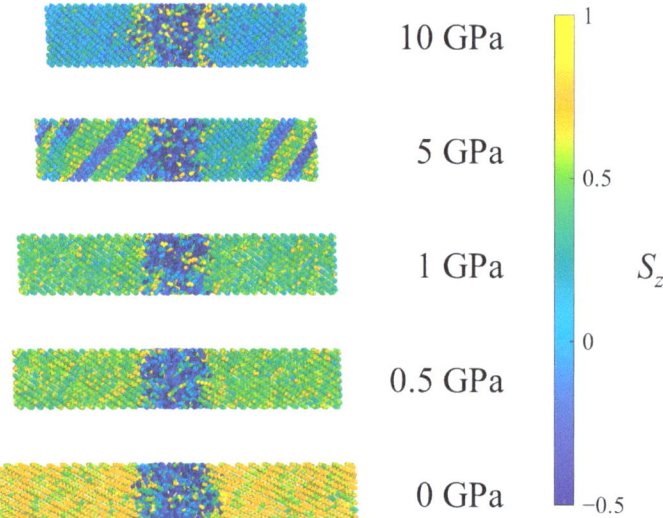

Figure 6. Example SCPE81 system under ambient conditions (0 GPa) and after equilibration at several pressures in Hugoniostatted shock simulations. Atoms are colored according to the orientational parameter S_z.

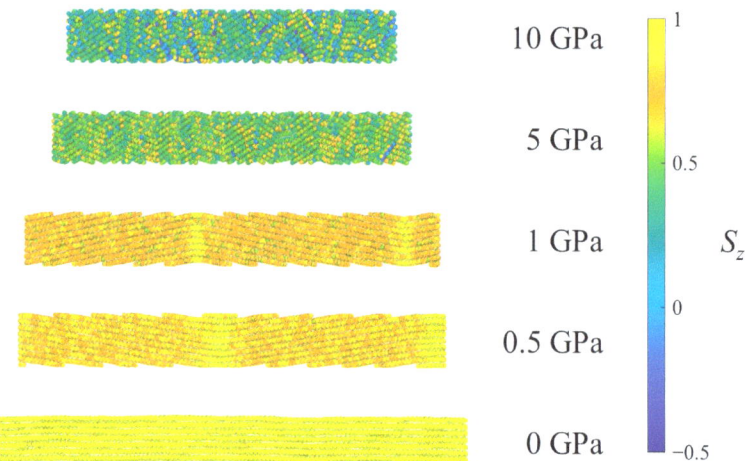

Figure 7. Example CPE system under ambient conditions (0 GPa) and after equilibration at several pressures in Hugoniostatted shock simulations. Atoms are colored according to the orientational parameter S_z.

Figure 8 shows the mean S_z values ($\langle S_z^c \rangle$ as defined in Section 2.5.2, Equation (8)) of the crystalline cluster as a function of pressure for the three systems. Through construction, the CPE system at $P = 0$ GPa is almost perfectly aligned with the z-axis; however, the bond chords of this system tilt to form kink bands in response to the strain imposed by the compression. Fluctuations of UA coordinates about those of the perfect crystal may dictate the direction that the bond chords tilt—this direction is not always consistent among the different starting seeds, but the absolute value of the angle is fairly consistent. Also shown in Figure 8 is a theoretical prediction of $\langle S_z^c \rangle$ vs. P for CPE using Pastine's model [62], assuming that the strain in the crystal manifests entirely as tilt of the bond chords. Assuming that the strain in the crystal is

$$\epsilon^c = 1 - |\cos \phi|, \qquad (12)$$

where ϕ is the angle the bond chords make with the z-axis, $\langle S_z^c \rangle$ is calculated according to Equation (5) and $\epsilon^c(P)$ is determined from Pastine's theory. The theoretical prediction and the simulation data have a root-mean-square deviation of approximately 0.032; deviations between the data and model may be due to crystallographic strain caused by an excess compression in the a or b unit cell dimensions (beyond that caused by chain tilt) or differences between the crystallographic unit cell in the current work and that of Pastine.

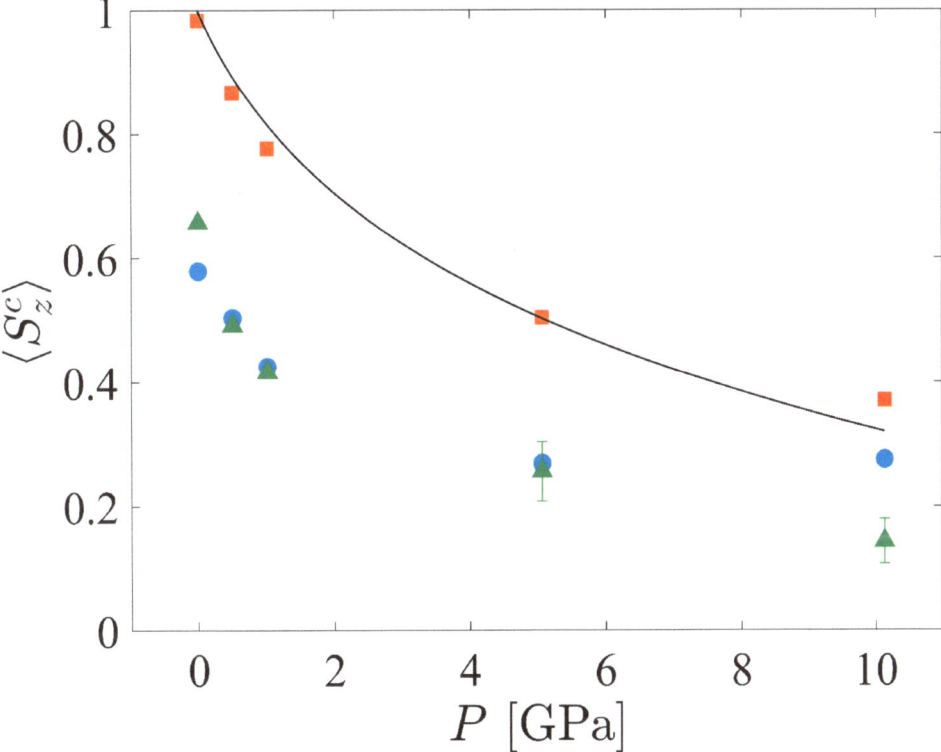

Figure 8. Mean orientational parameter, $\langle S_z^c \rangle$, vs. P for the crystalline populations of SCPE44 (blue circles), SCPE81 (green triangles), and CPE (orange squares). Also shown is a theoretical prediction of $\langle S_z^c \rangle$ vs. P for CPE based on the model of Pastine [62].

Figure 8 also shows the dependence of $\langle S_z^c \rangle$ on shock pressure for the crystalline domains of the SCPE systems. The orientational order within the crystalline domains for the two different SCPE systems are in close agreement up to 5 GPa but deviate at 10 GPa. The values for SCPE44 exhibit an upward shift after 5 GPa, while the values for SCPE81 appear to be "noisy" for 5 and 10 GPa. The non-monotonic decrease in the $\langle S_z^c \rangle$ values with increasing pressure for SCPE44 systems must be a result of interactions with the noncrystalline population because neither the theoretical nor simulation data for CPE show such features. An interchange of strain between the crystalline and noncrystalline regions is not likely because it would be reflected as non-monotonicity in v vs. P in Figure 2a, for example. Rather, it seems that, for $P \approx 10$ GPa, there is growth of the population of noncrystalline UAs at the expense of crystalline UAs (i.e., "melting"), thus eliminating some of the less crystalline UAs from the crystalline cluster and increasing the average orientation of the crystalline cluster. The kink boundaries observed for SCPE81 in Figure 6 result in the large error bars observed in Figure 8.

3.1.4. Crystallinity

To characterize the crystallinity of each of the systems, two different metrics are used. The first is the mass fraction of the crystalline population—χ^c, as defined in Section 2.5.2, Equation (9). χ^c alone is insufficient to fully characterize the crystalline order of the system when the crystalline region is imperfect (i.e., without perfect periodicity). For example, this definition always assigns a crystallinity of 100% to the CPE systems because these systems are characterized as a single cluster, even in the presence of kink bands (see the Supplementary Materials for further discussion). However, with increasing shock pressure, the nematic order of the CPE system decreases. Supplementary information is provided by the mean value of the nematic order parameter for the crystalline population, $\langle p_2^c \rangle$, calculated using Equation (8). Trends for both χ^c and $\langle p_2^c \rangle$ in all three systems are shown in Figure 9. Notice that $\langle p_2^c \rangle$ for SCPE81 at 5 GPa is less than the value for SCPE44 due to the formation of kink bands and the loss of the nematic order of UAs in between these kink bands. It appears that the crystallinity, χ^c, changes relatively little under the application of shock, but the nematic order within the crystalline lamellae, $\langle p_2^c \rangle$, decreases significantly.

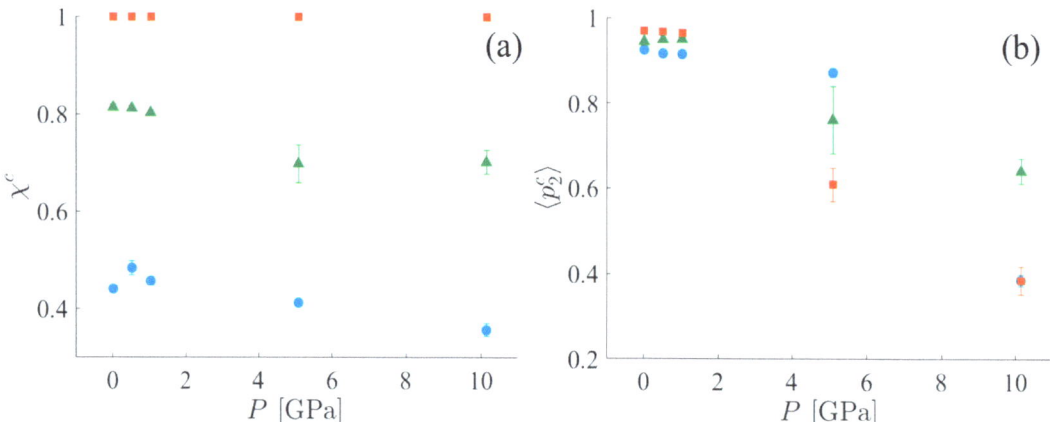

Figure 9. (**a**) χ^c and (**b**) $\langle p_2^c \rangle$ for SCPE44 (blue circles), SCPE81 (green triangles), and CPE (orange squares). Data are averages among the ten different starting configurations with error bars indicating three standard errors.

3.1.5. Potential Energy Contributions

The FCM clustering analysis is used to determine the mean potential energy contributions of both the crystalline and noncrystalline clusters of the systems separately. Potential energy contributions for the TraPPE-UA force field include nonbonded (pair) and bonded (bond, angle, and dihedral) contributions; these contributions are shown in Figure 10. The pair contribution, shown in Figure 10a, is lower for crystalline clusters than noncrystalline clusters because the former has a more stable chain packing arrangement. As pressure increases, the mean pair contributions of the crystalline clusters of both SCPE44 and SCPE81 increase nearly to the levels of the mean pair contributions of the noncrystalline clusters, indicative of decreasing order within the crystal cluster. In contrast, the mean bond contributions in Figure 10b are the same for both clusters within a single system.

The bond potential is by far the stiffest in the system, so the sensitivity of bond length displacements to the local environment (crystalline or noncrystalline) is negligible. Thus, the bond potential energy increase can mainly be attributed to the increase in temperature that accompanies increasing pressure along the Hugoniot curve. The angle energy, shown in Figure 10c, is the next stiffest mode in the systems. At low pressure, it is the same for both crystalline and noncrystalline clusters, similar to the bond energy. However, at elevated shock conditions it begins to differentiate for crystalline and noncrystalline clusters, indicating a sensitivity to the local environment. The mean angle contribution of the crystal cluster in SCPE81 closely tracks that of CPE, while the mean angle contribution of the crystal cluster in SCPE44 tracks more closely with that of the noncrystalline population for pressures as high as 10 GPa. Finally, the dihedral energy contribution is shown in Figure 10d; like the pair contribution, it is indicative of decreasing crystallographic order, especially for the case with the lowest crystallinity (SCPE44).

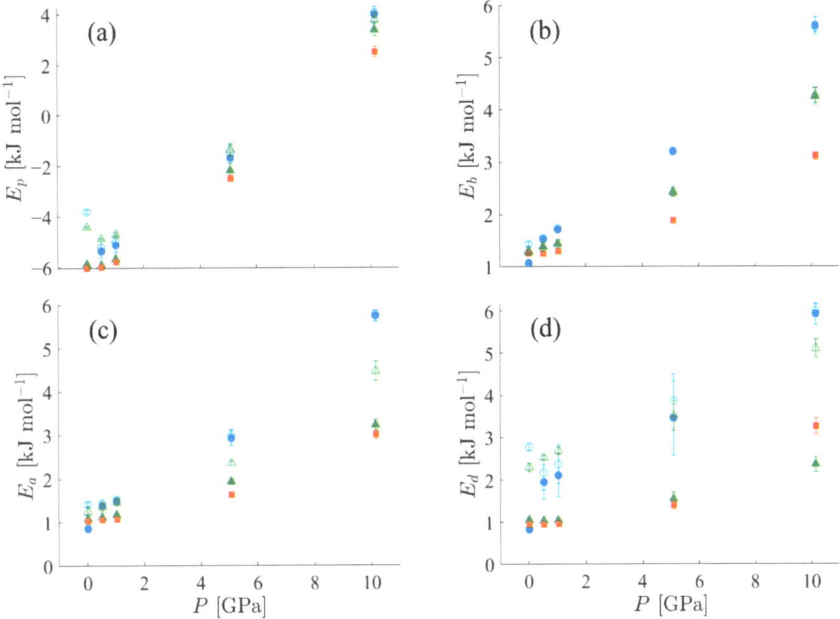

Figure 10. Potential energy contributions per UA for crystalline (filled symbols) and noncrystalline (empty symbols) populations of SCPE44 (blue circles), SCPE81 (green triangles), and CPE (orange squares, no non-crystalline population). The contributions are (**a**) pair (Van der Waals) energy, (**b**) bond energy, (**c**) angle energy, and (**d**) dihedral energy.

3.2. Hugoniostat Transient Evolution

Upon the initial application of the shock pressure, one observes two major regimes of compression, as illustrated by the evolution of local order parameters with logarithmic time in Figures 11–13. Figure 11 shows the spatial evolution of stress during the equilibration of shock. In the first, transient regime, which extends from approximately $t = 0.01$ to 10 ps, shear stress builds up in the crystalline domains of the systems, followed by compression of the crystalline and noncrystalline domains. The information computed by LAMMPS is $S_{\alpha\beta}$, where S is the negative of the per-UA stress tensor multiplied by volume, and the subscripts denote the components. To calculate per-UA pressure values, components of S are negated and then divided by per-UA volumes, calculated via Voronoi tessellation (see Section 2.4 for details). The shear stress for compression in the z-direction is computed as [8]

$$\tau = \frac{1}{2}\left(P_{zz} - \frac{P_{xx} + P_{yy}}{2}\right). \tag{13}$$

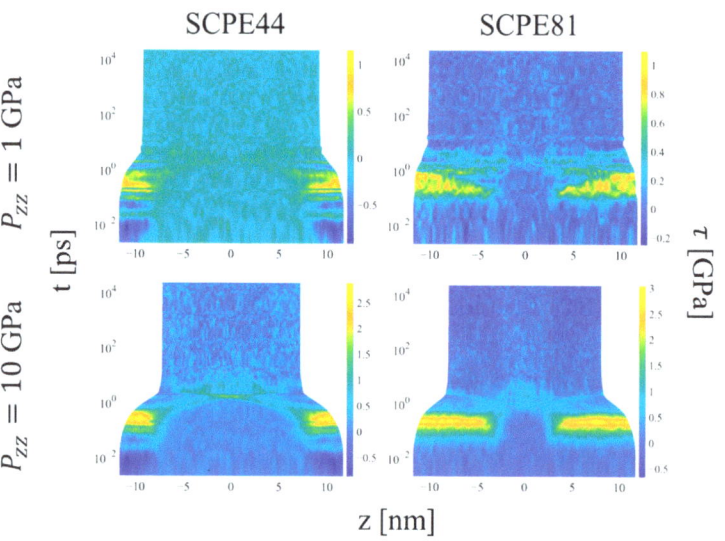

Figure 11. Heat plots for the shear stress as a function of position along the compression direction and time, for two systems and two applied pressures during Hugoniostatted simulations. The color scale is proportional to the shear stress. Plots for all systems and pressures are included in the Supplementary Materials.

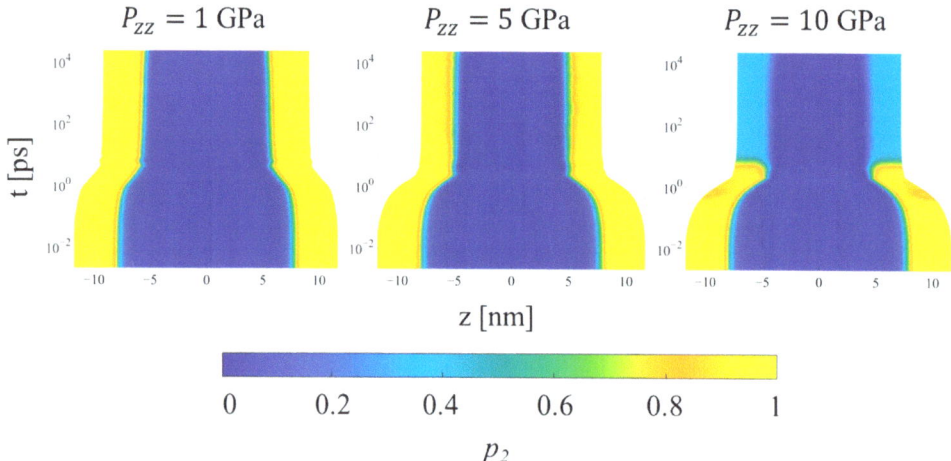

Figure 12. Heat plots of the p_2 order parameter for the SCPE44 as a function of position along the compression direction and time, for three different applied pressures in Hugoniostatted simulations. The color scale is proportional to p_2. Plots for all systems and pressures are included in the Supplementary Materials.

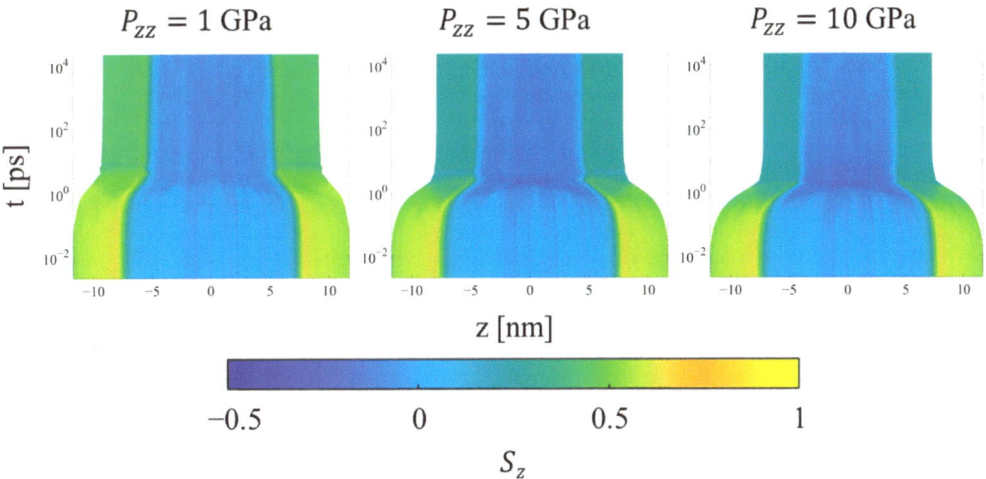

Figure 13. Heat plots of the S_z orientation parameter for SCPE44 as a function of position along the compression direction and time, for three different applied pressures in Hugoniostatted simulations. The color scale is proportional to S_z. Plots for all systems and pressures are included in the Supplementary Materials.

After a compression time on the order of picoseconds, the crystalline domains show a rise in shear stress to levels approximately half of the applied shock pressure in the z-direction. The low resistance to crystallographic slip in the UA models permits rotation of the chain stems in the crystalline domains in response to the shear stress, so that the compressive load is borne more by the softer nonbonded, intermolecular interactions and less by the stiffer bonded, intramolecular interactions; similar behavior was observed for small extensional strains (< 0.08) of UA SCPE models under isothermal uniaxial compression by Kim et al. [21]. At this point, the system experiences significant strain in both the crystalline and noncrystalline domains in response to the applied compressive stress. This behavior indicates that (1) there is a short delay during which the crystalline domain experiences the buildup of transverse and longitudinal stresses with respect to the shock direction that drive crystallographic slip, followed by (2) compression of the crystalline domain to equalize stress. The noncrystalline domain equalizes the stress in the different directions much more rapidly and thus never experiences a significant shear stress. The rise in shear stress would not be expected for compression of pure APE, due to the fast equalization of its stresses.

Once the final system volume is reached, the shear stress is nearly zero, indicating that the strain response serves to equalize the stress in all directions. The pressure also becomes uniform throughout the system, as a result of equilibration to the post-shock Hugoniot state. Figure 12 shows the local p_2 order parameter for several systems. The crystalline and noncrystalline domains remain clearly defined over the entire range of pressure. However, referring to Figure 9b, SCPE44 maintains high nematic order in the crystalline domain up to 5 GPa, although this order decreases dramatically at 10 GPa. SCPE81, on the other hand, exhibits a decrease in nematic order at 5 GPa, but with large variance; we hypothesize that this behavior is a consequence of the formation of kink bands in that system. Meanwhile, the orientational order parameter $\langle S_z^c \rangle$ confirms the tilting of chain stems in the crystalline domain away from the direction of applied load, as shown in Figure 11. One exception to this general trend is SCPE44, for which crystalline orientational order increases very slightly at high pressure, from 0.270 at 5 GPa to 0.275 at 10 GPa. This counterintuitive increase in $\langle S_z^c \rangle$ can be explained by a decrease in χ^c over the same range of pressure.

Finally, considering the trends for the CPE system, the uniformity of all metrics throughout the simulation is preserved during all simulations. The density increases uniformly during the transient regime, while the p_2 order parameter (not shown) decreases uniformly at higher pressures to values near 0.6 and 0.3 at 5 and 10 GPa, respectively. The orientational order parameter (Figure 14) decreases for all pressures in order to accommodate compression of the crystal region through buckling, kink band formation, and the mechanism of fine crystallographic slip. Interestingly, increasing the applied pressure from 5 to 10 GPa increases the amount of time required for the reorientation to complete, as shown by the intermediate values of S_z during the equilibration regime. In both cases, the decrease in S_z occurs gradually and monotonically even after the final system volume is achieved.

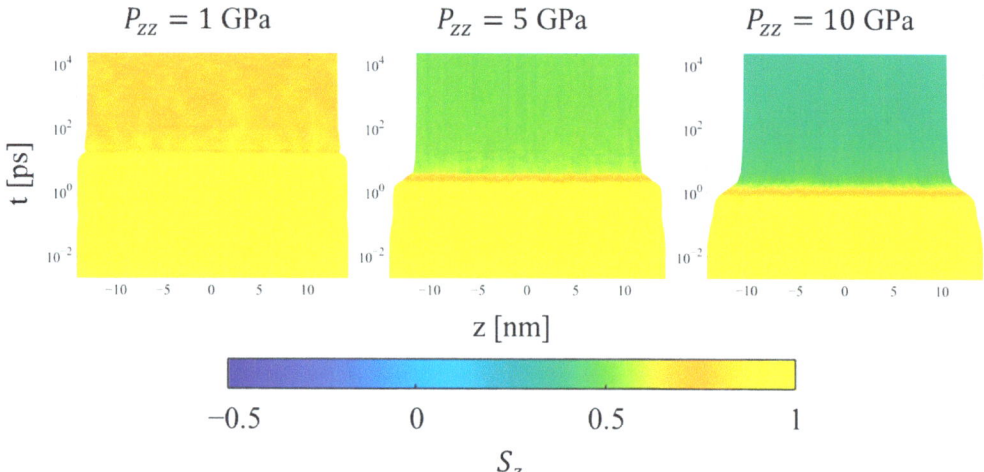

Figure 14. Heat plots of the S_z orientation parameter for CPE as a function of position along the compression direction and time, for three different applied pressures in Hugoniostatted simulations. The color scale is proportional to S_z. Plots for all systems and pressures are included in the Supplementary Materials.

4. Conclusions

This work analyzes the shock wave response of model SCPE and CPE systems for the purpose of understanding the changes to the configurational states associated with shocks of different pressures. Shock simulations were conducted using molecular dynamics with an equilibrium Hugoniostat called NP_{zz}Hug [8]. Clustering based on the FCM algorithm is introduced to allow adaptive clustering in response to changes in the distribution of the nematic order parameter, p_2, so that the trends of individual populations may be analyzed separately. The nematic order parameter is found to distinguish crystalline and noncrystalline UAs within each SCPE system with high fidelity and simplicity. We take advantage of this clustering to focus on the response of the crystalline populations in this work, because they exhibit a variety of deformation mechanisms depending on their initial degree of crystallinity and the applied shock pressure.

Examining the Hugoniostat trajectories, two potential energy storage mechanisms are identified: loss of nematic order within the crystal domain and change in the orientation of the crystal stems with respect to the crystalline–noncrystalline interface via crystallographic slip + compression. Both of these mechanisms increase the potential energy of the system and thus store energy of the shock wave by changing the system configuration. Additionally, for systems of sufficiently high crystallinity (or lamellar thickness), the formation of kink bands is observed within the crystalline region, as evidenced by the data for SCPE81 at

5 GPa in some instances, and for CPE at all shock pressures. The long period of these kink bands decreases with increasing shock pressure, consistent with a increasing energy buckling phenomenon. The formation of kink bands may be more prevalent in experiments than observed here, due to the ease with which crystallographic slip occurs in the UA model. Finally, at the highest pressure (10 GPa), kink bands apparently break up or disappear; we hypothesize that a higher temperature and lower nematic order lead to a decreased energy barrier for local slip and bend formation, and a degree of melting. For CPE, the angle of the chain tilt as a function of shock pressure is well-approximated by Pastine's theory [62].

At low shock pressures (up to about 1 GPa), all systems exhibit fine crystallographic slip. For the CPE systems, however, this slip is necessarily accompanied by kink band formation because of the lack of compliance imposed by the periodic boundary conditions. For the SCPE systems, on the other hand, the noncrystalline regions act as damping boundary conditions for the crystal chains, allowing them to tilt without kink band formation in response to the development of shear stress. At higher shock pressures, kink bands also begin to form in SCPE81 as further tilt of the crystal chains becomes energetically unfavorable. The SCPE44 systems do not form kink bands at any of the pressures simulated; instead, at 10 GPa and elevated temperature in the Hugoniot state, some of the crystalline population "melts" into the noncrystalline population in order to satisfy the geometric constraints caused by confinement while not altering the tilt angle. This melting is also observed in the convergence of the potential energy contributions of the crystalline populations to those of the noncrystalline populations in SCPE44 at 10 GPa, but not in SCPE81. This is also supported by the lower $\langle p_2^c \rangle$ values in SCPE44 compared to SCPE81 at 10 GPa. In fact, the $\langle p_2^c \rangle$ values in SCPE81 exceed those of the CPE systems at 10 GPa. This behavior suggests that the presence of some noncrystalline material can actually stabilize the crystalline domains against melting in an SCPE system by acting as something of a "shock absorber," but that too much noncrystalline material (which in this case correlates with lower crystallinity and thinner crystalline domains) can destabilize the crystalline domains with respect to melting. Further study of these systems may be able to more accurately determine an optimal combination of crystallinity and crystalline domain thickness to maintain the integrity of the crystal according to one of the aforementioned metrics.

Fundamentally, the deformation mechanisms observed in all of the simulated PE systems are consequences of the geometric confinement caused by the confined compression of the shock, the temperature increase in the Hugoniot state, and the atomic configuration in the initial state of the system. The modulus for intramolecular compression along the crystallographic c-axis (the chain axis) is an order of magnitude greater than the moduli for intermolecular compression along the a- or b-axes of the unit cell, so most of the deformation occurs through rotation of the unit cell to accommodate compression intermolecularly [62]. If the nematic order of the crystalline regions remains high, then the assumptions used by Pastine's model hold—namely, that there is at most a linear correction for compression along the c-axis and that the spatial arrangement of atoms remains periodic. Thus, we see for shock pressures of < 10 GPa that Pastine's prediction for stress as a function of strain in CPE can be used to predict the tilt of crystal stems leading to crystallographic slip + compression. These assumptions also approximately hold for the SCPE systems under the same pressure condition, as evidenced by the superposition of the SCPE curves in Figure 8; the downward shift of the SCPE results relative to the CPE result is due to the initial tilt of the crystal stems with respect to the direction of compression. Additionally, the SCPE systems exhibit some melting at high pressure, as indicated by a decrease in χ^c that is not observed for CPE or Pastine's model. This melting behavior accommodates a portion of the shock energy, as a result of which the nematic order within the crystal clusters does not decay as much for SCPE as it does for CPE.

Supplementary Materials: The following supporting information can be downloaded at: https://www.mdpi.com/article/10.3390/polym15214262/s1, Hugoniostat Transient Evolution, Figures S1–S3;

Evaluation of Clustering Methods by Silhouette Scores, Figures S4–S8. Reference [67] is cited in the Supplementary Materials.

Author Contributions: Conceptualization, G.C.R.; Methodology, J.P.M.; Formal analysis, J.P.M.; Writing—original draft, J.P.M.; Writing—review & editing, G.C.R.; Supervision, G.C.R. All authors have read and agreed to the published version of the manuscript.

Funding: This material is based upon work supported in part by the U. S. Army Research Office through the Institute for Soldier Nanotechnologies at MIT, under Collaborative Agreement Number W911NF-18-2-0048.

Institutional Review Board Statement: Not applicable.

Data Availability Statement: Publicly available datasets were analyzed in this study. This data can be found here: https://github.com/jpmikhail/Mechanisms_of_Shock_Dissipation_in_Semicrystalline_Polyethylene.

Acknowledgments: The authors gratefully acknowledge the helpful discussions and advice from our collaborators through the ISN and Army Research Laboratory, including Keith Nelson, Timothy Swager, Alex Hsieh, Jan Andzelm, and In-Chul Yeh.

Conflicts of Interest: The authors declare no conflict of interest.

References

1. Bradley, J.N. *Shock Waves in Chemistry and Physics*; Methuen: London, UK, 1962.
2. Courant, R.; Friedrichs, K.O. *Supersonic Flow and Shock Waves (Vol. 21)*; Springer Science & Business Media: Berlin/Heidelberg, Germany, 1999.
3. Varigonda, S.V.; Narayanaswamy, V. Investigation of Shock Wave Oscillations over a Flexible Panel in Supersonic Flows. In *AIAA Aviation 2019 Forum*; American Institute of Aeronautics and Astronautics: Reston, VA, USA, 2019; p. 3543.
4. Vermeer, D.J.; Meyer, J.W.; Oppenheim, A.K. Auto-ignition of hydrocarbons behind reflected shock waves. *Combust. Flame* **1972**, *18*, 327–336. [CrossRef]
5. Mansour, A.; Müller, N. A review of flash evaporation phenomena and resulting shock waves. *Exp. Therm. Fluid Sci.* **2019**, *107*, 146–168. [CrossRef]
6. Owens, F.J. EPR study of shock and thermally induced reaction in solid copper tetramine nitrate. *J. Chem. Phys.* **1982**, *77*, 5549–5551. [CrossRef]
7. Ortellado, L.; Vega, D.A.; Gómez, L.R. Shock melting of lamellae-forming block copolymers. *Phys. Rev. E* **2022**, *106*, 044502. [CrossRef] [PubMed]
8. Ravelo, R.; Holian, B.L.; Germann, T.C.; Lomdahl, P.S. Constant-stress Hugoniostat method for following the dynamical evolution of shocked matter. *Phys. Rev. B—Condens. Matter Mater. Phys.* **2004**, *70*, 014103. [CrossRef]
9. Reed, E.J.; Fried, L.E.; Joannopoulos, J.D. A method for tractable dynamical studies of single and double shock compression. *Phys. Rev. Lett.* **2003**, *90*, 235503. [CrossRef] [PubMed]
10. Evans, D.J.; Holian, B.L. The nose–hoover thermostat. *J. Chem. Phys.* **1985**, *83*, 4069–4074. [CrossRef]
11. O'Connor, T.C.; Elder, R.M.; Sliozberg, Y.R.; Sirk, T.W.; Andzelm, J.W.; Robbins, M.O. Molecular origins of anisotropic shock propagation in crystalline and amorphous polyethylene. *Phys. Rev. Mater.* **2018**, *2*, 35601. [CrossRef]
12. Liu, T.; Huang, A.; Geng, L.H.; Lian, X.H.; Chen, B.Y.; Hsiao, B.S.; Kuang, T.-R.; Peng, X.F. Ultra-strong, tough and high wear resistance high-density polyethylene for structural engineering application: A facile strategy towards using the combination of extensional dynamic oscillatory shear flow and ultra-high-molecular-weight polyethylene. *Compos. Sci. Technol.* **2018**, *167*, 301–312. [CrossRef]
13. Ronca, S. Polyethylene. In *Brydson's Plastics Materials*; Butterworth-Heinemann: Oxford, UK, 2017; pp. 247–278.
14. McKellop, H.A. The lexicon of polyethylene wear in artificial joints. *Biomaterials* **2007**, *28*, 5049–5057. [CrossRef]
15. Crist, B.; Fisher, C.J.; Howard, P.R. Mechanical properties of model polyethylenes: Tensile elastic modulus and yield stress. *Macromolecules* **1989**, *22*, 1709–1718. [CrossRef]
16. in't Veld, P.J.; Rutledge, G.C. Temperature-dependent elasticity of a semicrystalline interphase composed of freely rotating chains. *Macromolecules* **2003**, *36*, 7358–7365. [CrossRef]
17. Balijepalli, S.; Rutledge, G.C. Simulation Study of Semi-Crystalline Polymer Interphases. In *Macromolecular Symposia*; WILEY-VCH Verlag GmbH & Co. KGaA: Weinheim, Germany, 1998; Volume 133, pp. 71–99.
18. Strobl, G.R. *The Physics of Polymers: Concepts for Understanding Their Structures and Behavior*; Springer Science & Business Media: Berlin/Heidelberg, Germany, 2007.
19. In't Veld, P.J.; Hütter, M.; Rutledge, G.C. Temperature-dependent thermal and elastic properties of the interlamellar phase of semicrystalline polyethylene by molecular simulation. *Macromolecules* **2006**, *39*, 439–447. [CrossRef]
20. Lee, S.; Rutledge, G.C. Plastic deformation of semicrystalline polyethylene by molecular simulation. *Macromolecules* **2011**, *44*, 3096–3108. [CrossRef]

21. Kim, J.M.; Locker, R.; Rutledge, G.C. Plastic deformation of semicrystalline polyethylene under extension, compression, and shear using molecular dynamics simulation. *Macromolecules* **2014**, *47*, 2515–2528. [CrossRef]
22. Yeh, I.-C.; Andzelm, J.W.; Rutledge, G.C. Mechanical and structural characterization of semicrystalline polyethylene under tensile deformation by molecular dynamics simulations. *Macromolecules* **2015**, *48*, 4228–4239. [CrossRef]
23. Yeh, I.-C.; Lenhart, J.L.; Rutledge, G.C.; Andzelm, J.W. Molecular dynamics simulation of the effects of layer thickness and chain tilt on tensile deformation mechanisms of semicrystalline polyethylene. *Macromolecules* **2017**, *50*, 1700–1712. [CrossRef]
24. Kumar, V.; Locker, C.R.; in't Veld, P.J.; Rutledge, G.C. Effect of short chain branching on the interlamellar structure of semicrystalline polyethylene. *Macromolecules* **2017**, *50*, 1206–1214. [CrossRef]
25. Ranganathan, R.; Kumar, V.; Brayton, A.L.; Kroger, M.; Rutledge, G.C. Atomistic modeling of plastic deformation in semicrystalline polyethylene: Role of interphase topology, entanglements, and chain dynamics. *Macromolecules* **2020**, *53*, 4605–4617. [CrossRef]
26. Van Duin, A.C.; Dasgupta, S.; Lorant, F.; Goddard, W.A. ReaxFF: A reactive force field for hydrocarbons. *J. Phys. Chem. A* **2001**, *105*, 9396–9409. [CrossRef]
27. O'Connor, T.C.; Andzelm, J.; Robbins, M.O. AIREBO-M: A reactive model for hydrocarbons at extreme pressures. *J. Chem. Phys.* **2015**, *142*, 24903. [CrossRef] [PubMed]
28. Mattsson, T.R.; Lane, J.M.D.; Cochrane, K.R.; Desjarlais, M.P.; Thompson, A.P.; Pierce, F.; Grest, G.S. First-principles and classical molecular dynamics simulation of shocked polymers. *Phys. Rev. B* **2010**, *81*, 54103. [CrossRef]
29. Cochrane, K.R.; Desjarlais, M.; Mattsson, T.R. Density functional theory (DFT) simulations of polyethylene: Principal hugoniot, specific heats, compression and release isentropes. *AIP Conf. Proc.* **2012**, *1426*, 1271–1274.
30. Elder, R.M.; O'Connor, T.C.; Chantawansri, T.L.; Sliozberg, Y.R.; Sirk, T.W.; Yeh, I.-C.; Robbins, M.O.; Andzelm, J.W. Shock-wave propagation and reflection in semicrystalline polyethylene: A molecular-level investigation. *Phys. Rev. Mater.* **2017**, *1*, 43606. [CrossRef]
31. Bartczak, Z.; Galeski, A. Plasticity of Semicrystalline Polymers. In *Macromolecular Symposia*; WILEY-VCH Verlag: Weinheim, Germany, 2010; Volume 294, pp. 67–90.
32. Galeski, A.; Bartczak, Z.; Argon, A.S.; Cohen, R.E. Morphological alterations during texture-producing plastic plane strain compression of high-density polyethylene. *Macromolecules* **1992**, *25*, 5705–5718. [CrossRef]
33. Chen, G.Q.G.; Feldman, M. *The Mathematics of Shock Reflection-Diffraction and Von Neumann's Conjectures:(AMS-197)*; Princeton University Press: Princeton, NJ, USA, 2018; Volume 197.
34. Kazmierczak, T.; Galeski, A.; Argon, A.S. Plastic deformation of polyethylene crystals as a function of crystal thickness and compression rate. *Polymer* **2005**, *46*, 8926–8936. [CrossRef]
35. Brown, E.N.; Willms, R.B.; Gray, G.T.; Rae, P.J.; Cady, C.M.; Vecchio, K.S.; Flowers, J.; Martinez, M.Y. Influence of molecular conformation on the constitutive response of polyethylene: A comparison of HDPE, UHMWPE, and PEX. *Exp. Mech.* **2007**, *47*, 381–393. [CrossRef]
36. Jordan, J.L.; Rowland, R.L.; Greenhall, J.; Moss, E.K.; Huber, R.C.; Willis, E.C.; Hrubiak, R.; Kenney-Benson, C.; Bartram, B.; Sturtevant, B.T. Elastic properties of polyethylene from high pressure sound speed measurements. *Polymer* **2021**, *212*, 123164. [CrossRef]
37. Martin, M.G.; Siepmann, J.I. Transferable potentials for phase equilibria. 1. United-atom description of n-alkanes. *J. Phys. Chem. B* **1998**, *102*, 2569–2577. [CrossRef]
38. Bolton, K.; Bosio, S.B.M.; Hase, W.L.; Schneider, W.F.; Hass, K.C. Comparison of explicit and united atom models for alkane chains physisorbed on α-Al2O3 (0001). *J. Phys. Chem. B* **1999**, *103*, 3885–3895. [CrossRef]
39. EMC: Enhanced Monte Carlo A Multi-Purpose Modular and Easily Extendable Solution to Molecular and Mesoscale Simulations by Pieter J. in't Veld. (n.d.). Retrieved 22 December 2019. Available online: http://montecarlo.sourceforge.net/emc/Welcome.html (accessed on 27 September 2023).
40. Bassett, D.C.; Hodge, A.M.; Olley, R.H. On the morphology of melt-crystallized polyethylene-II. Lamellae and their crystallization conditions. *Proceedings of the Royal Society of London. A Math. Phys. Sci.* **1981**, *377*, 39–60.
41. Gautam, S.; Balijepalli, S.; Rutledge, G.C. Molecular simulations of the interlamellar phase in polymers: Effect of chain tilt. *Macromolecules* **2000**, *33*, 9136–9145. [CrossRef]
42. LAMMPS Molecular Dynamics Simulator. (n.d.). Retrieved 19 December 2019. Available online: https://www.lammps.org (accessed on 27 September 2023).
43. Thompson, A.P.; Aktulga, H.M.; Berger, R.; Bolintineanu, D.S.; Brown, W.M.; Crozier, P.S.; in't Veld, P.J.; Kohlmeyer, A.; Moore, S.G.; Nguyen, T.D.; et al. LAMMPS-a flexible simulation tool for particle-based materials modeling at the atomic, meso, and continuum scales. *Comput. Phys. Commun.* **2022**, *271*, 108171. [CrossRef]
44. Shinoda, W.; Shiga, M.; Mikami, M. Rapid estimation of elastic constants by molecular dynamics simulation under constant stress. *Phys. Rev. B* **2004**, *69*, 134103. [CrossRef]
45. Chantawansri, T.L.; Sirk, T.W.; Byrd, E.F.C.; Andzelm, J.W.; Rice, B.M. Shock Hugoniot calculations of polymers using quantum mechanics and molecular dynamics. *J. Chem. Phys.* **2012**, *137*, 204901. [CrossRef] [PubMed]
46. Marsh, S.P. (Ed.) *LASL Shock Hugoniot Data*; University of California Press: Berkeley, CA, USA, 1980; Volume 5.
47. Jorgensen, W.L.; Maxwell, D.S.; Tirado-Rives, J. Development and testing of the OPLS all-atom force field on conformational energetics and properties of organic liquids. *J. Am. Chem. Soc.* **1996**, *118*, 11225–11236. [CrossRef]

48. Brayton, A.L.; Yeh, I.-C.; Andzelm, J.W.; Rutledge, G.C. Vibrational Analysis of Semicrystalline Polyethylene Using Molecular Dynamics Simulation. *Macromolecules* **2017**, *50*, 6690–6701. [CrossRef]
49. Yi, P.; Rutledge, G.C. Molecular simulation of bundle-like crystal nucleation from n-eicosane melts. *J. Chem. Phys.* **2011**, *135*, 024903. [CrossRef]
50. Barber, C.B.; Dobkin, D.P.; Huhdanpaa, H. The quickhull algorithm for convex hulls. *ACM Trans. Math. Softw. (TOMS)* **1996**, *22*, 469–483. [CrossRef]
51. Nicholson, D.A.; Rutledge, G.C. Flow-induced inhomogeneity and enhanced nucleation in a long alkane melt. *Polymer* **2020**, *200*, 122605. [CrossRef]
52. Glielmo, A.; Husic, B.E.; Rodriguez, A.; Clementi, C.; Noé, F.; Laio, A. Unsupervised learning methods for molecular simulation data. *Chem. Rev.* **2021**, *121*, 9722–9758. [CrossRef] [PubMed]
53. Arthur, D.; Vassilvitskii, S. K-Means++ The Advantages of Careful Seeding. In Proceedings of the Eighteenth Annual ACM-SIAM Symposium on Discrete Algorithms, New Orleans, LO, USA, 7–9 January 2007; pp. 1027–1035.
54. Bezdek, J.C. *Pattern Recognition with Fuzzy Objective Function Algorithms*; Springer Science & Business Media: Berlin/Heidelberg, Germany, 1981.
55. Lloyd, S. Least squares quantization in PCM. *IEEE Trans. Inf. Theory* **1982**, *28*, 129–137. [CrossRef]
56. Bezdek, J.C. A Physical Interpretation of Fuzzy ISODATA. In *Readings in Fuzzy Sets for Intelligent Systems*; Morgan Kaufmann: Burlington, MA, USA, 1993; pp. 615–616.
57. Rousseeuw, P.J. Silhouettes: A graphical aid to the interpretation and validation of cluster analysis. *J. Comput. Appl. Math.* **1987**, *20*, 53–65. [CrossRef]
58. Tsu, R.; Gonzalez-Hernandez, J.; Chao, S.S.; Lee, S.C.; Tanaka, K. Critical volume fraction of crystallinity for conductivity percolation in phosphorus-doped Si: F: H alloys. *Appl. Phys. Lett.* **1982**, *40*, 534–535. [CrossRef]
59. Kavesh, S.; Schultz, J.M. Meaning and measurement of crystallinity in polymers: A review. *Polym. Eng. Sci.* **1969**, *9*, 331–338. [CrossRef]
60. Bowman, A.W.; Azzalini, A. *Applied Smoothing Techniques for Data Analysis: The Kernel Approach with S-Plus Illustrations*; OUP Oxford: Oxford, UK, 1997; Volume 18.
61. Hsieh, A.J.; Wu, Y.-C.M.; Hu, W.; Mikhail, J.P.; Veysset, D.; Kooi, S.E.; Nelson, K.A.; Rutledge, G.C.; Swager, T.M. Bottom-up design toward dynamically robust polyurethane elastomers. *Polymer* **2021**, *218*, 123518. [CrossRef]
62. Pastine, D.J. P, v, t equation of state for polyethylene. *J. Chem. Phys.* **1968**, *49*, 3012–3022. [CrossRef]
63. Ruoff, A.L. Linear shock-velocity-particle-velocity relationship. *J. Appl. Phys.* **1967**, *38*, 4976–4980. [CrossRef]
64. Agrawal, V.; Peralta, P.; Li, Y.; Oswald, J. A pressure-transferable coarse-grained potential for modeling the shock Hugoniot of polyethylene. *J. Chem. Phys.* **2016**, *145*, 104903. [CrossRef]
65. Lacks, D.J.; Rutledge, G.C. Simulation of the temperature dependence of mechanical properties of polyethylene. *J. Phys. Chem.* **1994**, *98*, 1222–1231. [CrossRef]
66. Olsson, P.A.; Schröder, E.; Hyldgaard, P.; Kroon, M.; Andreasson, E.; Bergvall, E. Ab initio and classical atomistic modelling of structure and defects in crystalline orthorhombic polyethylene: Twin boundaries, slip interfaces, and nature of barriers. *Polymer* **2017**, *121*, 234–246. [CrossRef]
67. Campello, R.J.; Hruschka, E.R. A fuzzy extension of the silhouette width criterion for cluster analysis. *Fuzzy Sets Syst.* **2006**, *157*, 2858–2875. [CrossRef]

Disclaimer/Publisher's Note: The statements, opinions and data contained in all publications are solely those of the individual author(s) and contributor(s) and not of MDPI and/or the editor(s). MDPI and/or the editor(s) disclaim responsibility for any injury to people or property resulting from any ideas, methods, instructions or products referred to in the content.

Article

Computational Requirements for Modeling Thermal Conduction in Polymeric Phase-Change Materials: Periodic Hard Spheres Case

Kevin A. Redosado Leon [1,*], Alexey Lyulin [2,3] and Bernard J. Geurts [1,2]

1 Mathematics of Multiscale Modeling and Simulation, Faculty EEMCS, University of Twente, P.O. Box 217, 7500 AE Enschede, The Netherlands; b.j.geurts@utwente.nl
2 Group Soft Matter and Biological Physics, Department of Applied Physics and Science Education, Eindhoven University of Technology, P.O. Box 513, 5600 MB Eindhoven, The Netherlands; a.v.lyulin@tue.nl
3 Multiscale Molecular Dynamics, Faculty EEMCS, University of Twente, P.O. Box 217, 7500 AE Enschede, The Netherlands
* Correspondence: k.a.redosadoleon@utwente.nl

Citation: Redosado Leon, K.A.; Lyulin, A.; Geurts, B.J. Computational Requirements for Modeling Thermal Conduction in Polymeric Phase-Change Materials: Periodic Hard Spheres Case. *Polymers* **2024**, *16*, 1015. https://doi.org/10.3390/polym16071015

Academic Editor: Patrick Ilg

Received: 4 March 2024
Revised: 2 April 2024
Accepted: 3 April 2024
Published: 8 April 2024

Copyright: © 2024 by the authors. Licensee MDPI, Basel, Switzerland. This article is an open access article distributed under the terms and conditions of the Creative Commons Attribution (CC BY) license (https:// creativecommons.org/licenses/by/ 4.0/).

Abstract: This research focuses on modeling heat transfer in heterogeneous media composed of stacked spheres of paraffin as a perspective polymeric phase-change material. The main goal is to study the requirements of the numerical scheme to correctly predict the thermal conductivity in a periodic system composed of an indefinitely repeated configuration of spherical particles subjected to a temperature gradient. Based on OpenFOAM, a simulation platform is created with which the resolution requirements for accurate heat transfer predictions were inferred systematically. The approach is illustrated for unit cells containing either a single sphere or a configuration of two spheres. Asymptotic convergence rates confirming the second-order accuracy of the method are established in case the grid is fine enough to have eight or more grid cells covering the distance of the diameter of a sphere. Configurations with two spheres can be created in which small gaps remain between these spheres. It was found that even the under-resolution of these small gaps does not yield inaccurate numerical solutions for the temperature field in the domain, as long as one adheres to using eight or more grid cells per sphere diameter. Overlapping and (barely) touching spheres in a configuration can be simulated with high fidelity and realistic computing costs. This study further extends to examine the effective thermal conductivity of the unit cell, particularly focusing on the volume fraction of paraffin in cases with unit cells containing a single sphere. Finally, we explore the dependence of the effective thermal conductivity for unit cells containing two spheres at different distances between them.

Keywords: conjugate heat transfer; high-fidelity simulation; effective thermal conductivity; OpenFOAM; resolution requirements; periodic systems

1. Introduction

The prediction of effective heat transfer is of great importance in a wide range of applications, particularly in the study of composite materials used for thermal energy storage (TES) [1]. TES refers to a system that stores heat energy for later usage and is based on the working principles of sensible and latent heat. A TES system is a sustainable energy solution that is commonly referred to as a 'heat battery'. These systems play a critical role in efficiently storing and releasing thermal energy, thereby resolving the problem that energy is often generated and consumed at different moments in time [2].

The properties of composite materials can vary greatly due to their heterogeneous nature. In fact, the desired properties for TES materials are a combination of high thermal conductivity and a significant sensible and latent heat capacity. In this context, paraffin emerges as a particularly compelling candidate material because of its latent heat properties (150–250 kJ kg^{-1}) [3,4] and widespread availability. Moreover, paraffin undergoes phase transitions within temperature ranges (0 to 90 °C) [3] that correspond closely to the requirements for domestic heating, thereby substantially enhancing its applicability [1]. However, the thermal conductivity of around 0.2 W m^{-1}K^{-1} in the solid state, and even 0.08 W m^{-1}K^{-1} [5] in the liquid state, would imply an impractically slow response to the loading and unloading of such a heat battery. Therefore, paraffin spheres encapsulated in polymer aerogels to prevent leakage [6], in combination with highly conductive nano-fillers, e.g., graphene [7,8], presents itself as a promising candidate composite material. With such materials, both a large storage heat capacity as well as high heat transfer rates may be achieved. These materials undergo a phase change as part of the heat storage process and will be referred to as phase-change materials (PCMs).

Accurately predicting the temperature distribution in structured heterogeneous materials is essential for understanding their functioning in heat batteries [9,10]. For system design, it becomes imperative to study the parameters governing heat transport on a small scale in a heat battery composed of paraffin for storage and nano-fillers for enhanced heat transport. We present the development of a fundamental simulation model and determine the spatial resolution requirements that should be met in order to achieve accurate predictions of the temperature distribution inside the material.

The analysis of the conjugate heat transfer (CHT) is challenging when aiming for an analytical temperature solution for complex heterogeneous media. In such cases, a numerical method is the only viable approach [11,12]. To understand the macroscopic heat transfer properties of a material, we will treat it as a continuum. Fourier first introduced the concept of thermal conductivity at a macroscopic scale [13], which led to the early development of Effective Medium Theory models by Maxwell-Garnett [14]. Numerous models to approximate the effective heat conductivity have been proposed since, which are either empirical, numerical or a combination of both [15]. However, these models often have unknown limitations. To overcome these uncertainties, our study uses a numerical approach to solve the complete underlying model formulated in terms of the governing partial differential equations.

Numerical discretization methods, including the finite volume method (FVM), finite element method (FEM) and finite difference method (FDM), are utilized to solve partial differential equations (PDEs). We adopt OpenFOAM [16,17], an open-source FVM computational fluid dynamics (CFDs) package. This simulation platform provides an option to add new solvers and post-processing to address specific problems specifically. This approach is suitable for the current heat battery study as it allows for proven numerical methods to be combined with tailored solutions to the problem of heat transfer in a configuration with multiple spheres in a temperature gradient.

This research focuses on numerically solving conduction-driven CHT in heterogeneous media composed of periodic configurations of stacked spheres of paraffin. The governing heat equations are discretized using OpenFOAM version 10 [16,17], where the use is made of the CHTMULTIREGIONFOAM solver, which provides the numerical solution to diffusive heat transport in domains containing various materials. The primary objective is to determine the accuracy with which the solution can be obtained and at what computational costs. In particular, the convergence of the solution upon refinement of the spatial grid is focused on. Ultimately, the rate of convergence obtained by OpenFOAM in multi-region simulations is determined and, correspondingly, the necessary spatial resolution needed to achieve a desired level of accuracy is quantified. By doing so, the reliability of the numerical approach for simulations of the heat transfer in stacked spherical particles is determined. This is crucial for the future investigation of heat conduction in genuinely complex systems, specifically focusing on the effective thermal conductivity (ETC) within spatially extended systems of randomly stacked spheres. The ETC of a one- and a two-sphere system per unit cell is determined to quantify the resolution requirements and specify the dependence of the effective conductivity on system parameters, such as the radius of the spheres and separation between spheres. The current investigations establish the feasibility and simulation conditions that should be adopted in simulations of more complex general configurations that are adopted to be determine the effective heat transfer in a so-called Representative Elementary Volume (REV), i.e., a unit cell for the periodic domain that contains a large number of spheres in an arbitrary configuration.

The numerical investigations have established that OpenFOAM can yield high-fidelity solutions, provided that a sufficient spatial resolution is employed. Achieving engineering accuracy, with errors in the temperature field within a few percent, necessitates approximately eight grid cells per diameter of the paraffin spheres. Nearly full convergence was attained with resolutions ranging from 32 grid cells per diameter and beyond. For these spatial resolutions, the approach displayed second-order convergence. A comparison between the ETC predicted by the Maxwell-Garnett model [14] and the numerical solution revealed close agreement up to a paraffin volume fraction of 30%. Beyond this volume fraction, gradual deviations were seen of 5–10%, e.g., at a 40–50% volume fraction. The successful modeling of the temperature field effects arising from the proximity of multiple spheres was achieved using OpenFOAM. The basic CHTMULTIREGIONFOAM solver was found to accurately predict the heat transfer and ETC in general unit cells with two paraffin spheres, including two overlapping as well as just touching spheres. This supports the potential extension of the simulation approach to configurations with multiple spheres per unit cell. The identified resolution requirement of eight grid cells or more per sphere diameter was seen to be sufficient even for configurations comprising multiple spheres.

The organization of this paper is as follows. In Section 2, the physical model and its mathematical formulation are introduced. The OpenFOAM implementation is described in Section 3. The simulation results specifying the temperature field and convergence upon grid refinement are discussed in Section 4. Finally, the concluding remarks are presented in Section 5.

2. Physical Model and Governing Equations

In this section, we first present the physical model of a heat battery in Section 2.1. The mathematical formulation of the governing equations is discussed subsequently in Section 2.2.

2.1. Physical Model of Heat Battery

The TES materials for phase-change materials (PCMs) exhibit diverse microstructures, designed to enhance heat transfer and facilitate rapid storage and release. The heat batteries that motivate the current study exploit a multiscale structure in which a large block of porous metal foam ($O(10^{-1}$ m)) is used to transfer heat quickly and over comparably large distances. Within the pores ($O((10^{-3}-10^{-2})$ m) [10] of this foam, spheres of paraffin of a typical radius $r = O((10^{-5}-10^{-4})$ m) are stacked in a random configuration, available for sensible and latent heat storage. Figure 1 provides a 2D representation illustrating the stacked spherical particles of different sizes within the foam.

In Figure 1, it is apparent that a single pore comprises three materials with distinct properties. The paraffin spherical inclusions within the pore exhibit a low thermal conductivity of approximately 0.2 W m^{-1}K^{-1} in the solid state and 0.08 W m^{-1}K^{-1} in its liquid state [5]. This is tremendously small compared to, e.g., the copper from which the metal foam is composed, which has a thermal conductivity of around 400 W m^{-1}K^{-1} [18]. Furthermore, the pore is saturated with still air, characterized by an even lower thermal conductivity of 0.0265 W m^{-1}K^{-1} [19]. The thermal conductivity ratio of solid and fluid phases (κ_s/κ_l) for air-saturated metal foams is over 8000, indicating that the contribution of heat transfer by air can be largely neglected [19].

Motivated by Figure 1, we will consider approximate configurations to develop reliable computational methods for the simulation of the temperature field that develops when such a configuration is subjected to a steady temperature gradient (Dirichlet boundary) in the Z direction. To that end, we consider spatially periodic systems in the X-Y directions generated by repeating a suitable unit cell (periodic boundaries). In this paper, we focus on periodic unit cells with one or two paraffin spheres of the same diameter inside to study the numerical capturing of the effect of such inclusions. This generic problem corresponds to an approximate stacking of the spheres and enables precise numerical investigations.

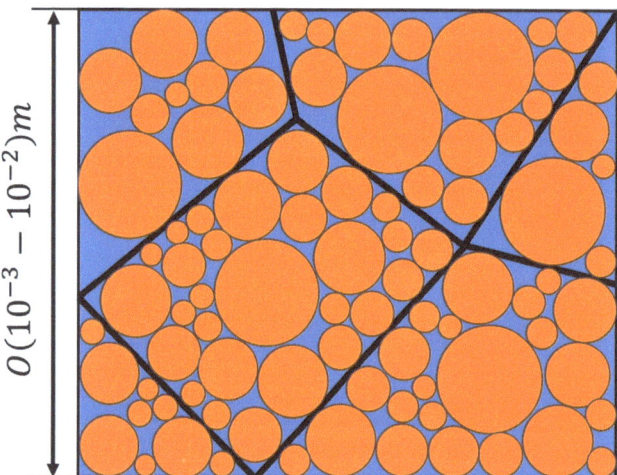

Figure 1. A two-dimensional representation of a TES microstructure composed of different size paraffin particles (orange) stacked in the pores of a foam (blue). The pores are bounded by a metal border, indicated symbolically by the thin black lines. In the actual porous metal foam, direct pathways connecting one pore with another are also contained—this is not included in the sketch.

After establishing the physical simulation domain, the PDE model for the heat transfer developing from a temperature gradient across the boundaries of the simulation box is specified next.

2.2. Mathematical Model

To accurately predict the thermal transport in composite domains, a CHT simulation provides a complete macroscopic model. This computational model enables the analysis of the contribution of conduction and convection mechanisms to the total heat transfer, consistently coupling all domains with appropriate interface conditions. We focus on process conditions that do not involve the melting of the paraffin—only heating and cooling are included at this stage. In this case, the heat battery problem considered needs to handle both the gaseous air (fluid domain) in the interstitial space left between the paraffin spheres, as well as the solid paraffin (solid domain). The system of partial differential equations (PDEs) governing the heat transfer in both the solid and fluid domains will be specified next. The numerical treatment of this model will be discussed in Section 3.1.

2.2.1. Conservation of Energy

The basic principle of conservation of energy can be expressed concisely in terms of the evolution of the specific total energy (per unit mass) e. Closely following [20], we can include all the relevant mechanisms for our problem. Taking into account the pressure and shear forces, as well as the force of gravity as a body force, we may apply the Reynolds transport theorem to the fundamental first law of thermodynamics [20] and arrive after some simplification at:

$$\frac{\partial(\rho e)}{\partial t} + \nabla \cdot (\rho \mathbf{u} e) = -\nabla \cdot \dot{\mathbf{q}}_S - \nabla \cdot (p\mathbf{u}) + \nabla \cdot (\boldsymbol{\tau} \cdot \mathbf{u}) + \rho \mathbf{g} \cdot \mathbf{u} + \dot{q}_V \qquad (1)$$

In this equation, $\mathbf{u}$ is the flow velocity, ρ denotes the material density, e stands for the specific total energy, p represents the pressure acting on the body, $\boldsymbol{\tau}$ is the viscous tensor and **g** represents gravity. Additionally, $\dot{q}_V$ accounts for the heat generated or destroyed per unit volume, while $\dot{\mathbf{q}}_S$ corresponds to heat transfer by diffusion, following Fourier's law:

$$\dot{\mathbf{q}}_S = -\boldsymbol{\kappa} \cdot \nabla T \tag{2}$$

Here, $\boldsymbol{\kappa}$ denotes the thermal conductivity matrix, and ∇T represents the gradient of the temperature field T. The thermal conductivity matrix $\boldsymbol{\kappa}$ is a material property that for anisotropic media takes the form [21]:

$$\boldsymbol{\kappa}(T) = \begin{pmatrix} \kappa_{11} & \kappa_{12} & \kappa_{13} \\ \kappa_{21} & \kappa_{22} & \kappa_{23} \\ \kappa_{31} & \kappa_{32} & \kappa_{33} \end{pmatrix} \tag{3}$$

To further understand the energy equation, we express the total energy in terms of the specific enthalpy, denoted as H, implying:

$$e = H - \frac{p}{\rho} + \frac{1}{2}\mathbf{u} \cdot \mathbf{u} \tag{4}$$

where **u** denotes the velocity field describing the motion of the medium. Enthalpy for Newtonian fluids in thermodynamic equilibrium can be considered a function of the pressure and temperature, i.e., $H = H(p, T)$, and can be evaluated using the standard equilibrium thermodynamic formula [22,23], which implies

$$dH = \left(\frac{\partial H}{\partial T}\right)_p dT + \left(\frac{\partial H}{\partial p}\right)_T dp = c_p dT + \left[v - T\left(\frac{\partial v}{\partial T}\right)_p\right] dp \tag{5}$$

Here, c_p represents the specific heat capacity at constant pressure, and v denotes the specific volume. Equation (5) describes a chemically inert system of fixed mass [23]. Combining Equation (1) with Equations (4) and (5), we can formulate the energy equation for an incompressible fluid as:

$$\frac{\partial}{\partial t}(\rho c_p T) + \nabla \cdot (\rho c_p \mathbf{u} T) = \nabla \cdot (\boldsymbol{\kappa} \cdot \nabla T) + \dot{q}_w \tag{6}$$

Equation (6) is a general equation where the source term $\dot{q}_w$ considers heating by shearing and pressure work. Equation (6) will be specified next for the solid and the fluid domain and solved later as a part of the total mathematical model.

2.2.2. Heat Transfer in Solid Domain

When dealing with solid domains, Equation (6) can be significantly simplified as there is no material flow and the density remains relatively constant with respect to the temperature. We can simplify the equation as follows:

$$\frac{\partial}{\partial t}(\rho c_p T) = \nabla \cdot (\boldsymbol{\kappa} \cdot \nabla T) + \dot{q}_V \tag{7}$$

The thermal conductivity of paraffin was examined in [5,24] and observed to be nearly isotropic and homogeneous, with variations in the heat conductivity of up to approximately 20% over a very wide temperature range of 300 to 650 K. Therefore, we make the simplifying assumption that the storage material can be treated as isotropic, homogeneous and with material properties that are independent of the temperature. This assumption enables expressing thermal conductivity as $\boldsymbol{\kappa} = \kappa \boldsymbol{I}$, where $\boldsymbol{I}$ is the identity matrix. The formulation in (7) can also be expressed in a non-dimensional form. In fact, upon introducing the reference time, length and temperature scales τ^*, L^* and T^* and assuming that ρ, c_p and κ are constant, we may write

$$\frac{\partial T^*}{\partial \tau^*} = \nabla^{*2}(T^*) + \dot{q}_V^* \tag{8}$$

in case the time scale and the forcing scale are chosen as

$$\tau^* = \frac{\kappa}{\rho c_p L^2} t \quad ; \quad \dot{q}_V^* = \frac{\kappa(T_{hot} - T_{cold})}{L^2} \dot{q}_V \tag{9}$$

where L is the given characteristic length. It is convenient to impose standardized temperature boundary conditions if one defines the dimensionless temperature as

$$T^* = \frac{T - T_{cold}}{T_{hot} - T_{cold}} \tag{10}$$

in terms of the temperatures T_{cold} and T_{hot} that define the temperature forcing of the system. Here, we use the same notation T for the dimensional and the non-dimensional formulation, as the difference is clarified by the context.

2.2.3. Heat Transfer in the Fluid Domain

Spheres arranged in a stack within a pore in the metal foam are enclosed by air. This air also contributes to the overall heat transfer. We approximate the air in the interstitial volume as an incompressible fluid and consider convection and diffusion as driving mechanisms. Correspondingly, the dynamics are governed by the continuity equation, the conservation of linear momentum and the temperature equation as specified above. Because the flow of air between the randomly stacked paraffin spheres is on a very small scale and subject to a modest temperature difference on the scale of the diameter of an individual sphere, the heat transfer is dominated entirely by diffusion. We substantiate this simplification next.

The problem of heat transfer by the air between the paraffin spheres is governed by the Rayleigh number Ra [22], which characterizes the phenomena of heat transfer for natural convection. For values below a critical Ra number, heat is transferred primarily through thermal conduction and the effects of natural convection are considered negligible. Ra is defined as

$$Ra = \frac{\beta g \Delta T L^3}{\nu \kappa} \tag{11}$$

Here, β, g, L, ν and κ represent the thermal expansion coefficient, gravitational acceleration, characteristic length, kinematic viscosity and thermal conductivity, respectively. Collecting typical values for these quantities [10], we observe the thermal expansion coefficient of air $\beta = 3.5 \times 10^{-3}$ K^{-1} [25]. Likewise, we recall that $g \approx 10$ m s^{-2} and take as the length scale for the interstitial air-filled domain the diameter of a paraffin sphere $L = 5 \times 10^{-6}$ m—this is likely to be an upper-bound for densely stacked spheres. The kinematic viscosity of air at room temperature is $\nu = 1.5 \times 10^{-5}$ m^2 s^{-1} and the thermal conductivity can be estimated at $\kappa \approx 2.6 \times 10^{-2}$ W m^{-1}K^{-1}. Adopting a very large temperature difference $\Delta T = 1$ K over a distance of the radius of a sphere of approximately 5 µm, we may estimate $Ra = O(10^{-14})$, i.e., we infer that only dissipative heat transport is of relevance here. Even in the case where one would consider an empty pore in the metal foam with a much larger characteristic length of $L = 5 \times 10^{-3}$ m, the Rayleigh number is found to be $Ra \approx O(10^{-5})$, i.e., much lower than the critical $Ra \approx O(10^2)$ for porous media [26]. Hence, the nonlinear convective transport is of little relevance here and diffusive transport dominates the heat transfer both in the solid and the fluid domain.

2.2.4. Interface and Boundary Conditions

The heat transfer problem we consider here is characterized by interface conditions that ensure (i) the continuity of the temperature and (ii) the continuity of the heat flux across the interface. We discuss these conditions in more detail next:

(i) Continuity of temperature: There is no temperature jump at the interface, meaning that the temperature when approaching the interface from one side is equal to the temperature when approaching the interface from the other side, i.e.,

$$(T)_{ij}(x^*) = (T)_{ji}(x^*) \qquad (12)$$

where $(T)_{ij}$ is the temperature at any point x^* on the interface between regions i and j, when approaching the interface from region i. Likewise, $(T)_{ji}$ is the temperature when approaching the same interface point x^* from region j.

(ii) Continuity of temperature flux: This condition ensures that the total heat flux density is continuous when crossing the interface between regions i and j, at any location x^*. This condition takes into account the thermal conductivity of each region:

$$\left((\kappa \cdot \nabla T) \cdot n\right)_{ij}(x^*) = \left((\kappa \cdot \nabla T) \cdot n\right)_{ji}(x^*) \qquad (13)$$

expressing the continuity of the normal component of the heat flux density at any location on the interface between regions i and j, irrespective of whether the interface is approached from region i or region j. Here, n denotes the normal vector on the interface at x^*.

2.2.5. Summary of Mathematical Model

The temperature distribution inside the domain consisting of air and paraffin is dominated by conduction in the parameter regime considered here. In the remainder of this paper, we will not consider explicit source terms. Hence, for region i, the problem is governed by:

$$\frac{\partial}{\partial t}(\rho c_p T)_i(x) = \nabla \cdot (\kappa \cdot \nabla T)_i(x) \qquad (14)$$

Periodic boundary conditions are used in the xy directions and the temperature is prescribed on the top and bottom of the unit cell in the z direction. Interface conditions with domain j are given by:

$$(T)_{ij}(x^*) = (T)_{ji}(x^*) \qquad (15)$$

for the continuity of the temperature and

$$\left((\kappa \cdot \nabla T) \cdot n\right)_{ij}(x^*) = \left((\kappa \cdot \nabla T) \cdot n\right)_{ji}(x^*) \qquad (16)$$

for the continuity of the heat flux across the interface.

3. Solver Description

In this section, we discuss the treatment of the governing heat equations in a finite volume framework as provided by OpenFOAM (Section 3.1). Moreover, the treatment of the adaptive meshing used for the accurate resolution of the finer details in the solution is presented (Section 3.2).

3.1. OpenFOAM Finite Volume Framework

OpenFOAM [16] is an open-source simulation platform for continuum mechanics. It utilizes the finite volume method (FVM) to discretize partial differential equations representing a wide range of physical phenomena [17]. We adopt OpenFOAM (version 10) in this study. OpenFOAM finds applications in diverse fields, including fluid dynamics, heat transfer and computational physics. By providing a comprehensive platform for numerical simulations, OpenFOAM enables the analysis of complex problems across various scientific and engineering disciplines.

The finite volume method (FVM) discretizes the computational domain into discrete control volumes, each representing a finite region within the domain. This numerical approach involves calculating fluxes across the faces of these control volumes and subsequently updating the values of the variables within each volume. In Figure 2, the conservation laws of a discrete volume V_c and fluxes (f_i) crossing through the discrete element walls are illustrated. This process ensures adherence to conservation laws at the discrete level. Consideration of a conservation equation for a general scalar variable ϕ is expressed as:

$$\underbrace{\frac{\partial(\rho\phi)}{\partial t}}_{\text{transient term}} + \underbrace{\nabla \cdot (\rho \mathbf{U} \phi)}_{\text{convective term}} = \underbrace{\nabla \cdot (\Gamma^\phi \nabla \phi)}_{\text{diffusion term}} + \underbrace{Q^\phi}_{\text{source/sink term}} \qquad (17)$$

where Γ^ϕ represents the diffusion coefficient of the ϕ property. Dropping the transient term in Equation (17) to simplify our discussion on how the FVM discretizes, and integrating over the element the volume V of an element C, yields:

$$\int_{V_C} \nabla \cdot (\rho \mathbf{U} \phi) dV = \int_{V_C} \nabla \cdot (\Gamma^\phi \nabla \phi) dV + \int_{V_C} Q^\phi dV \qquad (18)$$

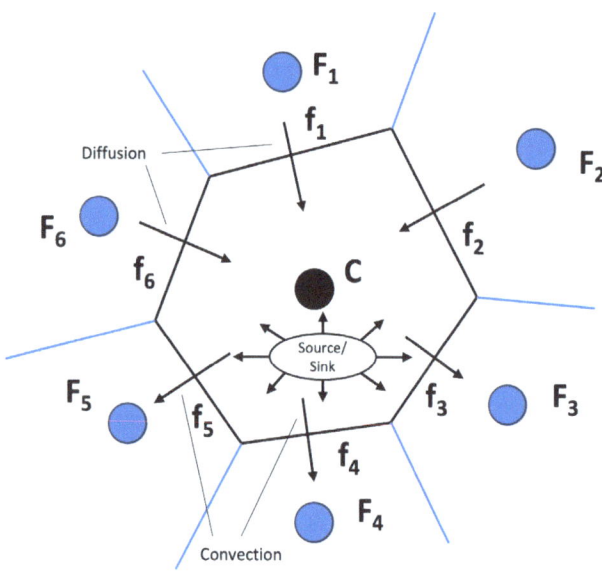

Figure 2. Conservation of a general scalar variable in a discrete element C of volume V_c.

Applying the divergence theorem to Equation (18) for both the convection and diffusive term and discretizing yields:

$$\sum_{faces(V_c)} \int_f (\rho \mathbf{U} \phi) \cdot d\mathbf{S} = \sum_{faces(V_c)} \int_f (\Gamma^\phi \nabla \phi) \cdot d\mathbf{S} + \int_{V_C} Q^\phi dV \qquad (19)$$

where $\mathbf{S}$ is the surface vector. Equation (19) expresses the conservative nature of the method, emphasizing that a surface integral must be resolved along the faces that constitute the volume V_c and a volume integral for the source Q^ϕ. In the FVM, a Gaussian quadrature is employed to numerically evaluate the surface integral with the fluxes crossing and the volume integral.

Figure 3 presents a grid cell in a structure, distinguishing between a uniform and an irregular example. The temperature Equation (8) is discretized on this mesh arrangement with the key variables stored in cell centers and at cell interfaces. For simulations involving stacked spherical particles in a domain, an unstructured mesh is well suited for accurately representing the geometry and capturing the physics of the system.

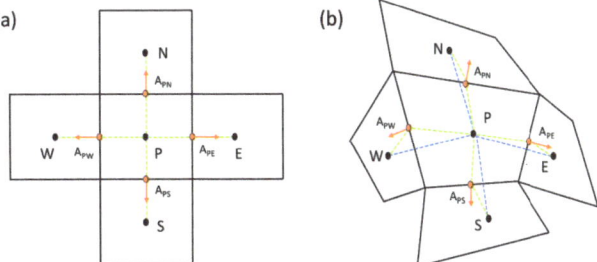

Figure 3. Cell in a structured uniform grid (**a**) and in a structured irregular grid (**b**).

The presence of multiple paraffin particles can be modeled in OpenFOAM by treating it as a multi-region case. OpenFOAM's chtMultiRegionFoam solver is purpose-built for simulating CHT problems involving general configurations. This solver enables the accurate modeling of heat transfer phenomena across different materials and regions, irrespective of whether these contain a fluid or a solid.

In structured grids, cells are arranged regularly, often in a Cartesian fashion, simplifying the identification of neighboring cells and facilitating interpolation and flux calculations. This regularity is particularly advantageous for simulations involving simple geometries and where a regular mesh can be easily generated. In contrast, unstructured grid cells lack a regular arrangement, offering flexibility in mesh generation but requiring more sophisticated algorithms for interpolation and flux calculations due to irregular cell shapes. Unstructured grids are highly recommended for complex geometries and situations where mesh generation may be challenging with a structured approach, such as the case of paraffin inclusions. However, a noteworthy consideration in unstructured grids is errors in the calculations in cases of highly irregular cell shapes.

3.2. Meshing

This study focuses on the heat transfer simulation in a stack of spherical particles. The fidelity of the simulation results depends on the spatial resolution and the quality of the meshing of the domain. To achieve high-quality meshing, the SNAPPYHEXMESH mesh generation tool in OpenFOAM was adopted. This tool is designed specifically for generating hexahedral (hex) and prismatic (wedge) meshes, suitable for subsequent numerical treatment with any of the OpenFOAM solvers [16].

In the context of mesh generation using SNAPPYHEXMESH, an initial background mesh comprising hexahedral cells covering the entire computational domain is established. To maintain simplicity, a structured mesh serves as the starting background, offering a well-defined foundation. Following this, a complex geometry surface within the computational domain is added, along with its corresponding boundaries. The SNAPPYHEXMESH utility then identifies the features on the specified surface, initiating an iterative process where the cells surrounding the surface gradually conform to its shape. It refines the mesh around specified geometries and according to the size of gradients, as illustrated in Figure 4. The process starts with a base mesh, and refinement is applied around the edges of the given geometry.

During this iterative process, systematic mesh refinement is applied to the newly defined geometry region, guided by the number of cells in the background mesh. If necessary, additional refinement may be introduced by splitting cells in specified regions. Subsequently, the geometry region is removed from the background mesh and introduced as a new region with independent properties. This comprehensive process ensures the generation of a high-quality hexahedral mesh that accurately represents the complex geometry within the computational domain. Parameters governing the adaptation process include cell size, surface feature refinement level, and surface curvature. Properly setting these parameters is crucial for achieving an optimal balance between mesh resolution and computational efficiency.

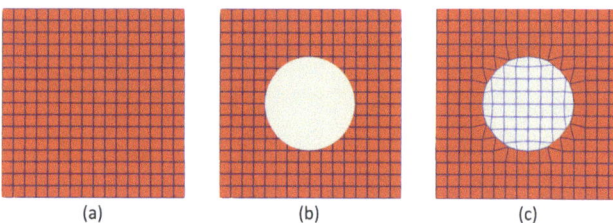

Figure 4. (**a**) Computation domain of structured cells constituting one region for the background mesh. (**b**) Structured background mesh with spherical inclusion surface. (**c**) New computational domain after SNAPPYHEXMESH consisting of two regions.

4. Convergence of Temperature Predictions upon Grid Refinement

In this section, we explore the resolution requirements that need to be met for an accurate simulation of heat transfer in periodic domains. Such a system is generated by repeating a unit cell in all three directions. We consider two types of periodically extended systems: (a) containing a single sphere per unit cell and (b) containing two spheres per unit cell. Apart from this difference, we may also consider periodic systems with different volume fractions of spheres in the system. For a single sphere per unit cell, we assess the convergence of the temperature field in terms of temperature profiles across selected lines and the corresponding convergence of the $L1$-norm of the error. For systems with two spheres per unit cell, we investigate in addition the resolution needed to resolve the total temperature field even in the case of spheres separated by very small distances and even overlap.

4.1. Convergence Study Setup

For an accurate simulation of conjugate heat transfer in configurations of stacked particles, a spatial resolution analysis is essential to determine the number of grid cells per unit cell needed to achieve a certain accuracy level. The periodic setting of stacked spheres mimics more general configurations of such spheres within the pores of a metal foam, as described in the Introduction.

We detail the simulation setup in a 2D representation for clarity next. The actual simulations are all conducted in 3D. In Figure 5a, a 2D sketch of the simulation domain is shown, divided into a large number of grid cells. Each grid cell is of size h^3 where the grid spacing is taken uniformly as $h = L/n$, with L the size of the periodic unit cell and n the number of grid cells along each coordinate direction. Figure 5b presents a single paraffin sphere with a diameter D embedded in air, and Figure 5c depicts two paraffin spheres separated by a specified distance H. In the case of two spheres per unit cell, quite general configurations are possible. To limit the convergence study, we consider two spheres directly above each other, sensing maximal thermal gradients. Configurations with the spheres at general relative placement will have somewhat smaller thermal gradients, making these less demanding for our resolution study—these are therefore omitted here. Future work is planned on several spheres in general relative configurations—preliminary investigations show that the current OpenFOAM approach is capable of addressing these problems as well.

The vertical direction will be identified as the z axis. In the z direction, a temperature difference is imposed, characterized by a temperature difference ΔT over the distance L. The steady temperature field T in the entire domain, including the spheres, is simulated and the accuracy of the predictions is quantified. For this study, we refine the spatial resolution using $n = 2^k$ grid cells per direction with $k = 1, 2, \ldots, 7$.

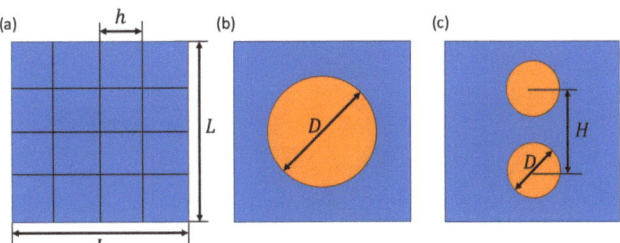

Figure 5. A two-dimensional representation of a periodic domain showing the following: (**a**) n^2 cells of h size composing a L^2 simulation box. (**b**) A single sphere with a diameter of D embedded within the simulation box. (**c**) Two spheres, each with a diameter D, whose centers are separated by a distance of H.

4.2. One Sphere per Unit Cell

The microstructure in the studied TES systems consists of multiple spherical particles of various sizes, densely packed together. To approximate the effective heat conductivity of such composite systems, we consider spatially extended periodic systems upon which a temperature gradient is imposed. In this subsection, we investigate periodic systems containing a single sphere per unit cell. This is a generic problem that is suitable to investigate numerical requirements for reaching a desired accuracy level.

The spatial resolutions that we consider are labeled in terms of the number of grid cells M that cover the diameter D of the sphere, i.e.,

$$M = \frac{D}{L/n} = \frac{D}{h} \tag{20}$$

in terms of the domain size L and number of grid cells n. Figure 6 illustrates a cross-cut displaying various resolutions at different M values. A temperature gradient directed from top to bottom is imposed. The steady temperature fields show a clear qualitative convergence with increasing M with coarse structures recognizable even as $M = 2$, as illustrated in Figure 7. This impression of convergence is quantified next.

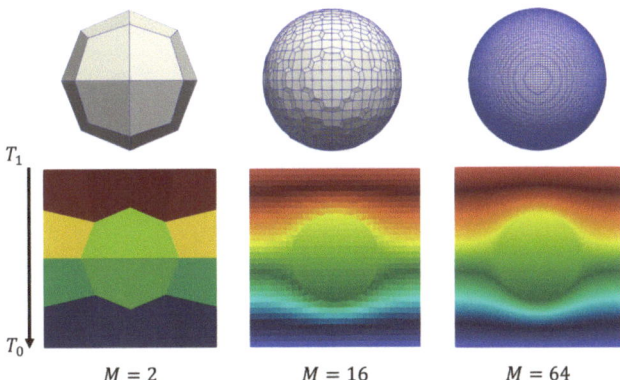

Figure 6. An embedded sphere in a periodic domain at different resolutions denoted by $M = D/h$. The predicted temperature fields T are shown below in terms of $T^* = (T - T_0)/(T_1 - T_0)$ in which T_0 and T_1 are the imposed temperatures on the bottom and top, respectively.

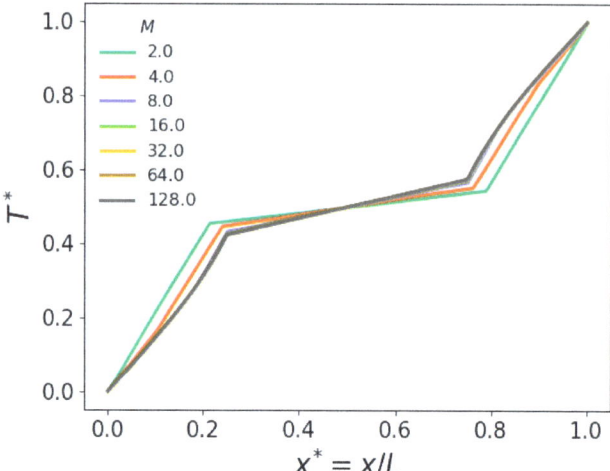

Figure 7. Temperature profile of sphere embedded in a cubic domain at different resolutions $M = D/h$.

To assess the convergence of the temperature predictions, we compare as a function of M the dimensionless, scaled temperature

$$T^* = \frac{T - T_0}{T_1 - T_0} \qquad (21)$$

where T_0 and T_1 are the imposed temperatures on the bottom and top of the periodic unit, respectively. We evaluate T^* along a vertical line through the middle of the sphere. The corresponding temperature profiles are depicted in Figure 8 showing quantitative convergence with nearly grid independence in the case $M \geq 8$.

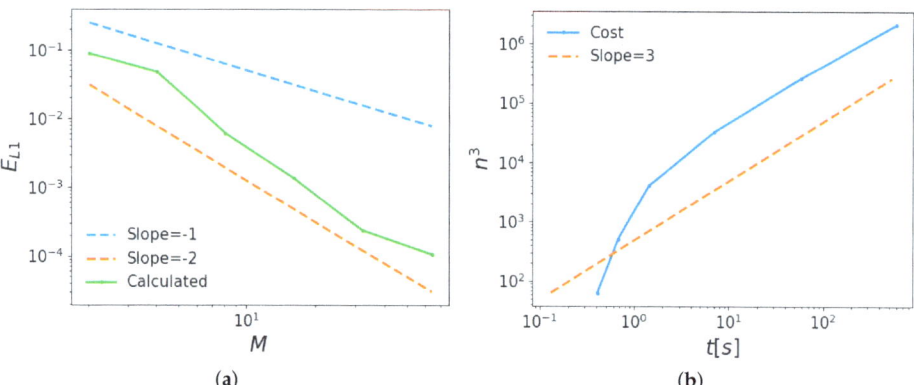

Figure 8. (**a**) The L_1—error for different spatial resolutions M. (**b**) The computational cost of the solver and SNAPPYHEXMESH time against the total number of cells.

To facilitate a further quantitative analysis of the convergence, we assess the L_1—error in the predicted temperature profiles using the result with $M = 128$ as the reference. We define the error as

$$E_{L_1}(M) = \frac{1}{n(M)} \sum_{i=1}^{n(M)} |e_i(M)| \tag{22}$$

where $n(M)$ denotes the total number of grid cells across the domain L at $M = D/h$. Moreover, the local error

$$e_i(M) = T^*(z_i(M)) - T^*(z_{i^*}(M_{ref})) \tag{23}$$

where $z_i(M)$ is the i-th grid point in the vertical grid corresponding to a selected value M and $z_{i^*}(M_{ref})$ is the corresponding grid point at $i = i^*$ in the $M_{ref} = 128$ grid in this study.

Figure 8a displays the convergence, indicating that indeed, for values of $M \geq 8$, convergence assumes asymptotic scaling equal to that of a second-order method. This was expected from the spatial discretization adopted in OpenFOAM employing Gaussian integration, which interpolates values from cell centers to face centers [16]. Finally, Figure 8b shows the increase in computational cost, which is defined as the time it takes for the complete mesh to be generated and the simulation to run, with increasing spatial resolution. We observe cubic scaling, verified by the power of three guiding lines.

In the next subsection, we consider the prediction of the temperature field in the case of two spheres per unit cell.

4.2.1. Effective Thermal Conductivity

In the preceding subsection, we introduced a parameter for the spatial resolution, denoted as M, and observed consistent second-order convergence beyond a value of $M = 8$.

The calculation of the effective thermal conductivity of the basic unit cell that contains the configuration of the paraffin spheres involves an examination of the overall thermal transport characteristics of the composite material, including its heterogeneity. In the numerical determination of ETC for a composite material, we assume steady conditions, which implies that the heat transfer across any plane at constant height z through the heterogeneous material remains constant. The total heat flow rate, denoted as $\dot{Q}$, flowing through a horizontal plane Γ at constant z is defined as follows:

$$\dot{Q} = \int_\Gamma dxdy \kappa(x,y,z) \partial_z T(x,y,z) \equiv \kappa_{eff} \frac{\Delta T}{L} A \tag{24}$$

Here, κ represents the local thermal conductivity, and $\partial_z T$ is the temperature in the z direction evaluated plane. This expression also introduces effective thermal conductivity, κ_{eff}, including the temperature difference ΔT across the vertical length of the simulation box L, and the simulation box plane area A. By rearranging this expression, we obtain for the ETC:

$$\kappa_{eff} = \frac{L}{A\Delta T} \int_\Gamma dxdy \kappa(x,y,z) \partial_z T(x,y,z) \tag{25}$$

In the steady state, the prediction of the ETC κ_{eff} is independent of the particular plane considered. We may also use this property to verify the numerical evaluation of the ETC. For convenience and accuracy, we exploit this definition only at planes that traverse the domain through air. Independence was also established for planes that traverse both air and paraffin.

Examining the dependence of κ_{eff} on the spatial resolution (M) is key for assessing what spatial resolutions are appropriate for reliable predictions. Figure 9 presents the temperature field (a) and the convergence of κ_{eff} for a paraffin sphere with a thermal conductivity of $\kappa_{paraffin} = 0.2$ W m^{-1}K^{-1} in the solid state [5] and surrounding air with $\kappa_{air} = 0.026$ W m^{-1}K^{-1} [27]. The observed accuracy of predicting ETC aligns well with the convergence of the underlying temperature field established earlier. In particular, also for κ_{eff}, sensible predictions are found beyond $M = 8$.

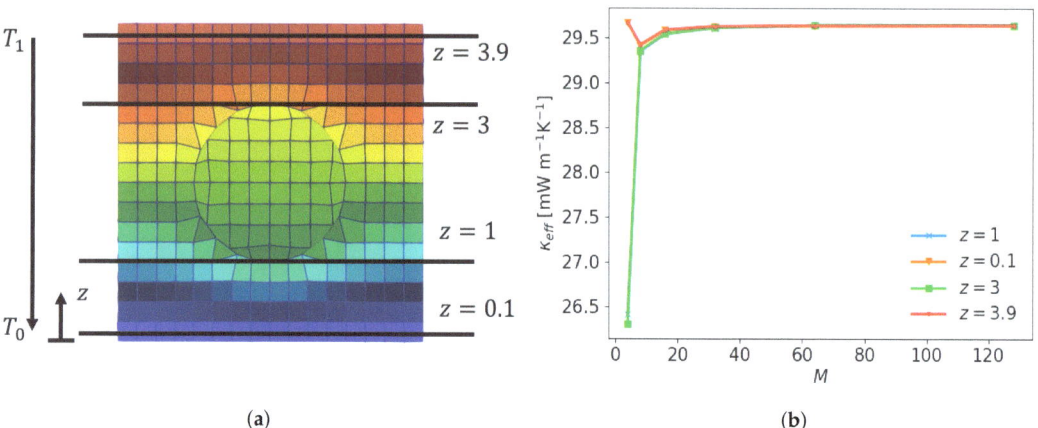

(a)   (b)

Figure 9. (a) Temperature field with an embedded sphere at $M = 16$ with different sample planes ($z = 0.1$, $z = 1$, $z = 3$ and $z = 3.9$) to determine κ_{eff} and (b) κ_{eff} at different spatial resolutions M evaluated on different sample planes.

4.2.2. Volume Fraction

Building upon the dependency of κ_{eff} on the spatial resolution M outlined in the preceding subsection, we consider the dependence of the effective thermal conductivity on the volume fraction of paraffin. Various constitutive (micro-mechanical) models [28] have been devised to explore the effective thermal conductivity of composite materials, with the Maxwell-Garnett [14] and Bruggeman [29] models as prominent examples. For unit cells containing a single sphere, effective heat conductivity is given by:

$$\kappa_{eff} = \kappa_m \left(1 + \frac{3\phi(\delta - 1)}{2 + \delta - \phi(\delta - 1)} \right) \quad (26)$$

The Maxwell-Garnett model, as depicted in Equation (26), offers a formulation for κ_{eff} [14]. Here, κ_m represents the thermal conductivity of the matrix material, while ϕ signifies the volume fraction of the filler material. The ratio between the thermal conductivity of the filler, denoted as κ_f, and that of the matrix is expressed as $\delta = \kappa_f / \kappa_m$. Notably, the Maxwell-Garnett model is tailored for spherical, non-overlapping particles, rendering it suitable for comparison with the simulation results.

Figure 10 presents a comparison between the numerical approximation and the Maxwell-Garnett model as a function of the volume fractions of the paraffin filler. Both approaches demonstrate a strong agreement, particularly up to a volume fraction of 30%. In these simulations, a constant spatial resolution of $M = 32$ per sphere diameter is upheld; note that this implies a growing computational cost with a reduced volume fraction—this posed no feasibility problem using OpenFOAM.

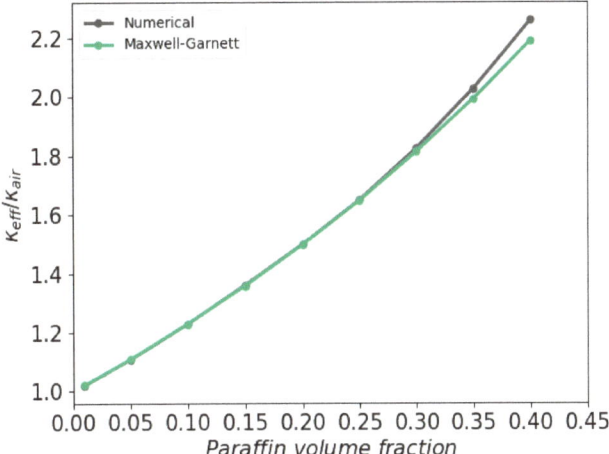

Figure 10. Effective thermal conductivity of binary mixture of still air and paraffin inclusions. The numerical approximation is compared with the Maxwell-Garnett model at a resolution of $M = 32$ per sphere diameter.

4.3. Two Spheres per Unit Cell

To further investigate the predictions for the temperature field when using a periodic model for extended systems, we next investigate unit cells that contain two spheres. These spheres can be in any relative configuration inside the basic unit cell, which poses different challenges to the numerical method. We consider two extreme situations:

1. Horizontal. If the two spheres are aligned horizontally, i.e., the line through the centers of the spheres lies in a constant z plane, the temperature gradient experienced by the spheres would be quite similar to the temperature gradient experienced by a single sphere. This is particularly true if the two spheres are separated far enough, making their mutual interactions diminish. This situation was already studied in the previous subsection.
2. Vertical. If the two spheres are aligned vertically, i.e., the line through the centers of the spheres is in the z direction, the temperature gradient experienced by each of the two spheres differs most from the single-sphere case. Moreover, the gradients seen in this configuration are the largest among the different configurations. Therefore, this configuration will be studied in this subsection.

Particles stacked within the unit cell may not always be well separated from each other. This mimics the situation when multiple spheres are stacked inside a pore of the metal foam in which a range of relative configurations may be expected. Therefore, we investigate the implications of different distances between the centers of the spheres on the predicted temperature field. We include overlapping, touching and separated configurations and consider the convergence of the corresponding solution upon grid refinement. Figure 11 illustrates four distinct cases, each characterized by different distance ratios $S = H/D$, measuring the distance between the centers of the spheres in units of the diameter of the spheres D, cf. Figure 5c. In terms of S, we observe that $0 \leq S \leq 1$ corresponds to partially overlapping spheres and $S > 1$ denotes separated spheres that, in principle, would allow for a grid fine enough to resolve the distance between the surfaces of the spheres with a number of grid cells.

Figure 12 illustrates two spheres separated by a small distance, corresponding to $S = 1.05$. In case a coarse mesh ($M = 4$) is used, SnappyHexMesh generates a computational grid that does not resolve the distance between the spheres but rather forms a dumbbell shape. As the spatial resolution increases, this artificial contact area diminishes until the gap is fully resolved at a sufficiently high resolution of $M = 32$. Although there are clear differences in the way the geometry is resolved at different resolutions, the main question of course is how such differences affect the prediction of the temperature field. We turn to this next.

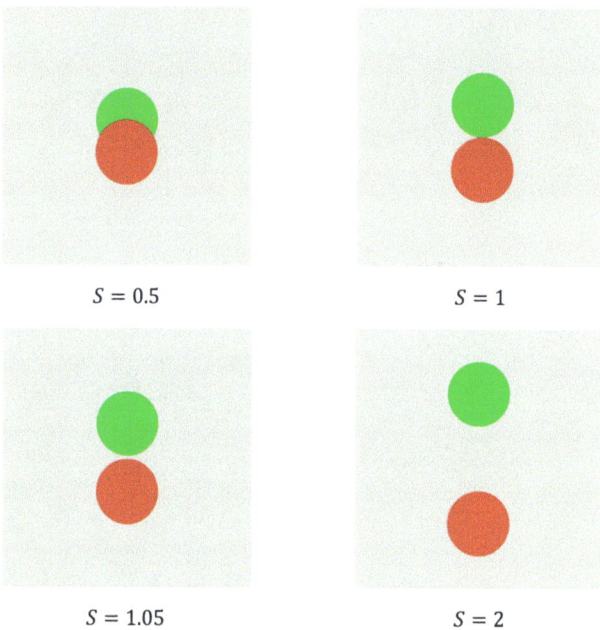

Figure 11. Two-sphere cases, illustrated as a red and green sphere of the same material, for different separations measured in terms of $S = H/D$, expressing the distance between the centers of the two spheres in units D.

Figure 12. SnappyHexMesh refinement for spheres, illustrated as a red and green sphere of the same material, separated by a small gap ($S = 1.05$) for different resolutions ($M = 4$, $M = 8$ and $M = 32$).

The cases depicted in Figure 11 ($S = 0.5, S = 1, S = 1.05$ and $S = 2$) have been simulated at various resolutions M. The corresponding temperature profiles are presented in Figure 13. We observe a characteristic convergence of the temperature profiles with an increasing resolution as already presented for unit cells containing a single sphere only. Again, for $M \geq 8$, good general agreement with the grid-independent solution is observed, where it is understood that the value of M refers to the number of grid cells across the diameter of a sphere. This value of M also appeared for the single-sphere case, suggesting that the interaction between the spheres in terms of the spatial temperature distribution is rather modest and no particularly strong gradients emerge in the two-sphere configuration. Finally, for the case $S = 1.05$ in which a small gap is present, even when not fully resolving the gap, the solutions are close to the fully resolved gap reference simulation.

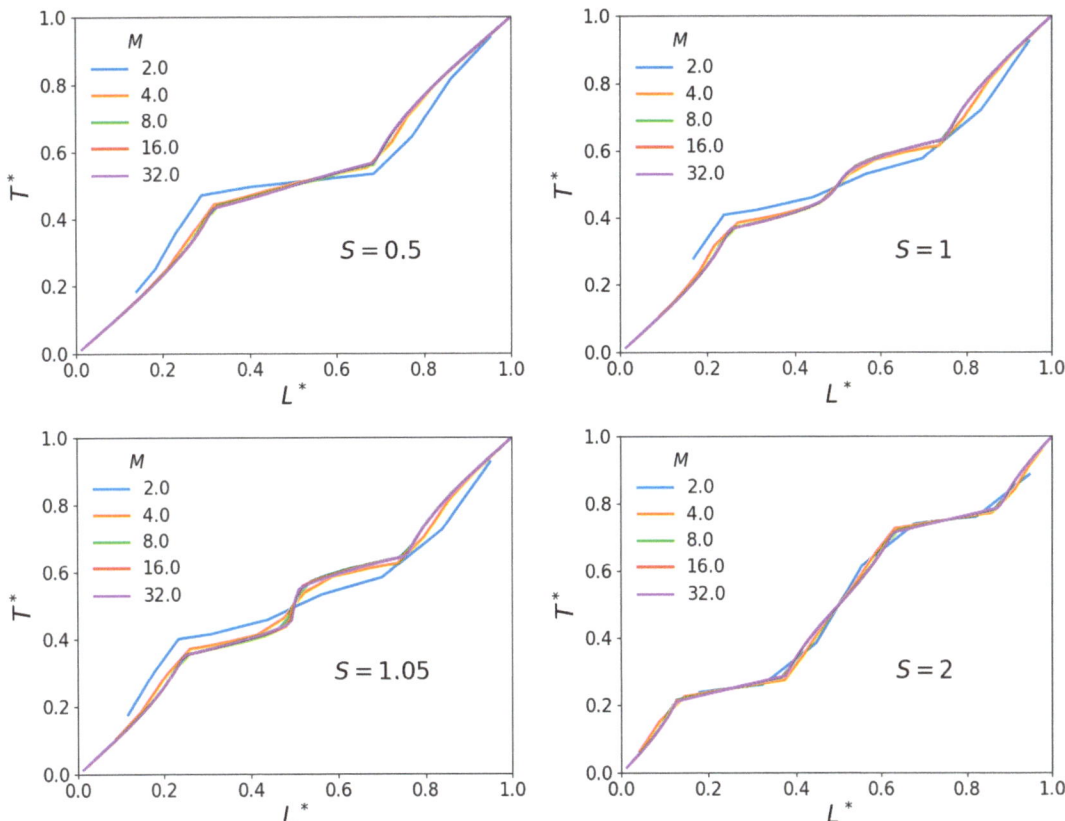

Figure 13. Vertical temperature profiles for various separations S and spatial resolutions M.

Effective Thermal Conductivity

The study of the ETC of a two-sphere paraffin system embedded in air is considered next as a function of the distance (S) between the sphere centers.

Figure 14 displays the dependence of κ_{eff} on the inter-sphere distance S. In this illustration, we used spheres of half the radius as used for the single-sphere case above. At $S = 0$, the predicted value therefore does not agree with the single-sphere case considered earlier. The two-sphere systems show a value of $\kappa_{eff} = 0.0263$ W m^{-1}K^{-1} in the fully overlapping case at $S = 0$. With increasing S, the effective heat conductivity increases and reaches a maximum near $S = 1$. In this configuration, the path along which the heat is transported is for a large extent contained in paraffin for which the heat conductivity is larger than in the surrounding air. For larger separations, the value of κ_{eff} reduces again to reach a plateau corresponding to two independent paraffin spheres. The simulation method appears to yield accurate predictions that can be used to define upscale theory as is considered in homogenization models.

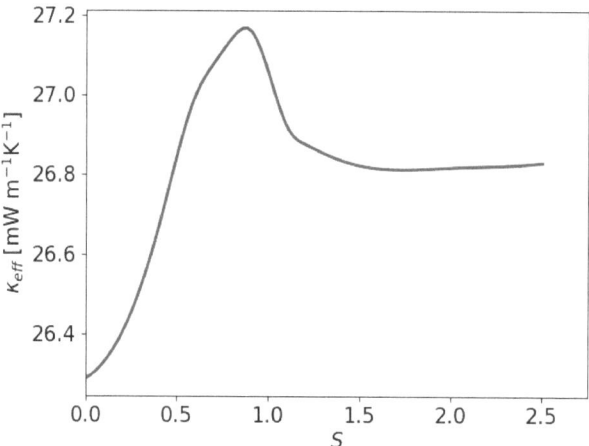

Figure 14. ETC of a vertically aligned two-sphere system at a spatial resolution of $M = 16$ as a function of the distance S between the sphere centers.

5. Conclusions

In this paper, we developed a simulation method with which heat transfer in structured heterogeneous media can be simulated. The heterogeneous medium is meant to represent in detail the working material in a future thermal battery. Specifically, one may think of spheres composed of paraffin, although the method developed here is general. The working material contains a system of spheres placed in a certain configuration, which is repeated periodically. The corresponding unit cell can have several such spheres inside to represent actual stacked spheres in a realistic domain.

The approach is implemented in OpenFOAM, using second-order finite volume discretization. The full conjugate heat transfer problem of a periodic system was addressed, in which a unit cell is repeated indefinitely in all three directions, subject to a steady temperature gradient. The heat transfer in the case of a unit cell with only one sphere inside was considered in a grid refinement study. Visually, rapid convergence was appreciated upon increasing the spatial resolution, which could be recognized in detail to be of second-order accuracy. In fact, on grids with $M \geq 8$ grid cells per diameter D of the spheres, good engineering accuracy was observed, yielding high-fidelity results upon further refinement. The computational effort was seen to scale as n^3 where n is the number of grid cells in each coordinate direction. The computational effort is sufficiently low to enable the simulation of extensive periodic models of the composite material.

Further examples of this problem were investigated by considering unit cells with two spheres inside. Grid refinement showed second-order convergence also in this case. Moreover, in terms of the separation parameter $S = H/D$, we simulated two-sphere problems with overlap ($0 \leq S \leq 1$) as well as without overlap ($S > 1$). Even in cases where possible small gaps between the two spheres would be smaller than the grid spacing h, the grid refinement showed continuous improvement upon increasing the resolution, with solutions that are very close to the eventual grid-independent solution. Hence, it appears that the under-resolution of tiny details in a complex stacking of spheres is not leading to large inaccuracies in the temperature field and the corresponding thermal transport.

The simulation method developed here was also illustrated in terms of the effective thermal conductivity κ_{eff}. We observed that at spatial resolutions $M \geq 8$ per sphere diameter, the effective conductivity can be computed reliably. This method can hence provide a basis for homogenization approaches to upscale the model to much larger systems. As an example, we calculated κ_{eff} as a function of the volume fraction of paraffin filler and compared this with the Maxwell-Garnett constitutive model. The numerical approximation closely mirrors the Maxwell-Garnett model up to a 30% volume fraction of paraffin filler. This correspondence diminishes for yet higher volume fractions, as the periodic boundaries imply that the temperature distribution around the paraffin spheres can no longer be described as independent of that around nearby spheres. Our method was also adopted to compute the κ_{eff} of a system of two spheres at different distances S. The two-sphere system revealed distinctive trends in the effective thermal conductivity. In fact, when going from overlapping to non-overlapping configurations, a peak ETC is observed slightly below $S = 1$, attributed to the longer paraffin thermal pathway with a higher heat conductivity compared to the embedding air. These findings contribute to the data-driven upscaling of heat transfer models in truly complex systems of polymer composite materials.

The new model developed in OpenFOAM will be extended to systems with coated spheres, with which it will become possible to further improve the heterogeneous material by assigning the coating to increase the overall heat transfer rate and increase the loading and unloading of the core of the multiple spheres storing heat effectively in large quantities. Specifically, paraffin spheres coated with graphene form an important example of such composite materials. This type of extension is currently under investigation—the results will be published elsewhere. The application of the new approach to extended systems requires high-performance computations, which is well possible based on OpenFOAM. In fact, a possible grid of $n = 1024$ and a resolution per diameter of $M = 8$–16 would enable simulations of extended systems with 128^3–64^3 spheres in a regular stack. This large-scale modeling forms the basis of future homogenization approaches that will enable the analysis of systems of realistic size and complexity.

Future research is devoted to effects due to variations in physical parameters, such as the volume fraction of the spherical inclusions. This aims to study the effect of changes in the physical system on the effective thermal conductivity. Additionally, the model will be extended by adding a thin coating composed of a material with very high thermal conductivity. A particular example would be the coating of paraffin spheres with graphene, thereby combining the fast and slow transport of heat in the system needed to realize particular designs for thermal batteries. Finally, after having established the resolution requirement for a single sphere and a pair of spheres, we will develop simulation methodologies that can handle large numbers of spheres (multi-spheres) touching each other. This would correspond closely to the situation motivated by Figure 1 and lead the way to realistic configurations. The detailed exploration and findings of these three studies will be presented elsewhere.

Author Contributions: Conceptualization, B.J.G. and K.A.R.L.; methodology, K.A.R.L.; formal analysis, K.A.R.L.; writing—original draft preparation, K.A.R.L.; writing—review and editing, B.J.G. and A.L.; visualization, K.A.R.L.; supervision, B.J.G. and A.L.; funding acquisition, B.J.G. and A.L. All authors have read and agreed to the published version of this manuscript.

Funding: This investigation is a part of the research initiative titled 'Nanofiller-enhanced wax for heat storage (Wax+)', identified by project number 18052 within the Open Technology Programme. The initiative receives (partial) funding from the Dutch Research Council (NWO), specifically under the Applied and Engineering Sciences (TTW) domain. Computations were performed at the Dutch national computing center Surf Sara, made possible by the 'Multiscale Modeling and Simulation' computing project, funded by NWO.

Institutional Review Board Statement: Not applicable.

Data Availability Statement: The raw data from the simulations are available from K.A. Redosado Leon.

Acknowledgments: The authors would like to thank H. Friedrich and L. Wijkhuijs from the Chemical Engineering and Chemistry Department and M. Boomstra from the Multiscale Simulations of Polymer Dynamics at TU/e for their fruitful collaboration.

Conflicts of Interest: The authors declare no conflicts of interest.

Abbreviations

The following abbreviations are used in this manuscript:

CFD	Computational Fluid Dynamics
CHT	Conjugate Heat Transfer
ETC	Effective Thermal Conductivity
FEM	Finite Element Method
FVM	Finite Volume Method
PDE	Partial Differential Equation
PCM	Phase Change Material
TES	Thermal Energy Storage
REV	Representative Elementary Volume

References

1. Du, K.; Calautit, J.; Wang, Z.; Wu, Y.; Liu, H. A review of the applications of phase change materials in cooling, heating and power generation in different temperature ranges. *Appl. Energy* **2018**, *220*, 242–273. [CrossRef]
2. Ahmad, T.; Zhang, D. A critical review of comparative global historical energy consumption and future demand: The story told so far. *Energy Rep.* **2020**, *6*, 1973–1991. [CrossRef]
3. Pereira Da Cunha, J.; Eames, P. Thermal energy storage for low and medium temperature applications using phase change materials—A review. *Appl. Energy* **2016**, *177*, 227–238. [CrossRef]
4. Sharma, A.; Tyagi, V.; Chen, C.; Buddhi, D. Review on thermal energy storage with phase change materials and applications. *Renew. Sustain. Energy Rev.* **2009**, *13*, 318–345. [CrossRef]
5. Soares, N.; Matias, T.; Durães, L.; Simões, P.; Costa, J. Thermophysical characterization of paraffin-based PCMs for low temperature thermal energy storage applications for buildings. *Energy* **2023**, *269*, 126745. [CrossRef]
6. Xie, L.; Wu, X.; Wang, G.; Shulga, Y.M.; Liu, Q.; Li, M.; Li, Z. Encapsulation of Paraffin Phase-Change Materials within Monolithic MTMS-Based Silica Aerogels. *Gels* **2023**, *9*, 317. [CrossRef] [PubMed]
7. Li, M. A nano-graphite/paraffin phase change material with high thermal conductivity. *Appl. Energy* **2013**, *106*, 25–30. [CrossRef]
8. Nazari, N.; Bahramian, A.R.; Allahbakhsh, A. Thermal storage achievement of paraffin wax phase change material systems with regard to novolac aerogel/carbon monofilament/zinc borate form stabilization. *J. Energy Storage* **2022**, *50*, 104741. [CrossRef]
9. Li, Z.; Yang, Y.; Gariboldi, E.; Li, Y. Computational models of effective thermal conductivity for periodic porous media for all volume fractions and conductivity ratios. *Appl. Energy* **2023**, *349*, 121633. [CrossRef]
10. Dinesh, B.V.S.; Bhattacharya, A. Comparison of energy absorption characteristics of PCM-metal foam systems with different pore size distributions. *J. Energy Storage* **2020**, *28*, 101190. [CrossRef]
11. Lopez Penha, D.; Stolz, S.; Kuerten, J.; Nordlund, M.; Kuczaj, A.; Geurts, B. Fully-developed conjugate heat transfer in porous media with uniform heating. *Int. J. Heat Fluid Flow* **2012**, *38*, 94–106. [CrossRef]
12. Lopez Penha, D.; Geurts, B.; Stolz, S.; Nordlund, M. Computing the apparent permeability of an array of staggered square rods using volume-penalization. *Comput. Fluids* **2011**, *51*, 157–173. [CrossRef]
13. Fourier, J.B.J. *Théorie Analytique de la Chaleur*; Chez Firmin Didot, Père et Fils: Paris, France, 1822.
14. Garnett, J.C.M.; Larmor, J. Colours in metal glasses and in metallic films. *Proc. R. Soc. Lond.* **1904**, *73*, 443–445. [CrossRef]
15. Wang, M.; Pan, N. Predictions of effective physical properties of complex multiphase materials. *Mater. Sci. Eng. R Rep.* **2008**, *63*, 1–30. [CrossRef]
16. Greenshields, C. *OpenFOAM v10 User Guide*; The OpenFOAM Foundation: London, UK, 2022.
17. Weller, H.G.; Tabor, G.; Jasak, H.; Fureby, C. A tensorial approach to computational continuum mechanics using object-oriented techniques. *Comput. Phys.* **1998**, *12*, 620–631. [CrossRef]
18. Meng, X.; Yan, L.; Xu, J.; He, F.; Yu, H.; Zhang, M. Effect of porosity and pore density of copper foam on thermal performance of the paraffin-copper foam composite Phase-Change Material. *Case Stud. Therm. Eng.* **2020**, *22*, 100742. [CrossRef]
19. Yang, X.H.; Bai, J.X.; Yan, H.B.; Kuang, J.J.; Lu, T.J.; Kim, T. An Analytical Unit Cell Model for the Effective Thermal Conductivity of High Porosity Open-Cell Metal Foams. *Transp. Porous Media* **2014**, *102*, 403–426. [CrossRef]
20. Moukalled, F.; Mangani, L.; Darwish, M. *The Finite Volume Method in Computational Fluid Dynamics: An Advanced Introduction with OpenFOAM® and Matlab*, 1st ed.; Number 113 in Fluid Mechanics and Its Applications; Springer International Publishing: Cham, Switzerland, 2016. [CrossRef]
21. Hahn, D.W. *Heat Conduction*, 3rd ed.; Wiley: Hoboken, NJ, USA, 2012.
22. Bird, R.B.; Stewart, W.E.; Lightfoot, E.N. *Transport Phenomena*, 2nd ed.; Wiley: New York, NY, USA, 2007.
23. Silbey, R.J.; Alberty, R.A.; Bawendi, M.G. *Physical Chemistry*, 4th ed.; Wiley: Hoboken, NJ, USA, 2005.
24. Li, Z.; Gariboldi, E. Review on the temperature-dependent thermophysical properties of liquid paraffins and composite phase change materials with metallic porous structures. *Mater. Today Energy* **2021**, *20*, 100642. [CrossRef]

25. Dixon, J.C. *The Shock Absorber Handbook*, 2nd ed.; Wiley-Professional Engineering Publishing Series; John Wiley: Chichester, UK, 2007.
26. Shiina, Y.; Hishida, M. Critical Rayleigh number of natural convection in high porosity anisotropic horizontal porous layers. *Int. J. Heat Mass Transf.* **2010**, *53*, 1507–1513. [CrossRef]
27. Thomas, L.C. *Heat Transfer*; Prentice Hall: Englewood Cliffs, NJ, USA, 1992.
28. Chen, H.; Ginzburg, V.V.; Yang, J.; Yang, Y.; Liu, W.; Huang, Y.; Du, L.; Chen, B. Thermal conductivity of polymer-based composites: Fundamentals and applications. *Prog. Polym. Sci.* **2016**, *59*, 41–85. [CrossRef]
29. Bruggeman, D.A.G. Berechnung verschiedener physikalischer Konstanten von heterogenen Substanzen. I. Dielektrizitätskonstanten und Leitfähigkeiten der Mischkörper aus isotropen Substanzen. *Ann. Phys.* **1935**, *416*, 636–664. [CrossRef]

Disclaimer/Publisher's Note: The statements, opinions and data contained in all publications are solely those of the individual author(s) and contributor(s) and not of MDPI and/or the editor(s). MDPI and/or the editor(s) disclaim responsibility for any injury to people or property resulting from any ideas, methods, instructions or products referred to in the content.

Article

A Precise Prediction of the Chemical and Thermal Shrinkage during Curing of an Epoxy Resin

Jesper K. Jørgensen [1,*], Vincent K. Maes [2], Lars P. Mikkelsen [1,*] and Tom L. Andersen [1]

1 Department of Wind and Energy Systems, Technical University of Denmark, 4000 Roskilde, Denmark
2 Bristol Composites Institute, University of Bristol, University Walk, Bristol BS8 1TR, UK; vincent.maes@bristol.ac.uk
* Correspondence: jkjjo@dtu.dk (J.K.J.); lapm@dtu.dk (L.P.M.)

Abstract: A precise prediction of the cure-induced shrinkage of an epoxy resin is performed using a finite element simulation procedure for the material behaviour. A series of experiments investigating the cure shrinkage of the resin system has shown a variation in the measured cure-induced strains. The observed variation results from the thermal history during the pre-cure. A proposed complex thermal expansion model and a conventional chemical shrinkage model are utilised to predict the cure shrinkage observed with finite element simulations. The thermal expansion model is fitted to measured data and considers material effects such as the glass transition temperature and the evolution of the expansion with the degree of cure. The simulations accurately capture the exothermal heat release from the resin and the cure-induced strains across various temperature profiles. The simulations follow the experimentally observed behaviour. The simulation predictions achieve good accuracy with 2–6% discrepancy compared with the experimentally measured shrinkage over a wide range of cure profiles. Demonstrating that the proposed complex thermal expansion model affects the potential to minimise the shrinkage of the studied epoxy resin. A recommendation of material parameters necessary to accurately determine cure shrinkage is listed. These parameters are required to predict cure shrinkage, allow for possible minimisation, and optimise cure profiles for the investigated resin system. Furthermore, in a study where the resin movement is restrained and therefore able to build up residual stresses, these parameters can describe the cure contribution of the residual stresses in a component.

Keywords: cure shrinkage; epoxy; finite element; UMAT; thermal expansion; volumetric shrinkage

Citation: Jørgensen, J.K.; Maes, V.K.; Mikkelsen, L.P.; Andersen, T.L. A Precise Prediction of the Chemical and Thermal Shrinkage during Curing of an Epoxy Resin. *Polymers* **2024**, *16*, 2435. https://doi.org/10.3390/polym16172435

Academic Editors: Alexey V. Lyulin and Valeriy V. Ginzburg

Received: 13 August 2024
Revised: 20 August 2024
Accepted: 24 August 2024
Published: 28 August 2024

Copyright: © 2024 by the authors. Licensee MDPI, Basel, Switzerland. This article is an open access article distributed under the terms and conditions of the Creative Commons Attribution (CC BY) license (https://creativecommons.org/licenses/by/4.0/).

1. Introduction

Residual stresses in cured thermosets like epoxies, polyesters, and polyurethanes are inherent to the curing process. Residual stresses can lead to unwanted warpage, which can lead to issues during assembly, for instance in wind turbine blade the root sections [1]. Furthermore, crack growth, tunnelling cracks, and delamination are affected by residual stresses in composites [2]. In some scenarios, reduced mechanical performance is observed, i.e., the fatigue behaviour is reduced [3,4].

These cure-induced residual stresses can be directly coupled to the cure-induced strains from combined thermal and chemical shrinkage [5]. The chemical shrinkage is known to be related to the volumetric change the thermoset undergoes due to the polymerisation [6]. Bogetti and Gillespie [6] proposed a linear relationship between the degree of cure and the volumetric shrinkage. This was experimentally confirmed by Shah and Schubel [7] and Khoun et al. [8], using a rheometer to quantify the shrinkage. Later, a non-linear model was proposed [9] allowing for a more complex shrinkage behaviour of thermosets. This non-linear shrinkage was later observed by measuring the volumetric shrinkage with a gravimetric/dilatometric setup by Li et al. [10]. Recently, experimental observations showed that the volumetric shrinkage determined using density measurements could be used to estimate the chemical shrinkage within reasonable accuracy, applying a linear fit [5]. Even though there has been a lot of research in this field, it remains unclear which approach yields the most accurate relationship between the degree of cure and the volumetric shrinkage.

Several phenomenological models exist to quantify the cure development, or the degree of cure as it is often referred to [11,12]. Some of these models capture only the kinetics and reaction patterns of the thermoset mixture [11], while others include the diffusion-controlled behaviour stemming from the influence of the glass transition temperature on the reaction rate [12]. The most well-established model for the evolution of the glass transition temperature in relationship with the degree of cure was coined by Dibenedetto [13]. As this transition occurs, thermosets are known to suffer significant changes in properties and behaviour [14], influencing the cure shrinkage.

Thermal shrinkage has also been a topic of substantial research over the years, with a common method for measuring thermal expansion being Thermal Mechanical Analysis (TMA) [8]. Previous studies have used this method to investigate glassy polymers, including both thermosetting and thermoplastic systems [8,15,16]. One study of an epoxy system using TMA [8] proposed a model for the non-linearity of thermal expansion approaching and crossing glass transition temperature for cured samples. The same study demonstrated that the thermal expansion is a function of the degree of cure above the glass transition temperature. Studies have shown that thermal expansion could also be measured with Dynamic Mechanical Analysis [15] and that glassy polymers can depend on heating and cooling [15,16]. Korolev et al. [16] showed that the difference in thermal expansion coefficient between heating and cooling could lead to a variation in the strain observed, thus demonstrating that thermal expansion in thermosets is complex.

This current work studies the cure shrinkage of a neat (i.e., without fibres) epoxy resin type typically used in wind turbine blade manufacturing. A cure kinetic and glass transition temperature model are used based on parameters found for a specific resin system investigated by Jørgensen et al. [17]. These models are used in experimental trials where the resin is free to contract and expand, with no external loads applied. The experimental method applied was proposed by Mikkelsen et al. [5]. This method uses fibre optic sensors with Fibre Bragg Gratings (FBG) similar to that used in other studies [3–5,18–20]. A conventional model describing the evolution of the chemical shrinkage with the degree of cure is investigated in terms of how it reflects the chemical shrinkage observed in experiments. In addition, a novel complex thermal expansion model is proposed to relate thermal effects to the measured cure shrinkage. This material behaviour is implemented into a simulation framework that considers the chemical and thermal shrinkage as the governing constituents used by Jørgensen and Mikkelsen [21]. In the end, a procedure will be delivered to accurately predict the shrinkage of thermosetting epoxy resins and allow for realistic minimisation in reducing residual stresses.

2. General Theory

This section describes the equations and models applied for both experimental and numerical aspects of this study.

2.1. Cure Kinetic Model

A cure kinetic model [12] accounting for the interaction between the glass transition temperature, T_g, and the evolution of the degree of cure, X, is used to account for the reduced molecular mobility effect by diffusion control into the cure predictions, resulting in a model defined as:

$$\frac{dX}{dt} = \frac{K(T)X^m(1-X)^n}{1+\exp[C(X-X_c(T))]}, \quad (1)$$

$$K(T) = A\exp\left(-\frac{e_a}{RT}\right),$$

$$X_c(T) = X_{cT}T + X_{c0}.$$

The model incorporates the Arrhenius reaction equation, $K(T)$, which consists of the preexponential factor, A, the activation energy, e_a, and the universal gas constant, R, and is commonly used to model the cure behaviour in epoxy resin systems. Furthermore, n and m are power law coefficients and in the denominator, the model is described by an exponential function which captures the reduction in rate of cure caused by the reduced molecular mobility at higher degrees of cure. This behaviour is governed by the diffusion constant, C, which captures how abruptly the cure reaction slows down and the critical degree of cure, X_c, which captures the degree of cure at which the polymer chains length and cross-links begin to prevent remaining reaction sites from meeting. This value depends on temperature through the glass transition temperature, T_g, as temperatures above T_g result in higher molecular mobility, which facilitates the meeting of reaction sites, and therefore, delays the drop in reaction rate. The critical degree of cure is computed using the baseline critical degree of cure at a temperature of zero, X_{c0}, and the increase in the critical degree of cure per degree increase in temperature, X_{cT}. The degree of cure X is calculated through numerical integration (2) in time as:

$$X = \int_0^t \frac{dX}{dt}dt. \quad (2)$$

The integrated X values, thus, depend on temperature, T, and time, t, relating the cure kinetic model to the specific curing profile applied to the studied epoxy resin.

2.2. Glass Transition Temperature

The DiBenedetto relation [13] in (3) relates X to the midpoint value of the T_g range. The relation involves the final T_g for a state of complete cure, $T_{g\infty}$, the initial T_g for a state of zero cure, T_{g0}, and a fitting parameter, ζ, and represents the ratio of the segmental mobility of the fully cured polymer to that of the initial monomers under the assumption of constant lattice energies [22] in the form:

$$T_g(X) = T_{g0} + \frac{\zeta X(T_{g\infty} - T_{g0})}{1-(1-\zeta)X}. \quad (3)$$

2.3. Cure-Dependent Load-Transferring Volumetric Shrinkage

In the experimental setup, shrinkage can only be measured if there is load transfer between the resin and the fibre optic sensor. The ability to carry a load is also needed for residual stresses to develop. Thus, the volumetric shrinkage of the resin in the liquid phase is ignored. The shrinkage model (4) only considers the load-transferring volumetric shrinkage, V_{sh}. To this end, the degree of cure at which the resin begins to transfer the load is denoted by X_σ, which is close to but not the same as the degree of cure at gelation, X_{gel}. The magnitude of X_σ is found experimentally based on a strain tolerance of $|\varepsilon| > 0.005\%$ from the optical fibre with FBG [5] and using the cure kinetic model with the thermal history measured using the thermocouple. This shrinkage model assumes that the shrinkage from the load transfer point until the end of the cure, X_{end}, results in the load-transferring volumetric shrinkage, V_{sh}^{end}. The shrinkage model is dependent on X through a second-order term Johnston [9] following the conditions:

$$V_{sh} = \begin{cases} 0, & X < X_\sigma \\ V_{sh}^{end}\left(\frac{X-X_\sigma}{X_{end}-X_\sigma}\right)^2, & X_\sigma \leq X < X_{end}. \\ V_{sh}^{end}, & X \geq X_{end} \end{cases} \quad (4)$$

The parameters used in the shrinkage model are determined in Section 6.

2.4. Complex Thermal Expansion

The thermal expansion model extends an earlier model [8] by segmenting the thermal expansion development into several transitional stages. This modification was motivated by the observed behaviour, which was judged to be best captured using a piece-wise linear curve. The expanded model also captures heat-up- and cooldown-dependent behaviour [15] not previously present. The structure of the thermal expansion transition is shown in Figure 1.

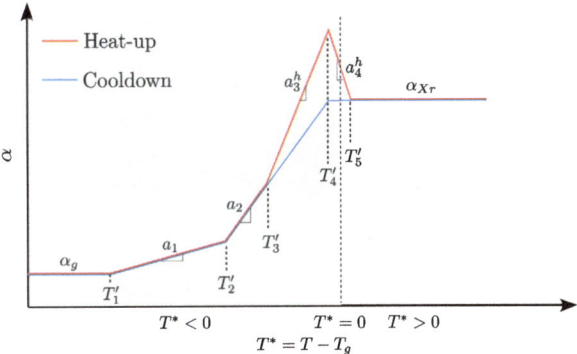

Figure 1. The thermal expansion model involving the transition for the difference between the instantaneous cure temperature T and the glass transition temperature, T_g.

The expanded model divides the thermal expansion into a thermal expansion for heat-up, α_{resin}^h, and one for the cooldown, α_{resin}^c. The expansion is related to the parameter T^*, which is the difference between the instantaneous temperature, T, and T_g. Before the load-transfer point, i.e., in the liquid phase, the thermal expansion of the resin is ignored (5) as it would not contribute to the cure-induced strain, and hence:

$$\alpha_{resin}^h = \alpha_{resin}^c = 0, \quad X < X_\sigma. \quad (5)$$

Figure 1 shows the behaviour below the glass transition $T^* < 0$ and after the glass transition temperature is passed $T^* > 0$. Far from the glass transition temperature, below T_1', the thermal expansion is constant, but above T_1' and through T_2', T_3', and T_4', the thermal expansion increases, still in the glassy regime as $T^* < 0$. Below T_3', the heat-up and cooldown increase in thermal expansion is described by a_1 and a_2. Above T_3', it becomes important to distinguish between heat-up and cooldown. The increase in thermal expansion during heat-up follows a_3^h. This is followed by a decrease in thermal expansion a_4^h across the glass transition temperature from T_4' to T_5'. During heat-up, the thermal expansion is constant relative to T_g above T_5', and similarly, during cooldown from T_4'. In both cases, the thermal expansion depends on the degree of cure when the glass transition is exceeded [8] and denoted α_{Xr}. The relationship with the degree of cure follows a parabolic development. The equations describing the thermal expansion development during heat-up are as follows:

$$\alpha_{resin}^h = \begin{cases} \alpha_g, & T^* < T'_1 \\ a_1(T^* - T'_1) + \alpha_g, & T'_1 \leq T^* < T'_2 \\ a_2(T^* - T'_2) + a_1(T'_2 - T'_1) + \alpha_g, & T'_2 \leq T^* < T'_3 \\ a_3^h(T^* - T'_3) + a_2(T'_3 - T'_2) + a_1(T'_2 - T'_1) + \alpha_g, & T'_3 \leq T^* < T'_4 \\ a_4^h(T^* - T'_4) + a_3^h(T'_4 - T'_3) + a_2(T'_3 - T'_2) + a_1(T'_2 - T'_1) + \alpha_g, & T'_4 \leq T^* < T'_5 \\ \alpha_{Xr}, & T^* \geq T'_5 \end{cases} \quad (6)$$

Similarly to the set of equations describing the heat-up (6), a set of equation prevail for the cooldown (7):

$$\alpha_{resin}^c = \begin{cases} \alpha_g, & T^* < T'_1 \\ a_1(T^* - T'_1) + \alpha_g, & T'_1 \leq T^* < T'_2 \\ a_2(T^* - T'_2) + a_1(T'_2 - T'_1) + \alpha_g, & T'_2 \leq T^* < T'_4 \\ \alpha_{Xr}, & T^* \geq T'_4 \end{cases} \quad (7)$$

The parabolic equation describes the cure-dependent thermal expansion α_{Xr} in (8) as:

$$\alpha_{Xr} = a_{X2}X^2 + a_{X1}X + a_{X0}. \quad (8)$$

The parameters describing the increase in the cure-dependent thermal expansion are a_{X2}, a_{X1} and a_{X0}. The thermal expansion parameters necessary in the proposed model (5)–(8) are fitted in Section 6.

3. Modelling Constituents

The following are the necessary constituents considered in the modelling used in this study.

3.1. Thermal Behaviour

The thermal behaviour applied (9) follows the energy balance equation [1]:

$$\Delta U = c_p \Delta T - H_T \frac{dX}{dt} \Delta t. \quad (9)$$

The incremental energy balance is described as ΔU for every time increment. The exothermal behaviour of the epoxy resin during curing is considered by including the total enthalpy of the reaction, H_T, multiplied by the cure rate and size of the time step. In addition, the resin density is as follows:

$$\rho_{resin} = \rho_{resin}^{init} X + (X_{end} - X)\rho_{resin}^{end}, \quad (10)$$

which is modelled using the rule of mixture between resin density in uncured state, ρ_{resin}^{init}, and cured state, ρ_{resin}^{end}, which were experimentally measured, see Section 6.

3.2. Mechanical Constituents

The constituents used for the mechanical behaviour are based on the constituents included in multiple studies [6,21]. The total, linear, cure-induced strain is taken as:

$$\Delta \varepsilon^{tot} = \Delta \varepsilon^{ch} + \Delta \varepsilon^{th}. \quad (11)$$

The incremental thermal strain, $\Delta \varepsilon_{th}$, develops according to:

$$\Delta \varepsilon^{th} = \alpha_{resin} \Delta T, \quad (12)$$

and the load-transferring incremental linear chemical strain, $\Delta\varepsilon_{ch}$, in the model is defined as the incremental isotropic change in the specific volumetric shrinkage following [6] with:

$$\Delta\varepsilon^{ch} = \sqrt[3]{1+\Delta V_{sh}} - 1. \tag{13}$$

Finally, the incremental volumetric shrinkage, ΔV_{sh}, is defined by the volume change of a cubic element normalised by its original volume and is thus unitless.

The development of the cure-dependent load transferring shrinkage over time can be related to the incremental chemical strain in (13) by deriving (4) to incremental form by differentiation for X and t as in (14). Giving the volumetric shrinkage in the incremental form for the modelling perspective:

$$\Delta V_{sh} = \frac{dV_{sh}}{dX}\Delta X\,;\quad \Delta X = \frac{dX}{dt}\Delta t. \tag{14}$$

4. Experimental Method

4.1. Material System

In the present study, an industrially available thermoset epoxy resin is investigated. The resin is a conventional diglycidyl ether of bisphenol-A (DGEBA). The hardener is a modified cyclo-aliphatic- and aliphatic-amine. A mixing ratio by weight used is base: hardener; 100:31, following supplier guidance.

4.2. Reaction Mechanics

The parameters for the cure kinetics model (1) are based on the work performed in a previous study [17]. The study included the fitting and analysis of DSC data from this specific resin system. It finalised a set of cure parameters given in Table 1 together with the total enthalpy of the reaction given later in Table 6. The parameters will predict the degree of cure from (1) and the midpoint value of the glass transition temperature from (3).

Table 1. Parameters used for the prediction of X and T_g of the specific resin system [17].

A [s^{-1}]	e_a [kJ/mol]	n [-]	m [-]	C [-]	X_{cT} [K^{-1}]	X_{c0} [-]	T_{g0} []	ξ [-]	$T_{g\infty}$ []
$2.50\cdot 10^5$	56.24	1.83	0.41	43.7	$5.27\cdot 10^{-3}$	-0.885	-42.0	0.487	89.0

4.3. Experimental Setup

The experimental setup applied in this study is equivalent to that used in a previous study [5] and similar to others [3,4,18–20]. The resin is in a stress-free state because there are no outer loads or constraining elements, allowing the resin to contract and expand and, therefore, to be considered unconstrained [5]. The setup shown in Figure 2 consists of a thin polymer bag, an optic fibre with Fibre Bragg Gratings (FBG), placed with a thermocouple inside the bag. The thermocouple monitors the temperature response during curing, and the FBG monitors the strain. The error of measurement from the optic sensor in a setup similar to this was discussed in the appendix of an earlier study [5]. Furthermore, the tail length of the optical sensor, from the FBG to the end of the sensor, is important for the accuracy of optical sensors [19]. It was found that the possible error from shear lag on the strain measured using this setup is negligible. As the tail length, l_f of the optical sensor was well above $420r_f$, where r_f is the optic fibre radius. This ratio was reported to give high sensibility even with low resin stiffness [19]. The resin is mixed, degassed, and injected into the polymer bag. Possible air entrapments during infusion are then removed from the bag. The dimensions of the neat resin after infusion are 150×150 mm and a thickness of 4 mm. The optic sensor and thermocouple are placed near the middle of the thickness.

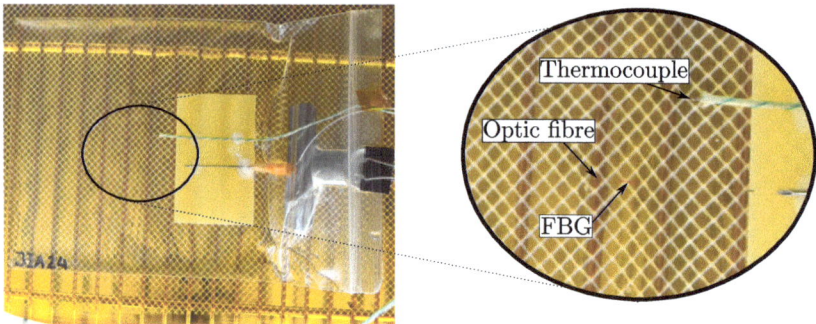

Figure 2. The experimental setup consists of a thin polymer bag. The specimen size is approx 150 × 150 mm and has an average thickness of 4 mm. A fibre optic sensor with an FBG and thermocouple is placed near the middle of the thickness.

5. Numerical Implementation

The numerical implementation was performed in the commercial finite element software Abaqus®2023. The applied material behaviour lies outside the boundaries of the built-in behaviour of the software. Therefore, the implementation used a user-defined material description through a FORTRAN programming-based subroutine offered by Abaqus®. This section gives a brief overview of the subroutines used and which parts of the models they were used to implement. A more detailed description and the actual subroutine can be obtained on request to the authors.

5.1. User-Defined Material Heat Transfer—UMATHT

The first part of the user-subroutine is the UMATHT, which handles the resin heat transfer and updates any changes in thermal properties. It is in this subroutine that the cure development is implemented. The main equations are the energy balance Equation (9), the degree of cure (1), the glass transition temperature (3), and the change in density (10).

5.2. User-Defined Expansion—UEXPAN

Coupled with the UMATHT, the subroutine UEXPAN is passed the necessary state variables from UMATHT to determine the thermal expansion (12) with framework from Figure 1 and the volumetric shrinkage (4) resulting in chemical strains (13). This results in the strain governed by (11).

6. Experimental Results

6.1. Cure Experiments

The cure profiles investigated with the setup explained in Section 4.3 are presented in Table 2. These profiles have been chosen to investigate the effects of different pre-cure temperatures and the effects of the length of the pre-cure on the resulting shrinkage. The notation of the naming follows that of a previous study [5]. The number refers to the cure temperature and the brackets [] denote the part of the cure profile considered the pre-cure, e.g., in $[40_L]80_L$, the [40] refers to 40 as the pre-cure isothermal temperature and 80 refers to an 80 isothermal post-cure. Additionally, the $()_L$ stands for a long cure time of 8 h or more, the $()_M$ stands for medium-length cure time, which is more than 2 h and less than 8 h, and $()_S$ stands for a short cure time of 2 h or less.

Table 2. Cure profiles for investigations of cure-induced strains for an unconstrained resin. The ramps for heating are 1 K/min for all cases. Cooling ramps are approximately −0.1 K/min for cases with cooling during pre-cure. Each investigated case's degree of cure results, and the final cure-induced strains measured at $T_{room} = 21$ are given.

Cure ID	1. Pre-Cure [h @]	2. Pre-Cure [h @]	Post-Cure [h @]	X_σ [%]	X_{pce} [%]	X_{end} [%]	$\varepsilon_{CI}^{21°C}$ [%]
$[40_L]80_L$	12 @ 40	-	10 @ 80	71.6	78.5	98.3	−0.500
$[50_L]80_L$	10 @ 50	-	10 @ 80	68.9	85.2	98.2	−0.536
$[50_M]80_M$	5 @ 50	-	6 @ 80	67.7	79.2	97.3	−0.603
$[50_S 70_S]80_M$	2 @ 50	2 @ 70	6 @ 80	66.8	90.0	97.5	−0.759
$[50_S 60_M]80_M$	2 @ 50	4 @ 60	6 @ 80	70.8	88.3	97.7	−0.629
$[50_S 60_L]80_M$	2 @ 50	8 @ 60	6 @ 80	72.8	90.3	97.8	−0.596
$[50_S 30_S]80_M$	1.5 @ 50	3 @ 30	6 @ 80	69.6	69.6	97.4	−0.791
$[50_S 30_M]80_M$	1.5 @ 50	6 @ 30	6 @ 80	69.7	72.9	97.4	−0.633

Figure 3 shows case $[50_S 70_S]80_M$ and the resulting temperature and cure-induced strain monitored over the duration of the cure profile. The degree of cure and, subsequently, the glass transition temperature are predicted based on the models in (1) and (3), respectively. A strain tolerance [5] determines the load transfer point. Based on the load transfer point, the time, the degree of cure, and the temperature at which load transfer occurs are found. The index $()_\sigma$ denotes the values at the load transfer point and in the plot it is denoted by narrow diamond-shaped points on the curves. Past the load transfer point, the points denoting the end of pre-cure $()_{pce}$, shown by wide diamond points, are plotted on the temperature and the degree of cure curves. The value of the degree of cure at the pre-cure end, X_{pce}, will be used to evaluate the effect of pre-cure length on the measured cure-induced strain. When the resin has cooled to room temperature $T_{room} = 21$ at the very end of the cure, the final cure-induced strain $\varepsilon_{CI}^{21°C}$ is found together with the final degree of cure. At this instance, the cure-induced strain is $\varepsilon_{CI}^{21°C} = -0.759\%$ and the corresponding degree of cure $X_{end} = 97.5\%$. The main results from the cases studied are compiled into Table 2 next to the cure profile parameters. The whole data figure set, like for the case illustrated in Figure 3, is available for download [23].

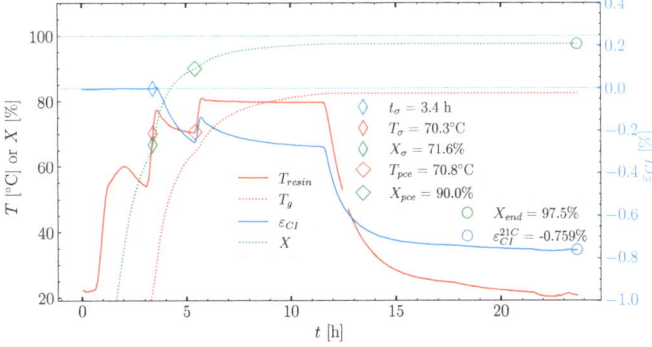

Figure 3. Cure experiment $[50_S 70_S]80_M$—strain measured with the optic FBG and temperature developing in the oven T_{oven}, the temperature recorded inside the resin T_{resin} by the thermocouple, as well as the degree of cure X and T_g predicted based on the recorded resin temperature. The blue dotted lines represent zero strain. the green dotted line represents a level of 100% cure.

Based on data analysis of all the cure experiments listed in Table 2, Figure 4 shows X_σ as a function of the temperature difference, ΔT, which is calculated as the difference between the temperature at the load transfer point, T_σ, and room temperature, T_{room}. Each case has a value of X_{pce}, which is colour-mapped across the investigated cases. This way, the plot demonstrates if the temperature and the length of pre-cure influence the degree of cure at the load transfer point. In the case of Figure 4, there is no obvious trend between ΔT and X_σ or X_{pce} and X_σ. Confirming that the parameter X_σ should be independent of the cure temperature and the length of the pre-cure. Thus, the overall behaviour agrees with a previous study [5]. The average degree of cure at load transfer was $X_\sigma = 69.7\%$, with a reasonably low variation.

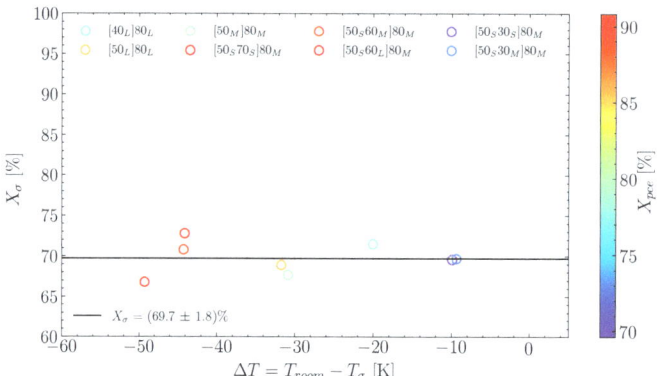

Figure 4. The degree of cure at the load transfer point, X_σ recorded for the experiments in Table 2. As a function of the difference in temperature between the load transfer point and room temperature, ΔT. Colour-mapped according to the degree of cure at the end of pre-cure.

The cure-induced strain has been plotted as a function of the temperature difference ΔT in Figure 5 for the cases. It is observed that there is a substantial scatter in the measurements. However, by using a linear fit, a trend between the temperature difference and the strain can be observed. The two measurements at $\Delta T = -10$ K have been excluded from the linear fit indicated by the grey line. This is due to these measurements seeming to be governed by other mechanisms. Therefore, the region from ΔT -20 K to 0 K is associated with some uncertainty. Hence, the grey-coloured trendline is used to demonstrate the region of uncertainty. The slope of 7.9×10^{-5} K^{-1} is similar to the slope found by a previous study for a similar unconstrained resin [5]. As there is a larger scatter around the linear fit than in the previous study, it is relevant to study the effect of X_{pce} for the different cases in Figure 5. The degree of cure at the pre-cure end, X_{pce}, seems to influence the cure-induced strain. If one observes the colour bar and the two measurements with $\Delta T \approx -10$, there is a significant difference in X_{pce} of around 5% reflected by the difference in colour. Similarly, for the two cases at $\Delta T \approx -30$, the difference in colour on the measurements lead to 4% difference in X_{pce}.

To better clarify the influence of X_{pce} on $\varepsilon_{CI}^{21°C}$, Figure 6, shows the cure-induced strain $\varepsilon_{CI}^{21°C}$ as a function of X_{pce} with a colour-map represented by the load transfer temperature T_σ. The figure shows that the strain observed differs even with the same T_σ, i.e., points with the same colour. However, the X_{pce} values on the horizontal axis differ for the same T_σ. Therefore, Figure 6 demonstrates a clear effect on the observed cure-induced strain with the evolution of cure past the load transfer point during pre-cure.

This pre-cure effect is investigated in more detail with Figure 7 for cases that lead to differences in cure-induced strain. Four cases have been selected: Figure 7a show the two cases $[50_S 30_S]80_M$ and $[50_S 30_M]80_M$ with $T_\sigma \approx 30$, and Figure 7b show two cases, $[50_L]80_L$ and $[50_M]80_M$ with $T_\sigma \approx 50$. The colours applied for each case follow that of the colour-mapping for X_{pce} applied in both Figures 4 and 5.

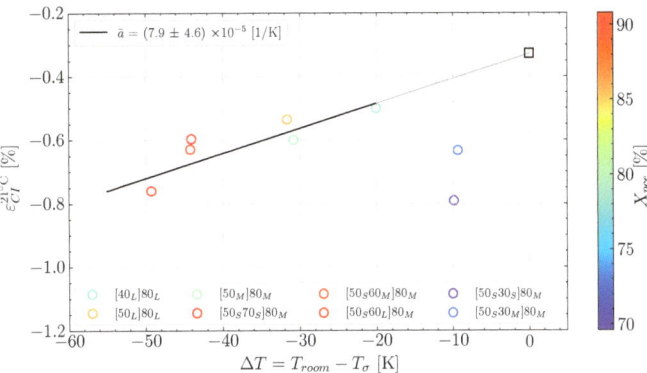

Figure 5. The final cure-induced strains measured at T_{room} as function of ΔT and colour-mapped according to X_{pce}.

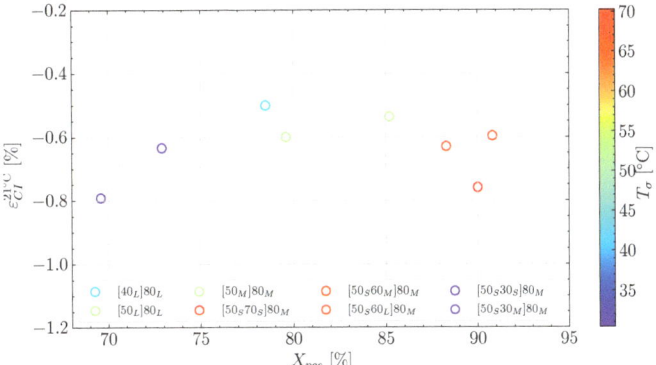

Figure 6. The relationship between $\varepsilon_{CI}^{21°C}$ and X_{pce}, demonstrating the effect of pre-cure length on the measured strain.

In Figure 7a, it can be seen that the added time in $[50_S 30_M]80_M$, and thereby higher X_{pce}, allows the resin to expand a little more than $[50_S 30_S]80_M$ in the heat-up followed by the pre-cure. This is reflected in the observed final cure-induced strain, $\varepsilon_{CI}^{21°C}$. The difference in expansion towards the post-cure between the two cases covers, for the most part, the difference in the cure-induced strain. Similarly, in Figure 7b, the longer pre-cure of the $[50_L]80_L$ case allows the resin to cure substantially more, resulting in a higher expansion relative to $[50_M]80_M$. Again, this results in a difference in the final cure-induced strain observed. Figure 7b also shows that the difference in pre-cure length affects the chemical strain, dominating the pre-cure and post-cure isothermals. For case $[50_L]80_L$, which has a substantially higher value of X_{pce}, the shrinkage at the end of the post-cure more or less has cancelled out, unlike $[50_M]80_M$, which is already in the negative strain regime at the end of post-cure. It is seen that the shrinkage during the pre-cure influences the total cure-induced strain by the end. This shrinkage is mostly chemical strain as the temperature changes are relatively small. The thermal expansion and the chemical shrinkage seem to relate to the magnitude of X_{pce} when the resin is heated up for post-curing. A similar effect, as demonstrated in Figure 7, has been observed previously [5]. However, the resin system studied then seemed to be less susceptible to the effect of pre-cure length.

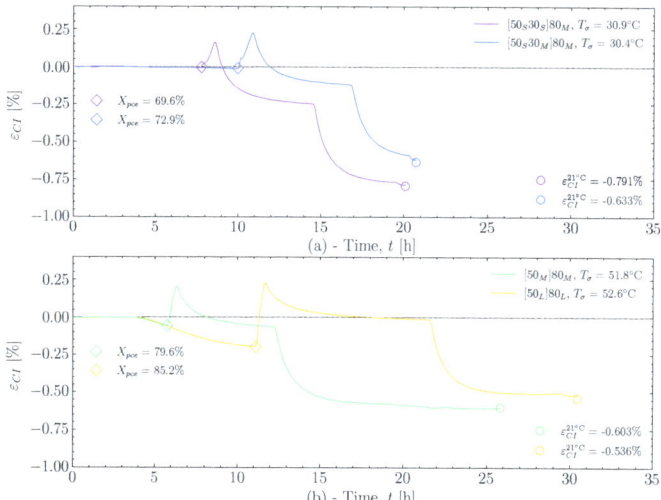

Figure 7. (a) The strain measured for case $[50_S30_S]80_M$ and $[50_S30_M]80_M$, demonstrating the effect of the length of pre-cure on these similar cases. (b) Strain measured for the cases $[50_L]80_L$ and $[50_M]80_M$ to show the effect pre-cure length.

6.2. Determining Volumetric Shrinkage

To evaluate the pre-cure effects in a simulation context, it is necessary to quantify the volumetric shrinkage related to the chemical strain and the thermal expansion behaviour of the resin. The following will quantify these shrinkages to create the necessary inputs for the model. For quantifying the volumetric shrinkage, the experiment $[50_M]80_M$ (see Figure 8) is used to fit the load-transferring linear chemical strain during the initial pre-cure hold at the constant temperature of 50 . The $[50_M]80_M$ case was ideal, as the temperature changes are small and the cure temperature is far away from T_g during the pre-cure, avoiding the vitrification effects from the glass transition.

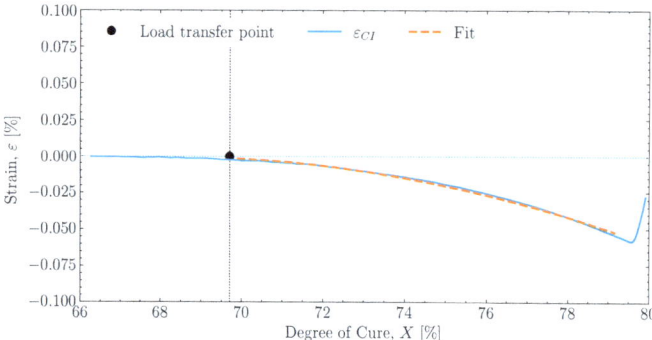

Figure 8. Fitting and extrapolation to determine the volumetric shrinkage based on strain measurements at pre-cure of $[50_M]80_M$. The blue line represents the strain measured, similar to Figure 3. The dashed orange line represents the fitted behaviour. The black dot represents the load transfer point, corresponding to X_σ.

The strain used to determine volumetric shrinkage is that measured from the load transfer point until just before the heat-up to the post-cure, which is combined with the predicted degree of cure using the cure kinetics model and the thermal history measured by the thermocouple. The temperature changes are so small in this region that the thermal contribution is assumed to have no influence. By applying the equation for the linear chemical strain (13) to the strain measured in this region as a function of degree of cure and substituting the incremental volumetric shrinkage with (4), the strain can be used to fit the evolution of the chemical shrinkage in the measured region. The parameters X_σ and X_{end} are input parameters, where X_σ is the average from Figure 4 and X_{end} is the maximum achievable, judged unlikely ever to exceed much more than 98%. Extrapolation from the fitted region can estimate the end value of the load-transferring volumetric shrinkage V_{sh}^{end}. By doing so, the extrapolated shrinkage is found to be -1.1%. This extrapolated value of V_{sh}^{end} together with the values of X_σ and X_{end} are found in Table 3.

Table 3. Volumetric shrinkage for the Johnston shrinkage model determined for $[50_M]80_M$ [9]. X_σ is the average taken from Figure 4 and X_{end} is upper realistic achievable bound.

Cure ID	X_σ [%]	X_{end} [%]	V_{sh}^{end} [%]
$[50_M]80_M$	69.7	98.0	-1.1

The total volumetric shrinkage from the liquid to the fully cured state is V_{sh}^{tot} of -5.2% based on density measurements. The resin density in liquid state ρ_{resin}^{init} was found using a liquid pycnometer, and the value was found to be $1088 \, \text{kg/m}^3$, based on an average of three measurements. The cured density ρ_{resin}^{end} was found using Archimedes principle on five samples cut from a cured panel with $X > 95\%$. The average value was found to be $1145 \, \text{kg/m}^3$ and is judged to be fairly independent of the cure conditions [5]. Judging by the magnitude of the total shrinkage V_{sh}^{tot}, the load-transferring part V_{sh}^{tot}, induced from X_σ until X_{end}, is considered reasonable, especially when compared with another study [19] estimating the load-transferring chemical shrinkage to be within -0.25% to -0.47%. The volumetric shrinkage fitted in this study would lead to a linear chemical strain of -0.36% when applying (13).

6.3. Fitting of Complex Thermal Expansion

A previously cured specimen, $[50_M]80_M$, was selected to fit the thermal expansion behaviour in a fully cured state. Before measuring the thermal expansion, the specimen was post-cured at 100 for 4 h to ensure no residual cure was left. According to a previous DSC analysis of this resin system, this should be sufficient to remove any residual cure of influence [17].

The conditions selected to measure the thermal expansion were 1 K/min and 3 K/min. Figure 9 shows the measured strain response of $[50_M]80_M$ by heating up and cooling down three times with the selected rates. The negative magnitude strains observed at the start before heat-up are the cure-induced strain of $[50_M]80_M$. Heating with two different rates was observed to have no significant effects on the measurements. Furthermore, the continuous heating and cooling of the sample at both rates showed no noteworthy hysteresis.

Based on the measured strain in Figure 9, the gradients can be found to fit the thermal expansion. The gradients are determined based on the 1 K/min data, segmented between heat-up and following cooldown. This is presented in Figure 10. The gradients of the measurements follow the behaviour described in Section 2.4 of the model proposed. The heat-up path is fitted to (6) and the cooldown to (7). The fitted thermal expansion in the glassy state α_g and the increase in thermal expansion a_1 to a_4^h are listed in Table 4 together with the values of T_1' to T_5'.

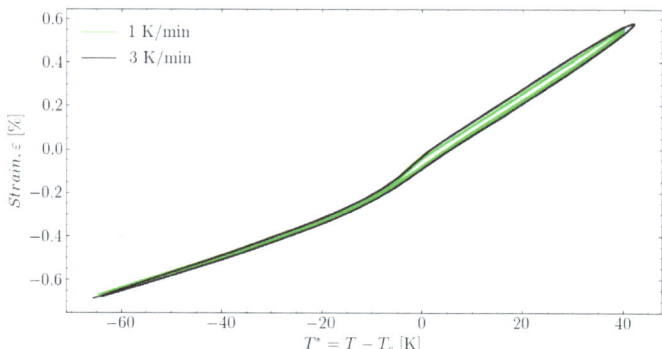

Figure 9. Strain measured based on a reheated $[50_M]80_M$ at both 1 K/min an 3 K/min after fully curing the specimen for 4 h at 100 .

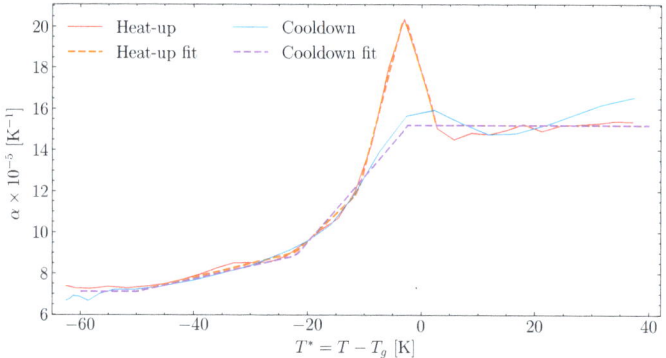

Figure 10. The thermal expansion evolution found by the derivative of the strain measured in Figure 9 for data measured at 1 K/min.

Table 4. Parameters fitted for the thermal expansion model (6) and (7) based on FBG measurements of reheated cured specimen.

α_g [K^{-1}]	a_1 [K^{-2}]	a_2 [K^{-2}]	a_3^h [K^{-2}]	a_4^h [K^{-2}]
7.11×10^{-5}	6.85×10^{-7}	2.67×10^{-6}	1.08×10^{-5}	-9.19×10^{-6}
T_1' [K]	T_2' [K]	T_3' [K]	T_4' [K]	T_5' [K]
-50	-22	-11	-3	2.5

For temperatures where $T^* > T_5'$, the cooldown and heat-up expansion are constant relative to the influence of T_g. However, as the resin is curing, thermal expansion changes during curing above T_g [8]. To determine the curing effect on thermal expansion, the cases listed in Table 2 have been measured during heat-up, past X_σ. The chemical shrinkage, based on the fit in Section 6.2, was subtracted from the measured cure-induced strain during the heat-up of the samples. The thermal expansion, α_{X_r}, was then fitted to the linear gradient observed from the point where $T^* = T_5'$ and until the heat-up ends. The fitted thermal expansion in the heat-up was correlated with the degree of cure, X, at the point where $T^* = T_5'$, ignoring possible changes in the degree of cure over the fitted interval. In Figure 11, the measured thermal expansion values for the different cases are plotted together with the fit of (8). The second-order fit of the cure-dependent thermal expansion above T_g seems to follow the measurements well. The fitted values are listed in Table 5. The function α_{X_r} will govern the cure-dependent thermal expansion during cooldown and heat-up, based on the little difference observed for the fully cured measurements in Figure 10.

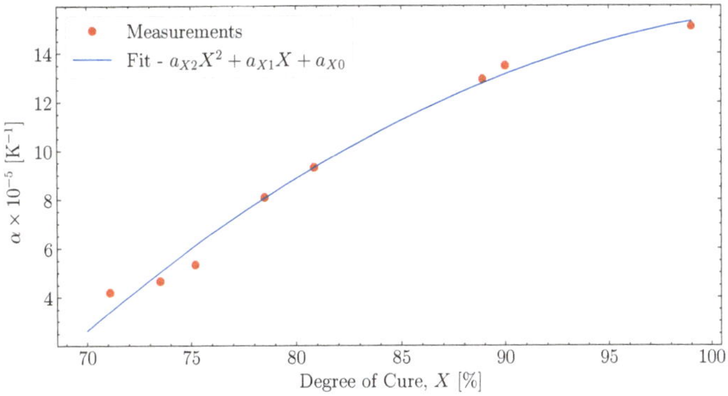

Figure 11. Evolution of thermal expansion for $T^* > T_5'$ as function of X, for heating and cooldown.

Table 5. The fitted parameters for the function α_{X_r} (8). This relation is only valid for values of $X > X_\sigma$.

a_{X0} [K^{-1}]	a_{X1} [K^{-1}]	a_{X2} [K^{-1}]
-96.0×10^{-5}	209.6×10^{-5}	-98.1×10^{-5}

7. Simulation of Cure Shrinkage

This section will predict the previously investigated cure-induced shrinkage with a simple 1D thermomechanical finite element model. The material models and behaviours described in Section 2 have been built into a modelling framework described in Section 5 and will be applied to elaborate on the material behaviour observed in Section 6.

Model for the Thermal and Cure-Induced Strain Predicitions

The model for predicting the resin shrinkage is based on a finite element framework [24] and is a simple 1D thermomechanical model. Earlier work has shown that cure-induced strain can be captured with a material point model [21]. In this study, a similar approach is applied. However, to predict the thermal behaviour, the through-thickness response of the resin is required. This allows for the additional effect of the exothermal release of heat during the curing. The model is illustrated with Figure 12. Here, the resin bulk is illustrated in an xy-plane, with y as the principal model direction through the thickness. The region presented in Figure 12 is a narrow cutout of a vast resin bulk. The length and width of the domain are much larger than the thickness of the observed area, and possible thermal effects from possible edges can be ignored. Then, by only considering the thickness, the model stretches from the surface of the resin, called boundary B, to boundary A, in the middle of the resin. Imposing the conditions listed below:

A $(u_y) = (0)$, $h = 0$ (symmetry of heat flow and displacement);
B $(u_y) = (free)$, $h = h_c$, $T = T_{oven}(t)$.

Boundary A is a symmetry condition for both the thermal and mechanical behaviour. As the resin is unconstrained, the model can contract in the y-direction. The heat flow h, from the surface of B, h_c, is the heat transfer coefficient enforced by the air movement possible from inside the oven. The temperature applied in this boundary, $T_{oven}(t)$, is the oven temperature measured for each case in Table 2. The thermal response of the simulation is monitored at boundary A, at the location corresponding to that of the thermocouple and the FBG sensor in the experiment. The strain produced in the simulation is evaluated at Boundary B.

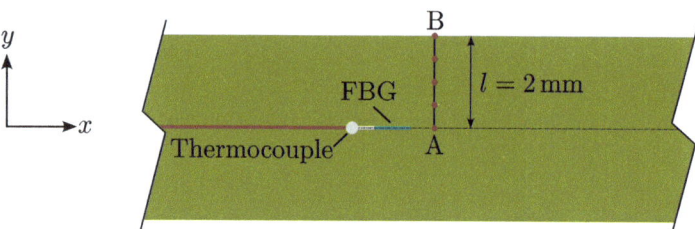

Figure 12. A cutout of the experimental setup showing the thermocouple and FBG. It illustrates how the 1D thermomechanical model is built to simulate the cure behaviour of the resin through the thickness.

For the prediction of the thermal behaviour, the necessary parameters are tabulated in Table 6. The total enthalpy of the reaction H_T has been measured with DSC for the specific resin system [17]. The densities in Table 6, as reported in Section 6, have been determined experimentally. The heat capacity $c_{p,resin}$ and conductivity k_{resin} were taken from [25,26], respectively. The convection coefficient for the air inside the oven has been taken from Carson et al. [27]. The cure-induced strain predicted by the model develops following the theory in Section 2. The primary components are the chemical and thermal strain, adding to the simulated cure-induced strain. In Figure 13, the simulation of the cure experiment $[50_S70_S]80_M$ is plotted, and the strain and temperature and strain from Figure 3 are included on top of the predicted strain and temperature by the model. In Appendix A, figures of the remaining cases for comparison based on the cases in Table 2 are compiled. To assist the description of the model behaviour, the following notation is used:

- ① 1. Isothermal— Pre-cure;
- ② 1. Ramp—Pre-cure;
- Ⓐ 2. Isothermal—Pre-cure;
- Ⓑ 2. Ramp —Pre-cure;
- Ⓒ Isothermal—Post-cure;
- Ⓓ Cooldown after Post-cure.

Table 6. Thermal properties for the simulations.

$c_{p,resin}$ [J/(kgK)]	k_r [W/(m²K)]	H_T [$\frac{J}{kg}$]	ρ_{resin}^{init} [kg/m³]	ρ_{resin}^{end} [kg/m³]	h_c [W/(m²K)]
1900 [25]	0.14 [26]	4.7×10^5	1088	1145	15 [27]

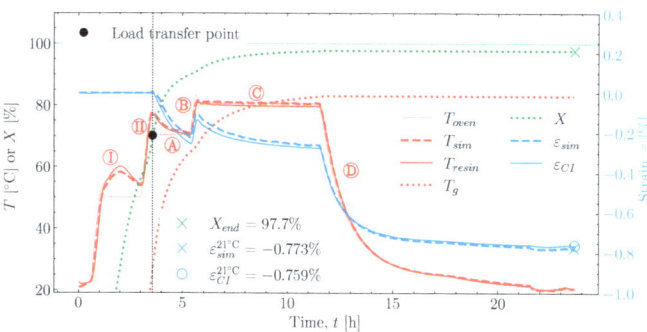

Figure 13. The predicted and measured values of temperature and strain over time for case $[50_S70_S]80_M$ as well as predicted X and T_g by the simulation. The final values of the predicted and measured strain are shown in the plot.

The predicted temperature in the simulation results from the temperature load $T_{oven}(t)$, the thermal boundary condition and the cure kinetic behaviour. This results in the development of the noticeable exotherm during the two parts of the pre-cure, ① and Ⓐ. The predicted temperature by the simulation matches well with the monitored temperature from the thermocouple inside the resin. This is also observed in the remaining eight cases studied, found in Appendix A.

The strain predicted depends on the thermal behaviour, as the temperature, corresponding degree of cure X, and glass transition temperature T_g are computed for every increment in the simulation. Once the load transfer point is reached, the incremental thermal and chemical strains develop. In Figure 13, the simulated cure-induced strain is predicted well. Both in terms of the shrinkage occurring during Ⓐ, which is influenced heavily by the thermal and chemical strain occurring simultaneously. Followed by the heat-up Ⓑ, then the post-cure Ⓒ and the cooldown Ⓓ, these also show good correlation between experiment and simulation. At the end of the cure, both the experimental observed cure-induced strain $\varepsilon_{CI}^{21°C}$ and the simulated $\varepsilon_{sim}^{21°C}$ are shown in Figure 13, as well as the final value of the simulated X. A comparison of the final simulated degree of cure with the final predicted one based on the thermocouple temperature monitored shown in Figure 3 is relevant. The differences are negligible; thus, the simulated cure development is accurate within the experimentally predicted. In terms of the differences observed between the simulated and measured strain, the deviation relative to the experiment was found to be within 2%. Hence, the simulation is overall satisfactory. The deviations and cure-induced strains observed for all the simulations and corresponding experiments are tabulated in Table 7. The overall deviation was found to be within 2–6% and the average deviation around 3%. With a simulation that matches the observed experimental behaviour well in all cases. The results from both experiments and simulations are available for download [23].

Table 7. End value of ε for experiments and simulations after the cure profiles evaluated at $T_{room} = 21$ and the deviation.

Cure ID	$[40_L]80_L$	$[50_L]80_L$	$[50_M]80_M$	$[50_S70_S]80_M$	$[50_S60_M]80_M$	$[50_S60_L]80_M$	$[50_S30_S]80_M$	$[50_S30_M]80_M$
$\varepsilon_{CI}^{21°C}$ [%]	−0.500	−0.536	−0.603	−0.759	−0.629	−0.596	−0.791	−0.633
$\varepsilon_{sim}^{21°C}$ [%]	−0.532	−0.511	−0.608	−0.773	−0.642	−0.583	−0.758	−0.609
Dev. [%]	6	5	2	2	2	2	4	4

To better clarify how the thermal strain prediction affects the model behaviour during the curing, the experimental and simulated cure-induced strain is plotted in Figure 14 as a function of temperature. The figure demonstrates that the model predicts the cooldown during Ⓐ well, although it underestimates the shrinkage somewhat in magnitude. During the following heat-up Ⓑ, the expansion observed in the experiment is parallel with the expansion simulated. Therefore, the simulation can capture the expansion and contraction observed experimentally while curing progresses. This is important as the contraction and expansion occurring during Ⓐ and Ⓑ both occur well above T_g. This means that the expansion and contraction should be influenced by curing as per the thermal expansion model applied in Section 2.4.

The final cooldown Ⓓ that occurs from Ⓒ and down to room temperature is unaffected by any significant changes in X and demonstrates that the model can also capture the cured contraction well from just below T_g and until far away from T_g. The simulated strain is plotted as a function of the degree of cure, X, against the experiment in Figure 15, where Figure 15a demonstrates the temperature development of the experiment T_{resin}, simulation T_{sim} and the oven temperature T_{oven} as a function of X. The temperatures of the experiment and simulation agree. There is a slight variation between the oven temperature and the resin temperatures.

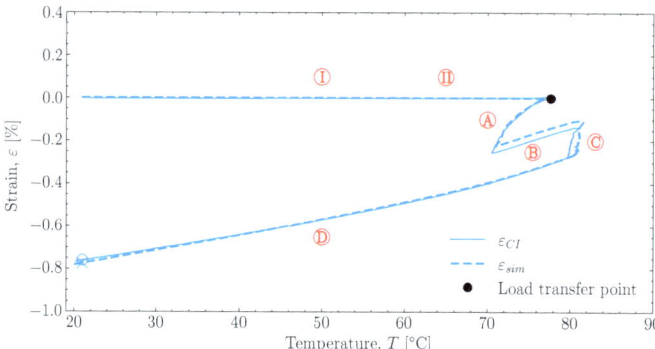

Figure 14. The predicted and measured values of strain as function of temperature for case $[50_S 70_S]80_M$.

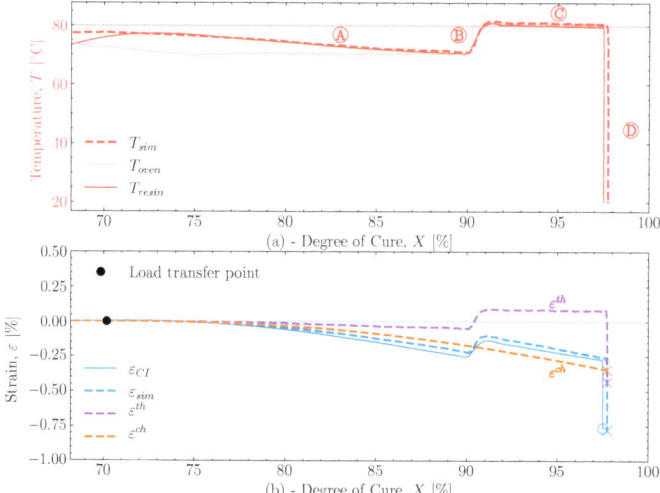

Figure 15. (a) The predicted and measured temperature in the resin as well as oven temperature as a function of the degree of cure. (b) The predicted and measured values of strain as a function of the degree of cure for case $[50_S 70_S]80_M$.

This lag appears due to the heat flow through the thickness of the sample. Figure 15b demonstrates the experimental and simulation strains as a function of X. This makes it easy to distinguish the thermal strain from the chemical strain observed in the simulation. As the temperature drops during the pre-cure Ⓐ, the simulated thermal strain is also observed to drop. The simulated chemical strain also decreases continuously as the degree of cure increases. It should be possible to check whether the chosen volumetric shrinkage model (4) adapted for the chemical strain matches the experimentally observed behaviour. The simulated strain ε_{sim} is seen to under-predict the shrinkage occurring during slightly Ⓐ, but follows in parallel with the experimental strain for the duration of Ⓒ. After that, the curing ends with the cooldown Ⓓ. Even though there generally is this slight offset between experimental and simulated, the offset does not increase or decrease slightly. Indicating that the proposed shrinkage behaviour follows the experimental behaviour well. The simulations are, therefore, quite capable of determining the effects observed experimentally.

To demonstrate this graphically across the range of cases, Figure 16 shows an extended version of Figure 5. The final simulated and measured values of cure-induced strains are plotted together, and the possible differences are shown. The trendline adapted in Figure 5 is not applied here as the pre-cure has shown a high dependency on cure-induced strains. The fact that a very low achieved ΔT for case $[50_S 30_S]80_M$ and $[50_S 30_M]80_M$ results in high cure-induced strain signifies the dominating effect due to pre-cure and, more precisely, X_{pce}. This is attributed mainly to the thermal expansion model applied. Stressing that even though ΔT influences the level of cure-induced strain observed it is necessary to consider the complex thermal expansion to determine the cure-induced strain accurately.

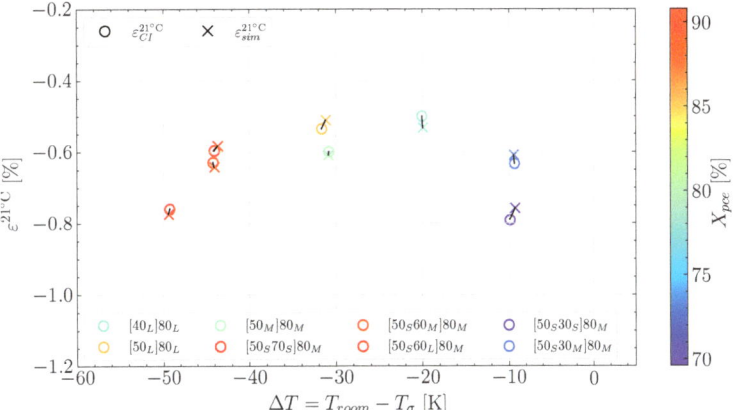

Figure 16. Comparison between the cure-induced strain from the experiments and simulations plotted together and linear and second-order tendencies plotted together. Circles indicate the measured shrinkages and the crosses represent the simulated shrinkage.

Adapted in the simulation, this cure-dependent thermal expansion results in a similar low expansion during the heat-up for the cases $[50_S 30_S]80_M$ and $[50_S 30_M]80_M$, as observed in Figure 17a. Demonstrating that the simulation can capture the complex thermal expansion behaviour observed. The low thermal expansion at a lower degree of cure results in more shrinkage being transferred at the cooldown after post-curing. As simulated for the two cases, $[50_S 30_S]80_M$ and $[50_S 30_M]80_M$ agreed well with the measured strain. This limits the ability to reduce the cure-induced strain for ΔT from -20 K to 0 K for this specific resin system. At the other end of the ΔT axis in Figure 16, the cases $[50_L]80_L$ and $[50_M]80_M$ as well as $[50_S 60_M]80_M$ and $[50_S 60_L]80_M$ show that an increase in cure-induced strain is present. However, the difference in X_{pce} between $[50_M]80_M$ and $[50_L]80_L$ is approximately 5% which, even though it is a more considerable difference than the 3.4% $[50_S 30_S]80_M$ and $[50_S 30_M]80_M$, the effect of ΔT is more dominating for higher values of X_{pce}. This is because the resin has cured significantly more; thus, the cure-dependent thermal expansion has developed much more, making the temperature at load transfer much more critical. This effect is reflected in the simulation and is due primarily to the implemented thermal expansion model. Therefore, the developed simulation can accurately predict the complex shrinkage observed over various experimental cases.

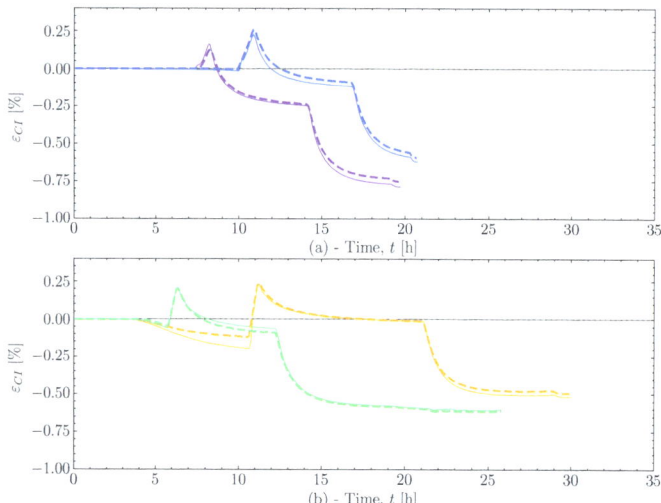

Figure 17. The comparison of pre-cure effects between the experiments with solid lines and simulations with dashed lines. The colours for the cases refer to the same colour bar for X_{pce} in Figure 16. (**a**) The experiments and simulations of case $[50_S 30_S]80_M$ $[50_S 30_M]80_M$. (**b**) The experiments and simulations of case $[50_L]80_L$ and $[50_M]80_M$.

8. Conclusions

A specific epoxy resin system has been studied to quantify the cure-induced strain expected to develop during the curing in an unconstrained experimental setup. The cure-induced strains arising from various experiments were rather complex for the cure profiles investigated. The thermal expansion during the heat-up at the end of the pre-cure was cure-dependent and dependent on the glass transition temperature. A novel complex thermal expansion model and a model for the load-transferring volumetric shrinkage related to chemical cross-linking of the resin were proposed. The governing factors, such as chemical and thermal shrinkage leading to the experimentally observed cure-induced strain, could be quantified by fitting experimental observations to the proposed models. This was performed to investigate the ability of the proposed models to capture the behaviour of the cure-induced strain seen experimentally.

A simulation method was proposed to simulate the cure-induced strain across various cases accurately. The simulations correlated well with the experiments and agreed with the experimental observations, thus validating the simulation method. The simulations showed that the complex thermal expansion and the conventional volumetric shrinkage models were necessary for accurately predicting the cure-induced strain. The behaviour of the resin studied depended on the load transfer temperature and the development of the degree of cure at the pre-cure stage. This makes the complex thermal expansion model especially essential for accurately predicting cure-induced strains in simulations.

To lower the cure-induced strain in an unconstrained system, like the one investigated, it is henceforth essential to consider the investigated effects to minimise potential residual stresses in a setup, constraining the resin behaviour, and thus, inducing residual stresses. In a compact sense, a list can be drawn of the parameters necessary to make precise predictions of cure shrinkage:

- Perform DSC analysis to characterise the cure behaviour and determine the parameters for the cure kinetic model, glass transition temperature evolution, and the enthalpy of the reaction;
- Determine the load transfer initiation in the resin and determine/estimate the load transferring part of the volumetric shrinkage;

- Define the complex nature of the thermal expansion of the specific resin.

The listed parameters are the key aspects necessary to define the field to potentially minimise cure-induced strains from the curing of thermosetting epoxy resins. The parameters and material behaviour presented in this work can be further utilised in experiments where epoxy is mechanically constrained by surroundings. Leading to accurate predictions of the thermal and chemical behaviour for the build-up of residual stresses.

Author Contributions: Conceptualisation, J.K.J. and L.P.M.; methodology, J.K.J. and L.P.M.; software, J.K.J. and V.K.M.; validation, L.P.M., V.K.M. and T.L.A.; laboratory resources, T.L.A.; data curation, J.K.J.; writing—original draft preparation, J.K.J.; writing—review and editing, J.K.J., L.P.M., V.K.M. and T.L.A.; supervision, L.P.M., V.K.M. and T.L.A. All authors have read and agreed to the published version of the manuscript.

Funding: This study acknowledges funding by Innovation Fund Denmark (Innovationsfonden) through the AIOLOS—Affordable and Innovative Manufacturing of Large Composites (0224-00003B) project. Acknowledgement to Horizon Europe, the European Union's Framework Programme for Research and Innovation, grant Agreement No. 101058054, towards tURbine Blade production with zerO waste (TURBO).

Institutional Review Board Statement: Not applicable.

Data Availability Statement: The data from the analysed experimental cure cases are made available, together with the model and scripts, building the analysis files for predicting the experimental behaviour investigated. The data are provided in '.csv' format and the script for ABAQUS simulations consists of a model file '.cae', a python script is provided making the necessary changes to the model '.py', and a script is provided for submitting several simulations in a shell script for Linux-based clusters '.sh' [23]. The subroutine for predicting curing is available [24].

Acknowledgments: Jonas Kreutzfeldt Heininge is acknowledged for technical assistance with the experiments. Nicolai Frost-Jensen Johansen is acknowledged for contributions to the data analysis concerning the experiments.

Conflicts of Interest: The authors declare no conflicts of interest.

Appendix A. Simulation Cases

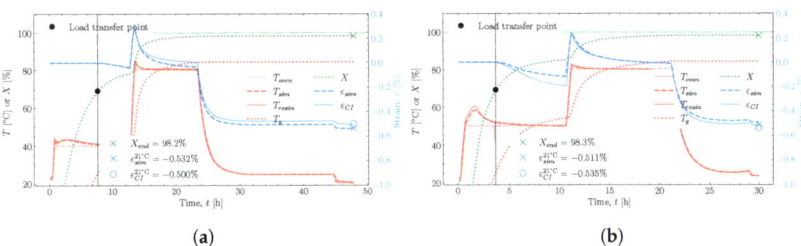

Figure A1. (a) Case $[40_L]80_L$ strain over time. (b) Case $[50_L]80_L$ strain over time.

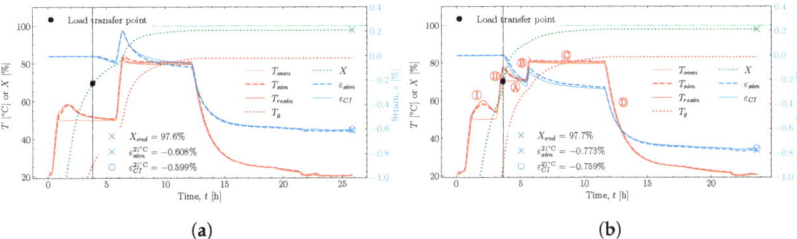

Figure A2. (a) Case $[50_M]80_M$ strain over time. (b) Case $[50_S70_S]80_M$ strain over time.

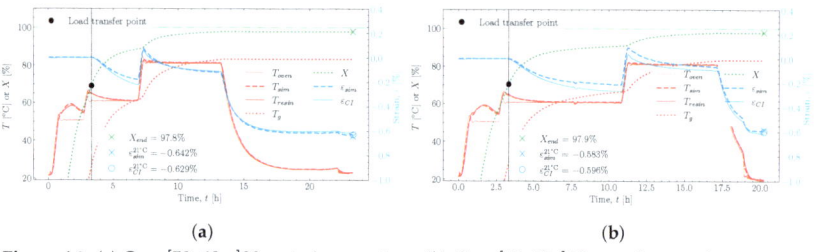

Figure A3. (**a**) Case $[50_S 60_M]80_M$ strain over time. (**b**) Case $[50_S 60_L]80_M$ strain over time.

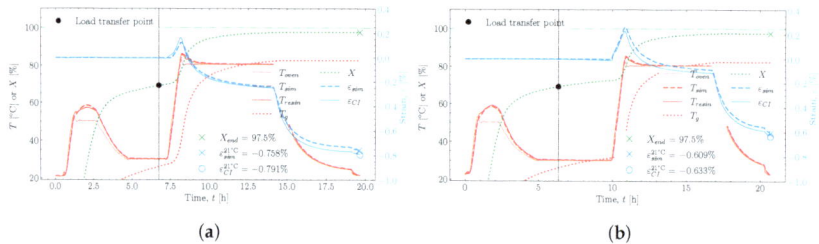

Figure A4. (**a**) Case $[50_S 30_S]80_M$ strain over time. (**b**) Case $[50_S 30_M]80_M$ strain over time.

References

1. Nielsen, M.W. Prediction of Process Induced Shape Distortions and Residual Stresses in Large Fibre Reinforced Composite Laminates: With Application to Wind Turbine Blades. Ph.D. Thesis, Technical University of Denmark, Kongens Lyngby, Denmark, 2013.
2. Jørgensen, J.B.; Sørensen, B.F.; Kildegaard, C. The effect of residual stresses on the formation of transverse cracks in adhesive joints for wind turbine blades. *Int. J. Solids Struct.* **2019**, *163*, 139–156. [CrossRef]
3. Mortensen, U.A. Process Parameters and Fatigue Properties of High Modulus Composites. Ph.D. Thesis, DTU, Roskilde, Denmark, 2019. [CrossRef]
4. Miranda Maduro, M.A. Influence of Curing Cycle on the Build-Up of Residual Stresses and the Effect on the Mechanical Performance of Fibre Composites. Ph.D. Thesis, DTU, Roskilde, Denmark, 2021.
5. Mikkelsen, L.P.; Jørgensen, J.K.; Mortensen, U.A.; Andersen, T.L. Optical fiber Bragg gratings for in-situ cure-induced strain measurements. *Zenodo* **2023**. [CrossRef]
6. Bogetti, T.A.; Gillespie, J.W. Process-induced stress and deformation in thick-section thermoset composite laminates. *J. Compos. Mater.* **1990**, *26*, 626–660. [CrossRef]
7. Shah, D.U.; Schubel, P.J. Evaluation of cure shrinkage measurement techniques for thermosetting resins. *Polym. Test.* **2010**, *29*, 629–639. [CrossRef]
8. Khoun, L.; Centea, T.; Hubert, P. Characterization methodology of thermoset resins for the processing of composite materials—Case study. *J. Compos. Mater.* **2010**, *44*, 1397–1415. [CrossRef]
9. Johnston, A.A. An Integrated Model of the Development of Process-Induced Deformation in Autoclave Processing of Composite Structures. Ph.D. Thesis, University of of British Columbia, Vancouver, BC, Canada, 1997.
10. Li, C.; Potter, K.; Wisnom, M.R.; Stringer, G. In-situ measurement of chemical shrinkage of MY750 epoxy resin by a novel gravimetric method. *Compos. Sci. Technol.* **2004**, *64*, 55–64. [CrossRef]
11. Kamal, M.R. Thermoset characterization for moldability analysis. *Polym. Eng. Sci.* **1974**, *14*, 231–239. [CrossRef]
12. Cole, K.C.; Hechler, J.J.; Noel, D. A New Approach to Modeling the Cure Kinetics of Epoxy Amine Thermosetting Resins. 2. Application to a Typical System Based on Bis[4-(diglycidylamino)phenyl] methane and Bis(4-aminophenyl) Sulfone. *Macromolecules* **1991**, *24*, 3098–3110. [CrossRef]
13. Dibenedetto, A.T. Prediction of the Glass Transition Temperature of Polymers: A Model Based on the Principle of Corresponding States. *Polym. Sci. Part Polym. Phys.* **1987**, *25*, 1949–1969. [CrossRef]
14. McKenna, G.B.; Simon, S.L. The glass transition: Its measurement and underlying physics. In *Handbook of Thermal Analysis and Calorimetry*; Cheng, S.Z.D., Ed.; Elsevier: Springboro Pike Miamisburg, OH, USA, 2002; Chapter 2, pp. 49–104.
15. Igor, N.; Shardakov, A.N.T. Identification of the temperature dependence of the thermal expansion coefficient of polymers. *Polymers* **2021**, *13*, 3035. [CrossRef]
16. Korolev, A.; Mishnev, M.; Ulrikh, D.V. Non-Linearity of Thermosetting Polymers' and GRPs' Thermal Expanding: Experimental Study and Modeling. *Polymers* **2022**, *14*, 4281. [CrossRef] [PubMed]
17. Jørgensen, J.K.; Mikkelsen, L.P.; Andersen, T.L. Cure characterisation and prediction of thermosetting epoxy for wind turbine blade manufacturing. *IOP Conf. Ser. Mater. Sci. Eng.* **2023**, *1293*, 012015. [CrossRef]

18. Hoffman, J.; Khadka, S.; Kumosa, M. Determination of gel point and completion of curing in a single fiber/polymer composite. *Compos. Sci. Technol.* **2020**, *188*, 107997. [CrossRef]
19. Hu, H.; Li, S.; Wang, J.; Zu, L.; Cao, D.; Zhong, Y. Monitoring the gelation and effective chemical shrinkage of composite curing process with a novel FBG approach. *Compos. Struct.* **2017**, *176*, 187–194. [CrossRef]
20. Khadka, S.; Hoffman, J.; Kumosa, M. FBG monitoring of curing in single fiber polymer composites. *Compos. Sci. Technol.* **2020**, *198*, 108308. [CrossRef]
21. Jørgensen, J.K.; Mikkelsen, L.P. Tailored Cure Profiles for Simultaneous Reduction of the Cure Time and Shrinkage of an Epoxy Thermoset. *Heliyon* **2024**, *10*, e25450. [CrossRef] [PubMed]
22. Pascault, J.P.; Williams, R.J. Glass transition temperature versus conversion relationships for thermosetting polymers. *J. Polym. Sci. Part Polym. Phys.* **1990**, *28*, 85–95. [CrossRef]
23. Jørgensen, J.K. Experimental and simulation results of cure-induced strains for a cure study of an conventional epoxy resin. *Zenodo* **2024**. [CrossRef]
24. Jørgensen, J.K. A thermomechanical finite element subroutine for Abaqus to precisely predict cure-induced strain of an epoxy thermoset. *Zenodo* **2024**. [CrossRef]
25. McHugh, J.; Fideu, P.; Herrmann, A.; Stark, W. Determination and review of specific heat capacity measurements during isothermal cure of an epoxy using TM-DSC and standard DSC techniques. *Polym. Test.* **2010**, *29*, 759–765. [CrossRef]
26. Wierzbicki, L.; Pusz, A. Thermal conductivity of the epoxy resin filled by low melting point alloy. *Arch. Mater. Sci. Eng.* **2013**, *61*, 22–29.
27. Carson, J.K.; Willix, J.; North, M.F. Measurements of heat transfer coefficients within convection ovens. *J. Food Eng.* **2006**, *72*, 293–301. [CrossRef]

Disclaimer/Publisher's Note: The statements, opinions and data contained in all publications are solely those of the individual author(s) and contributor(s) and not of MDPI and/or the editor(s). MDPI and/or the editor(s) disclaim responsibility for any injury to people or property resulting from any ideas, methods, instructions or products referred to in the content.

Article

Development of Fatigue Life Model for Rubber Materials Based on Fracture Mechanics

Xingwen Qiu [1], Haishan Yin [1,*], Qicheng Xing [1] and Qi Jin [2]

[1] College of Electromechanical and Engineering, Qingdao University of Science and Technology, Qingdao 266100, China; q936838338@163.com (X.Q.); qichengxing1997@126.com (Q.X.)
[2] Tongli Tire Co., Ltd., Jining 272100, China
* Correspondence: yinhais@163.com

Abstract: In this paper, the research on the fatigue damage mechanism of tire rubber materials is the core, from designing fatigue experimental methods and building a visual fatigue analysis and testing platform with variable temperature to fatigue experimental research and theoretical modeling. Finally, the fatigue life of tire rubber materials is accurately predicted by using numerical simulation technology, forming a relatively complete set of rubber fatigue evaluation means. The main research is as follows: (1) Mullins effect experiment and tensile speed experiment are carried out to explore the standard of the static tensile test, and the tensile speed of 50 mm/min is determined as the speed standard of plane tensile, and the appearance of 1 mm visible crack is regarded as the standard of fatigue failure. (2) The crack propagation experiments were carried out on rubber specimens, and the crack propagation equations under different conditions were constructed, and the relationship between temperature and tearing energy was found out from the perspective of functional relations and images, and the analytical relationship between fatigue life and temperature and tearing energy was established. Thomas model and thermo-mechanical coupling model were used to predict the life of plane tensile specimens at 50 °C, and the predicted results were 8.315×10^5 and 6.588×10^5, respectively, and the experimental results were 6.42×10^5, with errors of 29.5% and 2.6%, thus verifying the accuracy of thermo-mechanical coupling model.

Keywords: tire rubber; fatigue damage; numerical simulation

1. Introduction

1.1. Theory of Fatigue Research

Tires, seals, shock absorbers and other rubber-based materials will work to produce deformation, long-term exposure to alternating loads will lead to performance degradation, that is, fatigue damage. It is known that fatigue damage accounts for 80% of the failure of rubber products, so it is important to study the fatigue of rubber materials to improve their fatigue performance.

The fatigue damage of rubber materials is divided into two stages: the first stage is the gathering and sprouting of tiny defects inside the rubber to produce tiny cracks, i.e., the crack sprouting stage; the second stage is the expansion of the tiny cracks produced in the previous stage until the rubber material fails by fracture, i.e., the crack expansion stage [1]. Scholars classify the research methods into crack emergence and crack extension methods according to the two stages. The theory of fracture mechanics suggests that there are inherent defects in rubber materials, and when reinforcing materials are added to rubber products, the agglomeration of fillers also causes small defects inside the rubber, and this defect is the source of crack extension. Glanowski et al. [2] used X-ray computed microtomography to observe the fatigue of carbon filled natural rubber and came up with two damage mechanisms: one is the cavitation phenomenon at the poles of the agglomerates, and the other is fracture of the agglomerates, and Huneau et al. [3] studied the fatigue cracking of carbon filled natural rubber and found that the crack initiation mechanisms of carbon black aggregates and oxide aggregates were different, and only the cracks initiated by carbon black aggregates were accompanied by crack extension, because carbon black has stronger cohesion and adhesion to the matrix, and its cohesion is stronger than adhesion, so the crack budding is actually also the crack Therefore, it is of scientific significance to apply fracture mechanics to study the fatigue process of rubber materials.

According to fracture mechanics, when rubber cracks grow, a new surface will be generated, and the generation of a new surface will inevitably consume energy, that is, surface energy. When the mechanical energy storage consumed by the crack per unit area of expansion is greater than this re-sistance, the crack will expand [4]. The tearing energy is defined as the mechanical storage energy dU required to produce a crack per unit area dA [5], in which:

$$T = -\frac{dU}{dA} \quad (1)$$

where the negative sign indicates that the strain energy required to produce cracks in rubber materials decreases as the crack area increases.

The study of rubber fatigue life dates back to 1940, when Cadwell [6] studied the dynamic fatigue life of natural rubber, followed by Thomas [7], Lake [8], Lindley [9], Rivlin [10], and William V. Mars [11], which was developed over the years to form a whole set of theories, namely the crack extension theory based on fracture mechanics. They combined a large number of experimental data of crack extension under transverse loading, and divided the crack tip energy release rate-crack growth rate into four regions according to the maximum energy release rate in the cycle, as shown in Figure 1 [12], T_0 indicates the threshold tear energy, and the peak tear energy of stage I, $T_{max} \leq T_0$, at this stage the crack will not expand due to the external load; T_t is the transition tear energy, which indicates the transition from stage II to stage T_t is the transition tear energy, which indicates the tear energy corresponding to the transition from stage II to stage III; T_c indicates the critical tear energy, when $T_{max} \geq T_c$, the crack reaches stage IV, i.e., the destabilization expansion process. The approximate expressions for the crack growth rates of the four stages are also given, as follows

$$\begin{cases} \frac{da}{dN} = r_z & T_{max} < T_0 \\ \frac{da}{dN} = A_0(T_{max} - T_0) + r_z & T_0 \leq T_{max} < T_t \\ \frac{da}{dN} = BT_{max}^F & T_t \leq T_{max} < T_c \\ \frac{da}{dN} = \infty & T_{max} = T_c \end{cases} \quad (2)$$

where r_z is the crack extension rate when the tear energy is less than the threshold tear energy, and A_0, B, and F are the constants associated with the rubber material.

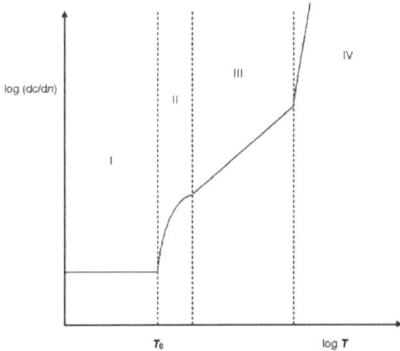

Figure 1. Crack expansion model of rubber material.

Among the above four stages, stage II and stage III belong to the stable expansion stage, and only the crack expansion rate of stage III satisfies the power exponential relationship with the peak tear energy, and this is also the stage in which the rubber products are often in when fatigue problems arise, so stage III can be used instead of the actual crack expansion process.

The visual representation can be obtained by taking the logarithm of both sides of the equation of stage III at the same time

$$\log \frac{da}{dN} = \log B + F \log T_{max} \tag{3}$$

Therefore, this equation becomes a straight line in logarithmic coordinates, and the influence of various factors on the crack extension of the rubber material can be shown more intuitively by the slope. Moreover, by establishing a fatigue life model to evaluate the crack expansion law, the data distortion caused by one experiment can be avoided.

After the crack extension rate is measured by the experimental platform built in this paper, it is also necessary to know the maximum tear energy density at the current strain level, and the equation of tear energy density is shown in Equation (4):

$$T_{max} = 2k(\lambda) l_0 E_0 \tag{4}$$

where is the strain energy density of the rubber specimen without pre-crack, is the length of the pre-crack, and is a function of the strain related to the rubber specimen, which is calculated as

$$k(\lambda) = \frac{\pi}{\sqrt{\lambda}} \tag{5}$$

where is the elongation of the test piece.

Substituting Equation (5) into (4) yields

$$T_{max} = \frac{2\pi l E_0}{\sqrt{\lambda}} \tag{6}$$

Combining Equations (3) and (6) yields

$$N = \int_{l_0}^{l_c} B^{-1} T^{-F} da = \frac{1}{B(F-1)\left(\frac{2\pi}{\sqrt{\lambda}} E_0\right)^F} \left(\frac{1}{l_0^{F-1}} - \frac{1}{l_c^{F-1}}\right) \tag{7}$$

where, is the length of the preset crack, is the length of the crack at fatigue failure. The Formula (7) is the formula for calculating the fatigue life of rubber materials.

1.2. The Effect of Temperature on Rubber Fatigue

Most rubber products are used in air and therefore most rubber fatigue studies are in air medium. Le-Gorju et al. [13] found that rubber exposed to air reacts with oxygen, especially at high temperatures where oxidation is more intense. Temperature is a crucial factor affecting the fatigue life of rubber, which is a temperature-sensitive material, and different types of rubber have different sensitivity to temperature; for example, the life of natural rubber at 110 °C is 1/4 of that at 0 °C, while the life of butadiene rubber decreases by 10,000 times [14]. Ruellan et al. [15] conducted uniaxial tensile fatigue experiments on filled natural rubber at different temperatures and found that strain crystallization was clearly observed on the fracture surface at room temperature, while at 90 °C, a significant decrease in crystallization could be observed on the fracture surface, and the crystallization disappeared completely at 110 °C. Therefore, high temperature affects the crystallization of natural rubber; Ngolemasango et al. [16] also found such a situation in their experiments. Rey et al. [17] studied the change in the properties of silicone rubber at different temperatures, and when the microstructure stable, the hardness of unfilled silicone rubber increased continuously with increasing temperature; while for filled silicone rubber, its hysteresis, stress relaxation and stress softening decreased continuously with increasing temperature. Haroonabadi et al. [18] thermally aged allene nitrile butadiene rubber (NBR) for 7 days and found that its crosslink density increased while its tensile and tear strength decreased. Chou et al. [19] thermally aged EPDM rubber for 6 months and then conducted fatigue experiments, which showed that the rubber life was reduced regardless of whether it was filled with carbon black. This is due to the fact that high temperature accelerates the thermal oxygen reaction, which continuously degrades the cross-linked network, so the increase in aging temperature and time irreversibly reduces the fatigue life of the rubber [20]. Luo et al. [21] conducted fatigue experiments on hourglass-type rubber specimens, in which real-time monitoring of the surface temperature revealed that the surface temperature remained in a stable interval for a long time, which accounted for most of the fatigue life. Then, the surface temperature increases sharply when the specimen is close to destruction, so the change in temperature can be used to determine when the specimen is reaching its fatigue limit. Then, the relationship between the steady-state temperature rise and the maximum principal strain is determined so as to determine the fatigue life of rubber.

The most important feature of natural rubber that distinguishes it from synthetic rubber is the crystallization phenomenon, which includes strain-induced crystallization and low-temperature-induced crystallization, and a large number of studies have shown that the fatigue life of natural rubber is substantially enhanced under non-relaxation loading (i.e., the ratio of minimum loading to maximum loading > 0) [22]. The crystallization phenomenon is considered to be the reason for the high fatigue resistance of rubber materials. Rubber materials exhibit different properties at different temperatures, for example, below the glass transition temperature, rubber is in a glassy state, and above that temperature, it is in a highly elastic state, indicating that rubber materials are highly temperature-sensitive materials. Most rubber products work in environments ranging from ambient temperature to 100 °C. Federico et al. [23] showed that microdefects in rubber materials at ambient temperature mainly originate from cavities generated by the separation of agglomerates from the rubber matrix, while at high temperature, they mainly originate from the fracture of agglomerates. Schieppati et al. [24] studied the fatigue properties of NBR, and found that the higher the temperature, the faster the crack growth rate, while the crack growth rate changed slightly at 25 °C~40 °C. Liu Xiangnan et al. [25] conducted uniaxial tensile fatigue tests on dumbbell-shaped natural rubber specimens, and found that the fatigue life was dispersive under the same conditions, so three probability distribution models, namely normal distribution, lognormal distribution and Weibull distribution were used to quantify the life distribution. Demiral et al. [26] made a finite element simulation of a bonded joint and used the user-defined UMAT subroutine of ABAQUS/Standard to link the static damage and fatigue damage models of the bonded zone to express the response of the bonded layer. Fang Yunzhou et al. [27] adopted the Arrhenius model, introduced the high temperature aging factor into the fatigue model with the engineering strain peak as the damage parameter, and accurately predicted the fatigue life of rubber bushing using finite element analysis. Weng et al. [28] subjected natural rubber to a high temperature (85 °C) and cyclic loading at the same time. SEM images confirmed the appearance of nano-scale cracks and cavities under the combined action of high temperature and cyclic loading. Unlike the fatigue loading conditions at room temperature, the cracks were caused by the nucleation effect of dissolved steam and gas in the low-molecular-weight domain of NR. The appearance of a low-molecular-weight domain is caused by thermal degradation products. Zhang et al. [29], based on the uniaxial tensile fatigue data of dumbbell-shaped rubber specimens, used the least square method to fit the functional relationship between the strain energy and temperature, established the high-temperature life-prediction formula and achieved good results. Luo et al. [30] studied the static tearing behavior of carbon-black-filled rubber at different temperatures and measured the critical tearing energy. The results showed that the critical tearing energy decreased exponentially with the increase of temperature.

1.3. Research Significance

The fatigue performance of rubber is affected by many factors, among which the temperature is the most significant. Many factors that affect the fatigue performance of rubber actually affect its fatigue by changing the temperature of rubber. For example, frequency itself has no effect on fatigue performance, and high frequency increases the temperature of rubber, thus affecting its fatigue life. Seichter et al. [31] reached the same conclusion when studying the effect of frequency on rubber fatigue. Therefore, the study of rubber fatigue should be based on temperature and establish the analytical relationship between temperature and fatigue life. At present, there are few studies on fatigue prediction of rubber based on crack propagation method, and no unified conclusion has been formed.

The main purpose of this paper is to build a fatigue experiment platform and design a fatigue experiment method based on the crack propagation method, so as to form a set of evaluation methods for studying rubber fatigue at variable temperature, and then establish a rubber thermal-mechanical coupling fatigue prediction model according to a large number of experimental results, which provides valuable reference for rubber fatigue research. Figure 2 is the technical route of research.

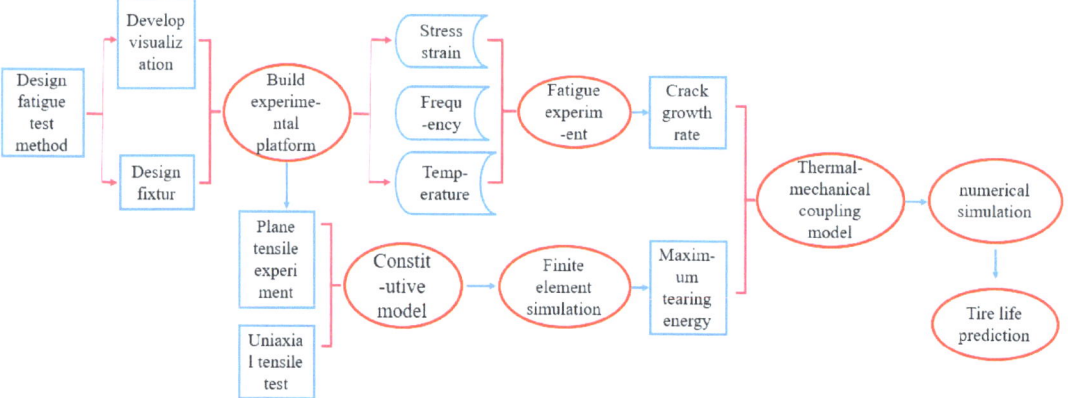

Figure 2. Research Route.

2. Construction of Fatigue Test Rig and Parameter Acquisition

2.1. Mullins Effect Study

When the rubber undergoes such a cyclic loading of stretching-unloading-restretching, the stresses of unloading and restretching are smaller than those of stretching, which is the Mullins effect of rubber, also known as stress softening phenomenon, and is a special mechanical property in rubber, especially in filled rubber, which has an important influence on the acquisition of stress-strain experimental data. This experiment was conducted to study the Mullins effect at different stretching speeds, starting from 2 mm/min and increasing to 500 mm/min, which were divided into low-speed stretching and high-speed stretching in order to distinguish them. In order to minimize the effect of speed during unloading in cyclic loading experiments, the unloading speed was unified to 2 mm/min when low-speed experiments were conducted, and the unloading speed was set to 50 mm/min under high-speed experiments. low-speed experiments were set to stretch at 2 mm/min, 5 mm/min, 10 mm/min, 20 mm/min; high-speed experiments were set to stretch at 50 mm/min, 100 mm/min, 500 mm/min.

2.1.1. Low-Speed Stretching (≤50 mm/min), the Effect of Stretching Speed on the Mullins Effect

In this experiment, experimental tests were conducted to stretch 12.5 mm, 25 mm, 37.5 mm, 50 mm, and 62.5 mm, i.e., strain levels of 50%, 100%, 150%, 200%, and 250% to study the effect of different stretching speeds on the Mullins effect, and the experimental data for 50% and 150% strains were selected in this case, as shown in Figures 3 and 4. It can be seen that the faster the stretching speed is, the higher the required stress is when stretching to the maximum strain, whether it is 50% or 150% strain; and it can be seen that the stretching curve for the larger stretching speed is above the curve for the lower speed.

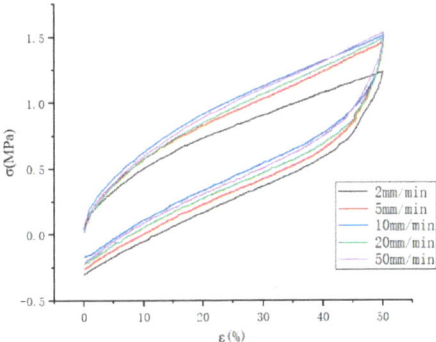

Figure 3. Tensile curves at different tensile speeds with strain levels of 50%.

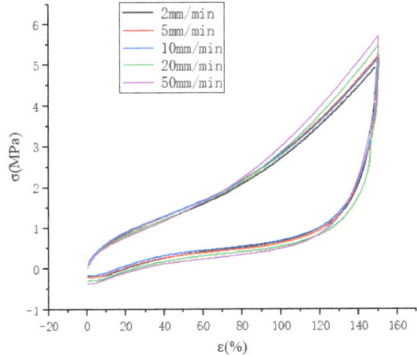

Figure 4. Stretching curves at different stretching speeds with 150% strain level.

In this experiment, experimental tests were conducted to stretch 12.5 mm, 25 mm, 37.5 mm, 50 mm, and 62.5 mm, i.e., strain levels of 50%, 100%, 150%, 200%, and 250% to study the effect of different stretching speeds on the Mullins effect. The experimental data for 50% and 150% strain were selected in this case, as shown in Figures 5 and 6. It can be seen that the faster the stretching speed is, the higher the required stress is when stretching to the maximum strain, whether it is 50% or 150% strain. It can also be seen that the stretching curve for the higher stretching speed is above the curve for the lower speed.

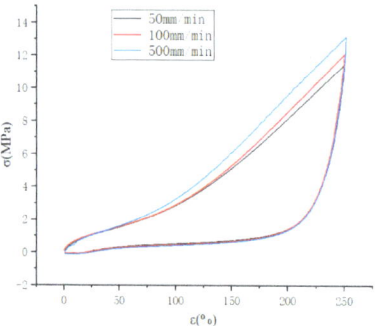

Figure 5. Cyclic curves for different stretching speeds at 250% strain level.

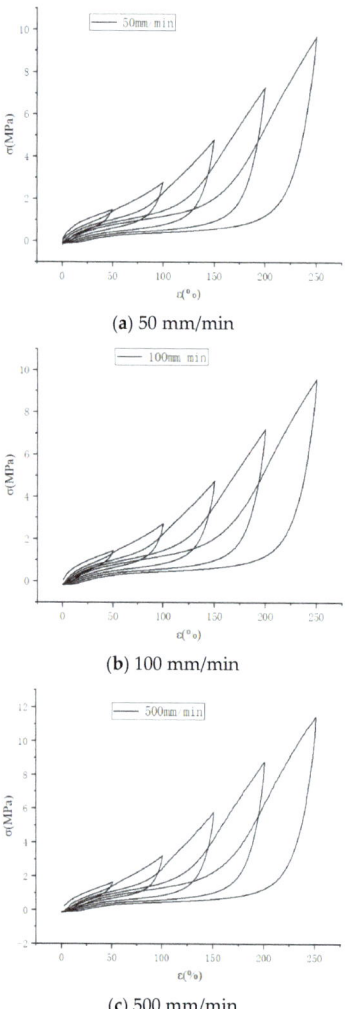

(**a**) 50 mm/min

(**b**) 100 mm/min

(**c**) 500 mm/min

Figure 6. Uniaxial cyclic stretching curves at different stretching speeds.

This can be explained by the high elasticity deformation theory of rubber [32]. The molecular structure of rubber creates a special property of high elasticity, where many elongated molecular chains are either adsorbed on the filler particles or in a free curl, with a large amount of flexibility and mobility. When the rubber is subjected to tensile load, the molecular chains are gradually stretched from the initial free-curl state so that the molecular chains are oriented and therefore the conformational entropy of the molecular chains decreases. From the thermodynamic point of view [33], the tensile load transmits mechanical energy to the rubber molecules, and this energy is converted into the thermal motion of the rubber molecules, which induces the rubber molecular chains to return from the stretched state to the free-curl state, increasing the conformational entropy. This is why the rubber material can return to its original length after intense deformation. The greater the stretching speed, the greater the strain rate, the higher the frequency, the more intensified the thermal movement of molecules, and the more high-speed stretching in a very short period of time. This means that the high-speed stretching process is somewhat adiabatic, that is, the heat generated by the thermal movement of molecules cannot be dissipated and can only increase, so it accelerates the increase in conformational entropy.

2.1.2. High-Speed Stretching ($\geq$50 mm/min), the Effect of Stretching Speed on the Mullins Effect

As shown in Figure 5, the cyclic stretching curve at a high speed is similar to that at a low speed.

And it can also be seen that the curve has a tendency to rise suddenly at 500 mm/min stretching speed, which may be a sudden increase in stress caused by the strain crystallization of natural rubber.

As shown in Figure 6, the cyclic stretching curves are plotted for stretching speeds of 50 mm/min, 100 mm/min and 500 mm/min, respectively, and it can be seen that the rising curve (c) has a sudden upward trend at the beginning of loading, which also proves the possibility of strain crystallization of natural rubber as described above. At the same time, it can be seen that the unloading curve of each cyclic stretching is steeply decreasing, which indicates that the cohesive structure inside the rubber is destroyed in a large amount, and the stress required for unloading decreases after each destruction, and the loading curve of the latter stretching is suddenly increasing near the top of the loading curve of the previous stretching, which also reflects the continuous destruction of the cohesive structure inside the molecule. From the thermodynamic point of view, the loading process stretches the molecular chain to the straightened state at each cycle of stretching, and the conformational entropy decreases accordingly, and the rubber deformation decreases when unloading, but the molecular chain does not return to the curled state completely, and the conformational entropy does not decrease again from the equilibrium state when loading again, but decreases from below the equilibrium state, so the required stress decreases when stretching again.

An explanation for the Mullins effect was given by Diani and Marckmann et al. [34,35], who explained the Mullins effect in terms of chain breakage, where they suggested that the rubber matrix reacts with the filler by cross-linking during vulcanization to form a cross-linked structure, and the applied load leads to the stretching and breaking of the molecular chains.

As shown in Figure 7, (a) is the state of the molecular chain when the rubber is not subjected to tensile load, and the molecular chains A, B and C are adsorbed on the filler particles through chemical adsorption and physical adsorption, and are in the bent state; (b) is the state of the rubber specimen when it is stretched. When the stretching experiment is carried out, the internal molecular chains of the rubber start to be stretched, firstly the shorter molecular chain C is stretched from the bent state to the straightened state, at this time the straightened molecular chain bears most of the stress and continues to stretch, the longer molecular chain A is also gradually stretched from the bent state to the straightened state, while the shorter molecular chain C is pulled off due to overload. This is the reason why the stress-strain curve of the re-stretching in the above cyclic stretching experiment is always lower than the stress-strain curve of the first stretching.

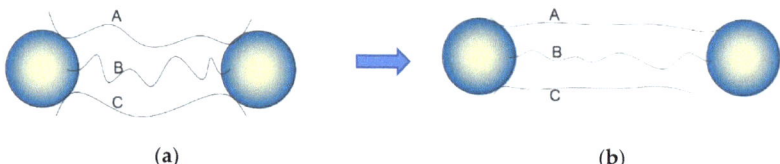

Figure 7. Rubber molecular chain breakage model. (**a**) the chain is not stretched. (**b**) the chain is stretched. A, B, and C are molecular chains.

The Mullins effect has an impact on the acquisition of rubber material parameters, so it is necessary to eliminate the Mullins effect of rubber materials when testing material parameters, as shown in Figure 8. The parameters of the rubber materials were first stretched in five cycles, and the data of the sixth stretching were taken.

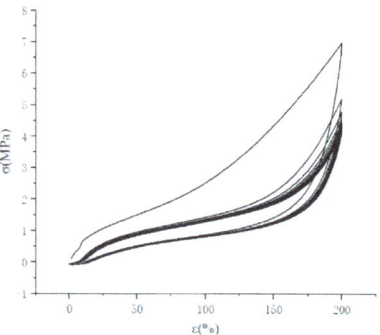

Figure 8. Graph of cyclic tensile stress-strain curve of rubber material.

From the above experiments on the effect of stretching speed on the Mullins effect, it is concluded that the damage to the internal structure of the rubber material is different when stretched to different strain levels, and when subjected to smaller strains, the filler network inside the rubber is damaged first, and when the strain level continues to increase, the reinforcement network generated by the rubber matrix and the filler starts to be damaged, and when stretched to a larger strain level, the interpolymer The network structure between the polymers also starts to be destroyed when the strain level is increased [36]. At the same time, this experiment also leads to another conclusion that the tensile experiments of rubber materials cannot be simply classified as static or dynamic, because the boundary of distinction is blurred, and this experiment was conducted by varying the tensile speed, and it was concluded that as the tensile speed increases, the stress required to stretch the rubber material to the same strain level also increases, which is due to the fact that the main unit of motion of rubber in the high-elastic state is "When increasing the stretching speed, the strain rate of rubber molecules increases, and the smaller "chain segments" are first subjected to tension and are the main units of motion, so the stress required for stretching is greater.

2.2. Fatigue Test Bench Construction

From the previous section, it is known that the crack expansion stage can be divided into four stages, and the third stage is the focus of calculation. To accurately calculate the fatigue life, accurate material parameters should be obtained, and as a polymer material capable of withstanding intense deformation and with nonlinear characteristics, an accurate description of its constitutive model is the key to calculate its fatigue life. From previous studies, it is necessary to fit the experimental data of uniaxial tensile, equiaxial tensile and plane tensile to describe the hyperelastic constitutive model of rubber, and the loading schematic is shown in Figure 9. These three experiments characterize the mechanical behavior of rubber materials in uniaxial tension, uniaxial compression, and pure shear states, respectively, and the alternating loads on rubber products in actual use can be represented by the coupling of these three force states [37].

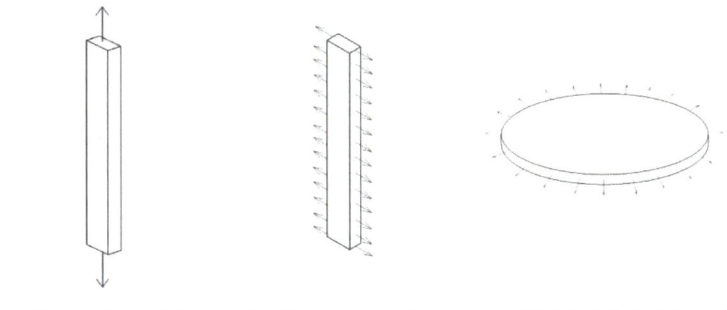

(a) Uniaxial stretching   (b) Planar stretching   (c) Isobiaxial stretching

Figure 9. Loading diagram of uniaxial tension, plane tension and equal biaxial tension.

From the mechanical point of view, The fatigue test in this paper is shear fatigue, and there is no equal biaxial tensile stress field. Only the experimental data of uniaxial tension and planar tension are needed to fit the hyperelastic constitutive model of rubber more accurately.

From Figure 9, it can be seen that the stress field of the rubber material is different for different specimen shapes and tensile states, and the rubber specimen required for uniaxial tensile is a flat dumbbell specimen, and the flat tensile specimen is self-designed, as shown in Figure 10. Mechanically, when the length-width ratio of the specimen is greater than or equal to 10, it can be approximately considered that the stress state is in a pure shear state, so the size of the specimen is designed to be 140 mm long, the height of the working area is 10 mm, and the thickness is 2 mm. For fatigue test, it is necessary to preset a crack of 25 mm, as shown in Figure 11.

(a) Uniaxial tensile specimens   (b) Flat tensile specimen

Figure 10. Uniaxial tensile specimen and plane tensile specimen.

Figure 11. New fixture.

In the uniaxial tensile experiments, the Mullins effect has a greater impact on the acquisition of experimental data, which can be essentially eliminated after 5–6 times of stretching, so the stress–strain data from the sixth stretch is selected after five cycles of stretching first. However, it should be noted that the creep phenomenon also exists in rubber materials, so after five times of cyclic stretching, the equipment should be allowed pause for 1 min so that the creep phenomenon can be alleviated, then the extensometer should be reclamped to maintain the standard distance and straightened state to ensure the length of 25 mm, and then the sixth stretching can be carried out.

The equipment used for uniaxial stretching of rubber is the universal experiment machine of high speed rail and its own fixture. The plane stretching experiment requires self-designed fixture for the experiment because there is no national standard, and the fixture is shown in Figure 11. To ensure the accuracy of the experimental data, each test was conducted three times and the average value was taken. The process of obtaining tensile stress-strain data is shown in Figure 12, with a tensile speed of 50 mm/min first 5 cycles of tensile to eliminate the Mullins effect, and the sixth tensile to obtain stress-strain data after resting for 1 min.

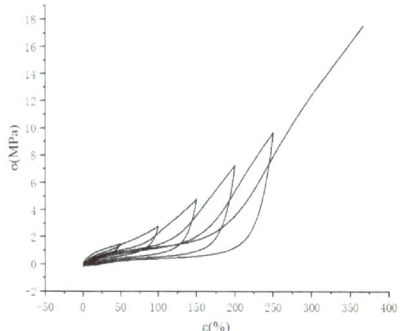

Figure 12. Stress-strain curve.

In this paper, based on the relationship between the crack expansion rate and tearing energy established by Thomas et al., the fatigue life calculation model of rubber materials was established in order to obtain the material parameters required for calculating the fatigue life of rubber using the crack expansion method. We designed the fixture, specimen and vulcanization mold used in the shear fatigue experiment of rubber materials and built the planar tensile and fatigue experimental platform according to the characteristics of fatigue experiments, as shown in Figure 13a. Since the conventional fixture is prone to screw loosening after a long period of fatigue experiments, which leads to the problem of the specimen clamping sliding and eventually distorting the stress–strain data during the test, the new fixture and specimen used in this paper can avoid the above problems. These two reasons are likely to lead to errors in the experimental results and affect the accuracy of the subsequent life calculation process. Therefore, after considering various factors, we designed our own fixture and specimen. The designed specimen is vulcanized into a cylindrical shape at the upper and lower ends, and the upper and lower two cylinders are embedded in the cylindrical groove of the designed fixture. The computer can display the tensile stress in real time, and adjust the stress to 0, that is, to reach the initial unstressed state of the specimen, after which the experiment can begin. This fixture can avoid the clamp loosening caused by the screw becoming loose accurately capture any quick changes in any part of the specimen, and also avoid errors caused by the inaccurate height of the single measurement work area. When carrying out fatigue test, a 25 mm crack is preset on one side of the specimen, as shown in Figure 13b, According to fracture mechanics, when the crack length is greater than or equal to twice the height, the influence of edge effect on the shear fatigue test can be ignored. Additionally, the accuracy when measuring the crack expansion length affects the accuracy when constructing the fatigue life calculation model, so this test bench is designed with a vision system for measuring the crack expansion length, as shown in Figure 13b.

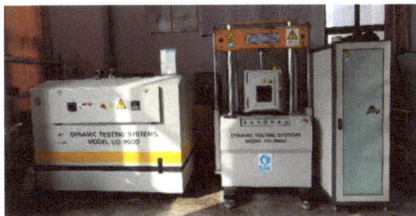

(a) Fatigue test bench.

(b) Visual system.

Figure 13. Dynamic fatigue analysis system.

2.3. Selection of Constitutive Model

This experiment is conducted as a shear fatigue experiment, and its stress field is a combination of uniaxial and planar tension, so fitting the intrinsic parameters requires the use of stress-strain data from uniaxial and planar tension, however, some literature uses only data from uniaxial tension experiments for fitting, and the accuracy of the fitting results needs to be considered. Assuming that the rubber deformation is within the Gaussian chain and conforms to the Gaussian distribution function, the strain energy function equation [38],

$$W = \frac{1}{2}G\left(\lambda_1^2 + \lambda_2^2 + \lambda_3^2 - 3\right) \qquad (8)$$

where is the strain energy of the rubber, is the shear modulus, and is the elongation in the x, y, and z directions, respectively.

In uniaxial stretching, the elongation in the three directions are $\lambda 1 = \lambda$, $\lambda 2 = \lambda 3 = \lambda - 1/2$, so the (8) equation becomes

$$W = \frac{1}{2}G\left(\lambda^2 + \frac{2}{\lambda} - 3\right) \tag{9}$$

Derivation of Equation (9) yields

$$\sigma = G\left(\lambda - \frac{2}{\lambda^2}\right) \tag{10}$$

Equation (10), which is the equation of state for uniaxial stretching of rubber materials; Similarly, the equation of state for plane stretching is obtained for the state of $\lambda 1 = \lambda$, $\lambda 2 = 1$ and $\lambda 3 = \lambda - 1$.

$$\sigma = G\left(\lambda - \frac{1}{\lambda^3}\right) \tag{11}$$

Isobiaxial stretching with $\lambda 1 = \lambda 2 = \lambda$ and $\lambda 3 = \lambda - 2$ gives the equation of state for isobiaxial stretching

$$\sigma = 2G\left(\lambda - \frac{1}{\lambda^5}\right) \tag{12}$$

From the state equations of the above three stress fields, it can be seen that the specimens in the same elongation ratio have the highest stress when subjected to equal biaxial stretching, followed by planar stretching, and the lowest is seen in uniaxial stretching. The gap between them can be seen more clearly from the experimental graph.

As shown in Figure 14 for the experimental data and simulation data of uniaxial stretching and plane stretching, it can be seen that there is a large gap between the stress-strain curves of uniaxial stretching and plane stretching, and the hyperelastic model needs to be used in calculating the strain energy density of rubber, so for the stress field of the combination of uniaxial stretching and plane stretching, the data of uniaxial stretching cannot be used only, which will cause large deviation, so it is necessary to The uniaxial tensile and planar tensile data need to be obtained simultaneously. In this paper, we use Ogden model to fit the experimental data with a high degree of overlap with the experimental data, which can meet the calculation requirements. The results are shown in Table 1.

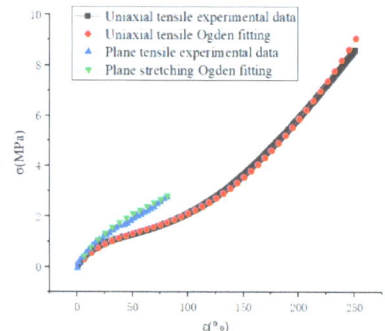

Figure 14. Uniaxial tensile and planar tensile experimental data and fitted data.

Table 1. Material parameters fitted by Ogden model.

No.	MU1	MU2	MU3	ALPHA1	ALPHA2	ALPHA3
Parameters	1.485	−5.016	10.038	1.704	12.501	−24.999

Note: The unit of each parameter is MPa.

The non-working area of the specimen in this experiment is cylindrical, which is likely to cause convergence problems for the finite element simulation, and deleting this part during the simulation will not affect the simulation results, so the non-working area is removed from the simulation. As shown in Figure 15, the strain energy density is required in the calculation model of ABA fatigue life, and the strain energy density is calculated by finite element simulation in this paper. The size of the rubber fatigue specimen in this experiment is 140 mm × 30 mm × 2 mm, and the size of the working area is 140 mm × 10 mm × 2 mm. As the finite element model is established in QUS, the material parameters are set and submitted for calculation. The maximum strain energy density of the working area is obtained. As shown in Figure 15, the grid division diagram of the finite element model is shown, and Figure 16 shows the loading mode.

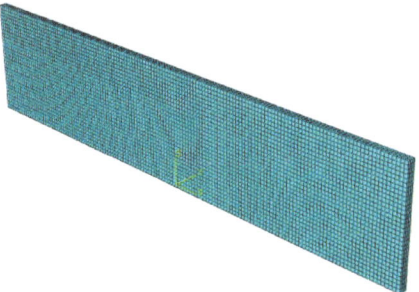

Figure 15. Finite Element Model Grid Diagram.

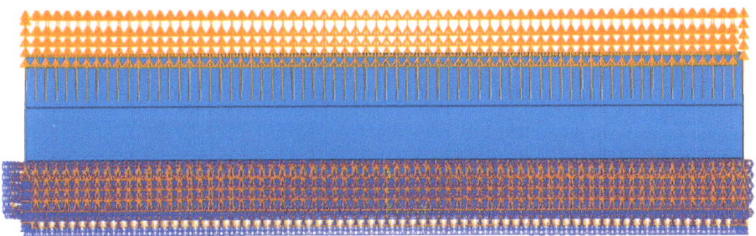

Figure 16. Loading Mode.

There is an edge effect in the calculation of the finite element model, as shown in Figure 17b, which needs to be removed manually, so the maximum strain energy density should be selected after removing the value of the fixture edge. The strain energy densities at different strain levels are shown in Table 2. The maximum tear energy Tmax can be obtained by substituting the maximum strain energy density under different strains into Equation (4).

Table 2. Strain energy density at different strain levels.

Strain	30%	50%	80%	100%
strain energy density	0.745	1.385	2.311	3.370
Tearing energy	102.637	177.633	270.572	374.313

Note: The unit of strain energy density is mJ/mm^3.

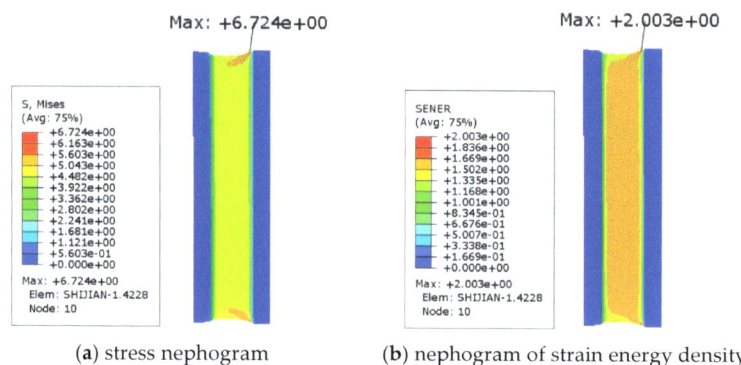

(a) stress nephogram (b) nephogram of strain energy density

Figure 17. Simulation Results of Stress and Strain Energy Density.

3. The Relationship between Fatigue and the Variables

Rubber is a temperature-sensitive polymer material, and different working conditions will have different effects on its fatigue performance, especially the temperature. High temperature will make the rubber material soft, so the stress-strain curve at high temperature will also be different. This paper intends to investigate the effects of different factors on the fatigue performance of rubber by conducting fatigue experiments at different temperatures, different frequencies, different orientations and different loading methods.

3.1. Stress Ratio

The stress ratio R is the ratio of the minimum stress to the maximum stress that the rubber specimen is subjected to during cyclic loading. For different rubber materials, the effect of stress ratio on fatigue life is different. Scholars at home and abroad have conducted a large number of experiments to study the effect of stress ratio, and although some useful conclusions have been drawn, the overall law is not universal, so to understand the effect of stress ratio of the formulation used in this experiment, it needs to be studied by experiment.

Poisson et al. [39] conducted uniaxial tensile fatigue experiments using neoprene and showed that when the stress ratio $R \geq 0.2$, the fatigue life of the specimen increased with the increase of the stress ratio, which means that the crack expansion rate was decreasing. In the study of magneto-rheological elastomers, Yong [40] concluded that the life of magneto-rheological elastomers remained almost unchanged as the stress ratio increased, which he attributed to the fact that the fatigue of magneto-rheological elastomers belongs to high circumferential fatigue and the stresses applied during the experiments were much less than their fatigue limits.

While this experiment was conducted with different stress ratios, as shown in Table 3, it was found that the fatigue life of the rubber material surged when the stress ratio was applied for the experiment, and the fatigue life increased by an order of magnitude with the stress ratio, and the fatigue life of the rubber increased with the increase of the stress ratio, which indicates the possibility of crystallization of natural rubber in the tensile state, as observed by Beatty et al. [41] in their study Non-crystalline rubber in the stress ratio $R > 0$ was not observed to enhance the life; and the rubber material has creep phenomenon, the rubber material creep during long time clamping, as the experiment proceeds, the stress ratio R due to the creep of rubber will become smaller and smaller, so the change in rubber strain caused by the stress ratio R will be smaller and smaller, then the transfer to the crack interface, caused by the damage to the crack tip should be small. This may be the reason why the increase of stress ratio will cause the fatigue life to increase.

Table 3. Crack expansion rates at different stress ratios.

No.	R	Frequency	Waveform	Strain	dc/dN
1	0	4 Hz	Sine wave	80%	6.143×10^{-4}
2	1/3	4 Hz	Sine wave	80%	6.944×10^{-5}
3	1/2	4 Hz	Sine wave	80%	3.472×10^{-5}

Note: The unit of crack expansion rate is mm/c.

3.2. Loading Method

As shown in Figure 18, the effect of different loading methods on fatigue life is negligible, and the fitted function curves are

$$\log\left(\frac{dc}{dN}\right) = 3.117\log(T_{max}) - 10.571 \quad (13)$$

$$\log\left(\frac{dc}{dN}\right) = 3.114\log(T_{max}) - 10.561 \quad (14)$$

$$\log\left(\frac{dc}{dN}\right) = 3.137\log(T_{max}) - 10.614 \quad (15)$$

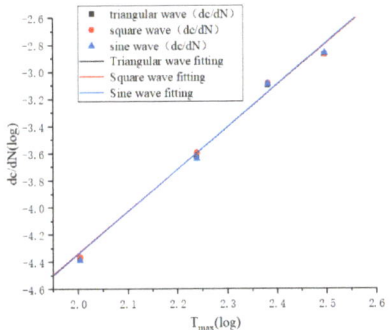

Figure 18. Fitting curves of tear energy and crack expansion rate under different loading methods.

The above Equations (13)–(15) are the fitted curves under triangle wave, square wave and sine wave loading respectively, and it can be seen that the slope and intercept difference of the three equations are very small, because the different loading methods affect the change of strain following stress, and the same material has the same kinematic unit, so the relaxation time of strain relative to stress change is unchanged at the same frequency; meanwhile At the same time, the temperature of the specimen will not rise when loading at low frequency, so it will not produce the change of temperature and thus affect the change of internal energy. These two reasons may be the reason why the loading method does not affect the fatigue life of rubber materials.

3.3. Frequency

The effect of frequency on the rubber material is reflected in the fact that high frequency increases the temperature of the rubber material, which affects the crack expansion rate. As shown in Figure 19, fatigue experiments were conducted at 1 Hz, 5 Hz, 8 Hz, 12 Hz, and 15 Hz, respectively, and it can be seen that the change in crack expansion rate at 1 Hz, 5 Hz, and 8 Hz is small, while the crack expansion rate increases significantly at 12 Hz and

15 Hz frequencies. The fitting equations for the five frequencies are shown in the following Equations (16)–(20).

$$\log\left(\frac{dc}{dN}\right) = 3.259\log(T_{max}) - 10.803 \qquad (16)$$

$$\log\left(\frac{dc}{dN}\right) = 3.283\log(T_{max}) - 10.845 \qquad (17)$$

$$\log\left(\frac{dc}{dN}\right) = 3.282\log(T_{max}) - 10.831 \qquad (18)$$

$$\log\left(\frac{dc}{dN}\right) = 3.389\log(T_{max}) - 10.948 \qquad (19)$$

$$\log\left(\frac{dc}{dN}\right) = 3.507\log(T_{max}) - 11.128 \qquad (20)$$

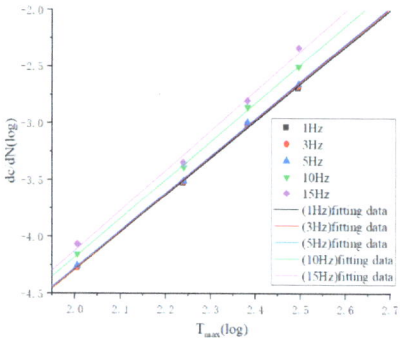

Figure 19. Fitting curves of crack expansion rate and tearing energy at different frequencies.

From the first three equations, the slope and intercept difference is very small, while the slope of the last two equations increases significantly, indicating the enhanced crack expansion rate. It indicates that the first three frequencies did not lead to a significant increase in the temperature of the rubber specimen, so the change in frequency did not have an effect on the fatigue life, while the high frequencies led to an increase in the temperature of the rubber, so the internal energy and conformational entropy of the rubber increased, making the generation of new surfaces easier and therefore accelerating the crack extension rate.

3.4. Effect of Orientation on Fatigue Life

As shown in Table 4, the fatigue life of the specimens under different orientations has a large difference, and the orientation refers to the direction of the rubber material taken during vulcanization of the rubber specimen. The polymer chain of the rubber material is a long chain linear structure, so the direction of the molecular chain will be affected by the process. Generally, the direction of the molecular chain is the same as the calendering direction, so the way of taking rubber during vulcanization is also very important. If the orientation of the vulcanized specimen is the same as the calendering direction, the modulus and strength of the specimen can be improved; when the direction of the preset crack of the fatigue specimen is the same as the calendering direction, which is equivalent to the tearing direction and the molecular chain direction is parallel to the direction of calendering, and the direction of the preset crack is perpendicular to the direction of calendering, then the molecular chain must be torn off if the crack wants to expand, so the crack expansion rate is lower when the orientation is perpendicular, i.e., the fatigue life is higher.

Table 4. Crack extension rate under different orientations.

No.	Orientation	Strain	Waveform	dc/dN
1	Vertical	50%	Sine wave	2.778×10^{-4}
2	Parallel	50%	Sine wave	3.543×10^{-4}

Note: The unit of crack expansion rate is mm/c.

3.5. Effect of Mullins Effect on Fatigue Life

From the crack expansion rates in Table 5, it can be seen that the Mullins effect has some influence on the fatigue life. Of course, the difference between the specimens with and without the Mullins effect eliminated from the crack expansion rate is not large, which can also be considered as being caused by the experimental error. The elimination of the Mullins effect will cause a certain degree of damage since the cyclic tension process will cause the destruction of the filler network, the debonding between the rubber matrix and the filler, etc., which can create internal micro-defects of different degrees in the rubber. However, it cannot be simply assumed that the Mullins effect causes a reduction in fatigue life because the 5–6 cycles of stretching required to eliminate the Mullins effect are negligible compared to the fatigue life of tens of thousands of cycles. But whether it is reasonable to stretch to a higher strain level to eliminate the Mullins effect or whether it has an effect on the fatigue life requires further systematic experiments to verify.

Table 5. Effect of Mullins effect on crack expansion rate.

No.	Mullins Effect	Strain	Waveform	dc/dN
1	Eliminated	50%	Sine wave	2.454×10^{-4}
2	Not eliminated	50%	Sine wave	2.315×10^{-4}

Note: Eliminating the Mullins effect means stretching the specimen to 200% strain for 5 times and then presetting the crack for the fatigue test.

3.6. Construction of a Thermodynamic Coupling Model for Fatigue Life of Rubber

The importance of temperature for rubber is self-evident, and scholars started to study the effect of temperature on the fatigue life of rubber materials a long time ago. Mars found that the life of butadiene rubber decreased by 10^4 and that the life of natural rubber decreased by four times when the temperature increased from 0 °C to 100 °C. Viscoelasticity is one of the characteristics of rubber, and its elastic energy storage is transformed into heat energy when subjected to alternating load for a long time, which increases the temperature of rubber, especially for tire products, and the internal temperature of tires can reach nearly 100 °C when rotating at high speed, so it is of great significance to study the effect of temperature on rubber.

In this experiment, the crack expansion behavior and static tearing behavior of rubber materials are studied by changing the ambient temperature, the critical tearing energy is determined by static tearing experiments at different temperatures, and the critical tearing energy is the criterion on which to judge whether the rubber crack expansion is destabilized. Firstly, the specimens with preset cracks are stretched at a speed of 50 mm/min until they tear, and the state of instantaneous fracture is the strain corresponding to the critical tear energy.

In order to prevent damage to rubber caused by excessive thermal aging, the specimens were preheated at a predetermined temperature for 12 min before each trial, the surface temperature of rubber was measured to reach the predetermined temperature, and then the experiment was started.

As shown in Figure 20, the fatigue life fitting curves at different temperatures are plotted, and the fitting functions are, respectively

$$\log\left(\frac{dc}{dN}\right) = 3.019\log(T_{max}) - 10.779 \quad (21)$$

$$\log\left(\frac{dc}{dN}\right) = 3.157\log(T_{max}) - 10.658 \quad (22)$$

$$\log\left(\frac{dc}{dN}\right) = 3.106\log(T_{max}) - 10.380 \quad (23)$$

$$\log\left(\frac{dc}{dN}\right) = 4.020\log(T_{max}) - 11.906 \quad (24)$$

$$\log\left(\frac{dc}{dN}\right) = 4.263\log(T_{max}) - 12.583 \quad (25)$$

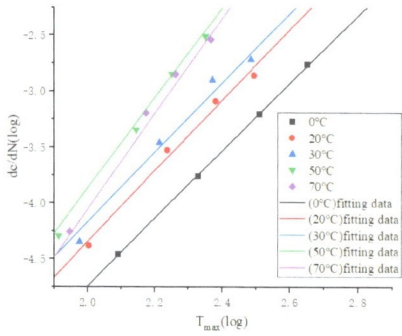

Figure 20. Fitting curves of crack expansion rate and tearing energy at different temperatures.

Equations (21)–(25) are the fitting functions at 0 °C, 20 °C, 30 °C, 50 °C and 70 °C, respectively. It can be seen that the slope of the curve gradually increases with the increase of temperature, but the slope at 20 °C is larger than that at 30 °C, which means that the crack expansion rate is higher at 20 °C. The possible reason is that the increase in temperature makes the rubber soft, and the stress is smaller when reaching the same strain, so the stress concentration of the crack tip is low, and the failure stress is not reached. This means that crack branching is more likely to occur and that there is more energy dissipation in the generation of new surfaces, therefore hindering the expansion of the crack. The phenomenon of crack branching was also observed in the experiment. Meanwhile, the fitted curve at 0 °C is lower than other that at temperatures, indicating that the rubber material is more prone to crack defects at higher temperatures, thus intensifying crack expansion. While continuing to increase the temperature, the crack expansion rate in the temperature range of 50–70 °C continues to increase, indicating that in this temperature range, the fatigue life of the rubber material decreases sharply due to the high temperature, and more micro-defects are generated inside the rubber when bearing the load. The long time spent bearing the alternating load makes the micro-defects gather and expand, while the temperature increase as the activation energy of the crack tip increases, causing the cracks expand rapidly. From above, it can be seen that high temperatures limit the strain-induced crystallization of rubber, so the life strengthening disappears. The above two aspects may be the reason for the poor fatigue resistance of rubber materials under high temperature.

Therefore, it can be concluded that the fatigue performance of rubber materials with the change of temperature is not a linear increase or decrease, but rather there are three stages: from 0 °C to about 20 °C, the crack expansion rate increases with the increase in temperature; while from 20 °C to 30 °C, temperature instead appears to increase the fatigue life; and when continuing to increase the temperature to between 50 °C and 70 °C, the fatigue life decreases sharply.

It can be seen that temperature has a great influence on the fatigue life of rubber materials, and the working environment of many rubber products is not room temperature, so it is necessary to take the influence of temperature into account when establishing the calculation model. From the above study, it is found that the tearing energy is related to temperature, so the tearing energy is considered as a function of temperature, and from the calculation formula of tearing energy:

$$T_{max} = 2k(\lambda)l_0 E_0 \tag{26}$$

$k(\lambda)$ is a function related to the strain of the rubber specimen, which is calculated as

$$k(\lambda) = \frac{\pi}{\sqrt{\lambda}} \tag{27}$$

Substituting (27) into (26) yields

$$T_{max} = \frac{2\pi l E_0}{\sqrt{\lambda}} \tag{28}$$

The tearing energy obtained at different temperatures is plotted in Figure 21, and fitting the experimental data points yields that the temperature and tearing energy satisfy a power-of-three relationship, as in Equation (29):

$$f(t) = a_1 t^3 + a_2 t^2 + a_3 t + a_4 \tag{29}$$

where t is the temperature, a_1, a_2, a_3, and a_4 are the relevant constants. Thus the equation for correcting the tearing energy, the

$$T_{max} = \frac{2\pi l E_0 f(t)}{\sqrt{\lambda}} \tag{30}$$

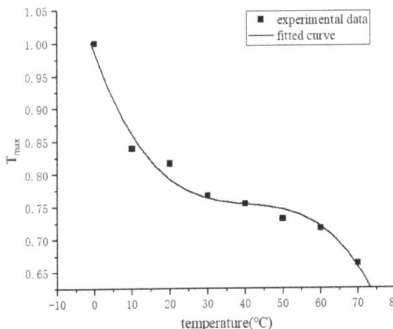

Figure 21. Tearing energy versus temperature.

The following are fitted in this paper at 30%, 50%, 80% and 100% strain with critical tearing energy respectively as shown below:

$$f(t) = -3.128e^{-6}t^3 + 3.810e^{-4}t^2 - 0.016t + 0.985 \tag{31}$$

$$f(t) = -3.251e^{-6}t^3 + 3.911e^{-4}t^2 - 0.016t + 0.982 \tag{32}$$

$$f(t) = -6.316e^{-6}t^3 + 7.214e^{-4}t^2 - 0.026t + 0.975 \tag{33}$$

$$f(t) = -7.501e^{-6}t^3 + 8.576e^{-4}t^2 - 0.031t + 0.970 \tag{34}$$

$$f(t) = -3.963e^{-6}t^3 + 4.430e^{-4}t^2 - 0.020t + 0.971 \tag{35}$$

Thus, the rubber thermodynamic coupling fatigue model can be obtained as

$$N = \int_{l_0}^{l_c} B^{-1}T^{-F}da = \frac{1}{B(F-1)\left(\frac{2\pi f(t)}{\sqrt{\lambda}}E_0\right)^F}\left(\frac{1}{l_0^{F-1}} - \frac{1}{l_c^{F-1}}\right) \tag{36}$$

An interesting phenomenon was also found in the experimental observation of crack extension paths. In the range of 30–100% of strain level, the crack paths of crack extension experiments at room temperature were more regular, basically expanding perpendicular to the direction of loading, while the extension paths at high temperature were more tortuous, usually zigzagging forward. The possible reason is that the temperature of the rubber crack surface is higher in high temperature environment, and the surface molecules are more easily activated, which accelerates the rate of crack expansion.

3.7. Validation of Fatigue-Thermal Coupling Model

The hyperelastic intrinsic model and fatigue life prediction model of rubber were established in the previous section. In order to ensure the accuracy of the subsequent tire simulation calculation, the prediction effect of the constructed model needs to be verified, so the fatigue life of the specimen at 50 °C was predicted by establishing the finite element model of the planar tensile specimen and compared with the Thomas model and experimental results to verify the prediction effect of the constructed thermodynamic coupling model, as shown in Figure 22 is the finite element model of the plane tensile specimen. Figure 23 shows the loading amplitude curve.

Figure 22. Finite element model of the planar tensile specimen.

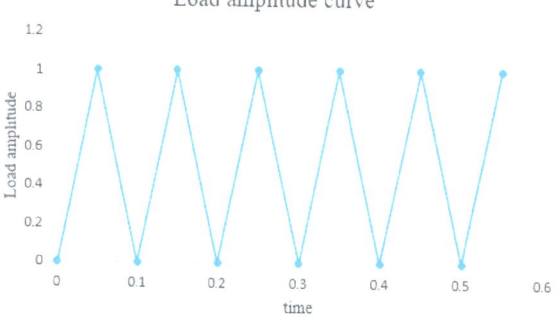

Figure 23. Loading amplitude curve.

The stress–strain clouds at 50% strain under planar tension are shown in Figure 24.

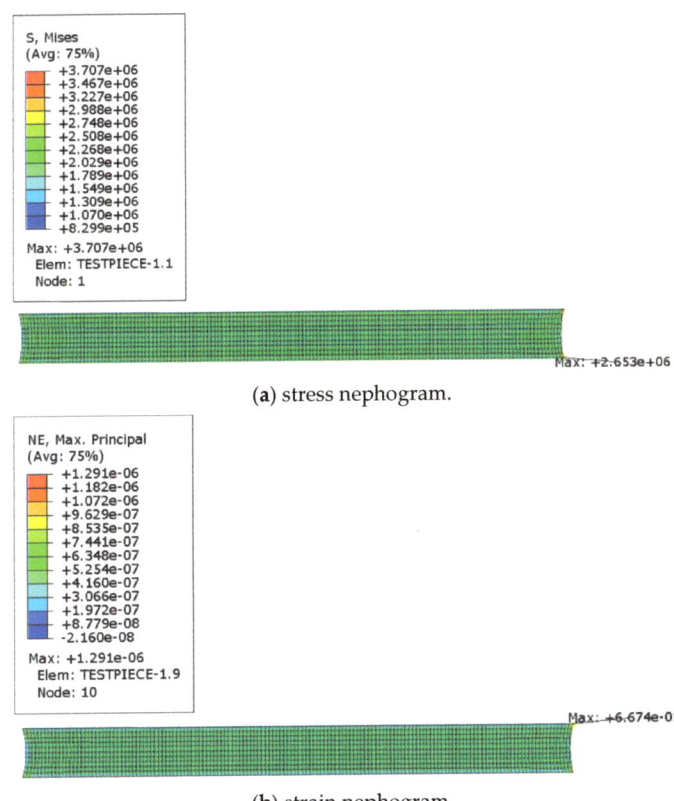

(a) stress nephogram.

(b) strain nephogram.

Figure 24. Stress-strain cloud of plane tensile specimen.

The calculated ODB file is imported into Endurica2020 software to calculate its fatigue life, and the Thomas model in the software library is selected for the calculation of life, and the finite element calculation result of life is shown in Figure 25, which is 8.315×10^5 times.

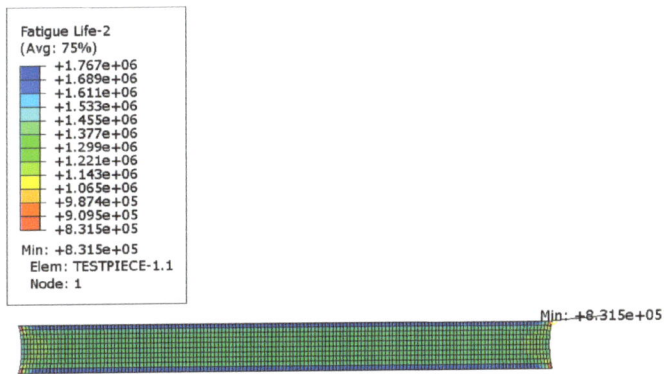

Figure 25. Life nephogram of plane tensile specimen.

As shown in Figure 26, the cloud plot of rubber specimen life calculated by using the thermally coupled fatigue model built in the previous section, the result is 6.588×10^5 times, and the experimental result is shown in Figure 27, which is measured as 6.42×10^5 times, as shown in Table 6, the deviation between the predicted result of Thomas model and the actual one is 29.5%, and the error between the predicted result of the thermally coupled fatigue model built in this paper and the experimental result is 2.6%. It indicates that the calculation accuracy of the proposed thermodynamic coupling fatigue model is high and can meet the requirements of the subsequent tire fatigue life calculation.

Figure 26. Nephogram of fatigue life of specimen under thermal-mechanical coupling.

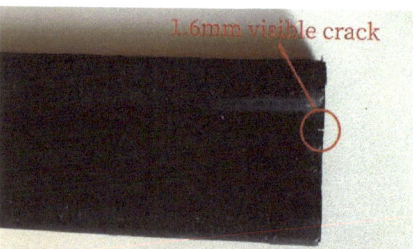

Figure 27. Fatigue specimen life.

Table 6. Comparison of life prediction effects.

Different Models	Thomas Model	Thermodynamic Coupling Model	Experimental Results
Fatigue life	8.315×10^5	6.588×10^5	6.42×10^5

4. Study of Fatigue Micromechanisms in Rubber

Many scholars, when studying fatigue, often characterize it by macroscopic features such as the length, type, and lifetime of cracks, while the damage of polymeric materials such as rubber often begins with the breakdown of the bonding between the atoms of polymer chains. And by Champy et al. [42], who studied the fatigue process of natural rubber, it was concluded that cracking starts with debonding of the agglomerates from the rubber matrix and cavities at the ends of the agglomerates, and Pérocheau et al. [43] also found the phenomenon of cavities at the ends of the agglomerates, and Le Cam, who studied [44] the fatigue life of carbon black-filled natural rubber, came to a similar conclusion that the agglomerates and rubber matrix debonded to produce cracks.

Therefore, it is known from the above studies that rubber products need a variety of fillers in the rubber matrix to meet different requirements for use, and the addition of fillers will inevitably form agglomerates, and when subjected to alternating loads, defects will be formed between the rubber matrix and the agglomerates, and these defects will gradually expand and gather to form microcracks, and the fatigue failure of rubber materials starts from these microcracks. From a large number of scholars' studies, it is known that the size of the initial microcrack has a large influence on the fatigue life of rubber materials, and the size of the initial crack mainly depends on the size of the agglomerates and the aggregates of inorganic fillers such as ZnO, and observing the scale of the agglomerates by scanning electron microscopy (SEM) is the best way at present. This chapter attempts to verify whether the use of the crack extension method is consistent with reality by studying the surface morphology of rubber materials before and after fatigue, and to explore the connection between the macroscopic and microscopic aspects of fatigue.

4.1. Experiments and Discussion of Results

4.1.1. Experiment

In order to visually observe the changes of surface morphology of rubber materials before and after fatigue, this experiment was first conducted using a fatigue test bench, and then the new surface generated by crack expansion of the experimental specimen was cut off; the specimen without fatigue test was also quickly cut with a blade to expose the surface, and then the prepared specimen was vacuum-treated and scanned by SEM instrument.

4.1.2. Discussion of the Results

Figure 28 shows the fast cut surface of the specimen without fatigue experiments. From the photo with 10,000 times magnification, it can be seen that the fillers are all well dispersed, in which the rubber matrix is the continuous phase and the other fillers are the dispersed phase. And it can be seen from the figure that the infiltration of the filler and the rubber matrix is relatively good, the vast majority of the filler is wrapped in the rubber matrix, and only a small part of the filler particles are not wrapped by the rubber, which is the source of the fatigue cracks in the rubber material.

Figure 28. SEM10000 times morphology of rubber surface.

Figure 29 shows the surface of the specimen after fatigue photographed with three-dimensional morphology, and it can be found that there are both relatively smooth and rough areas on the surface, and there are obvious fatigue streaks, which are obvious characteristics of ductile materials. These two kinds of areas were also observed when Yanhong Wang studied the fatigue phenomenon. This is due to shear fatigue experiments, the tearing energy used is much less than the critical tearing energy, so it has not reached the degree of instantaneous damage, and the fracture surface caused by this slow tearing is generally rough; and when expanding along the rough surface, the crack is not expanding uniformly in one direction, so it will cause stress concentration, and when the stress exceeds the range that the rubber can withstand, it will tear instantly thus producing a smooth area. Also from the figure, we can see that the longer the fatigue life, the more stripes and the more obvious the grooves are, which means the crack expansion rate is low.

Figure 29. Three-dimensional photos of fatigue surface.

Figure 30 is a 100-fold SEM photograph of the fatigue surface and the cut-off surface of the specimen without fatigue experiments, we can see that the surface before and after fatigue is very different, the surface without fatigue experiments is very smooth and has a beach-like ripple, which is a sign of cut-off; while the fatigue interface can be The surface without fatigue test is very smooth and has beach-like ripples, which is a sign of cut-off; while the fatigue interface can be seen to have peeling phenomenon, which is layer by layer overlapping, which is the rough surface formed by long time shear fatigue.

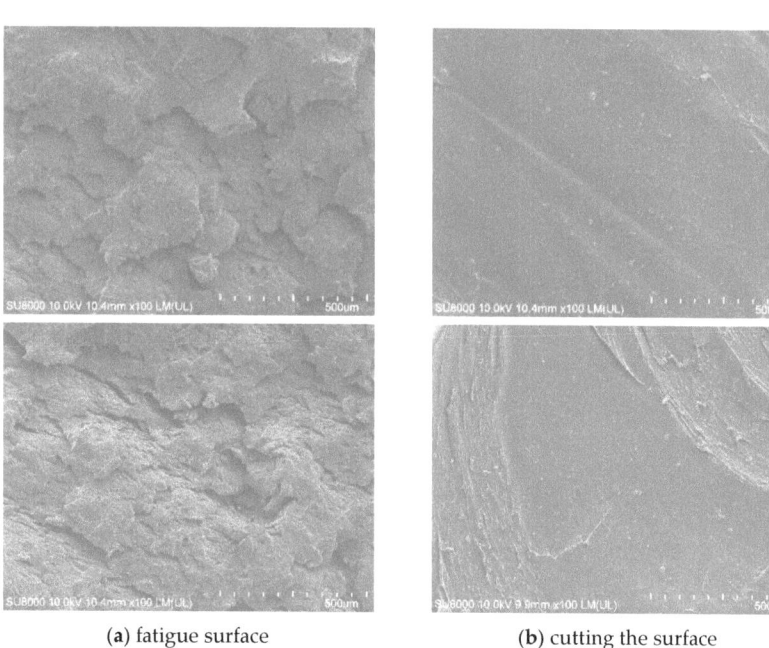

(a) fatigue surface   (b) cutting the surface

Figure 30. Morphology of Rubber Specimen under SEM100 times.

Figure 31 shows the SEM scanning photos of the unfatigued specimens. It can be observed that the fatigue-free interface is smooth, but there are pits. This is because the rubber specimen hardly stretches when it is cut off, so when the material is cut off, the filler sticks to the fracture surface.

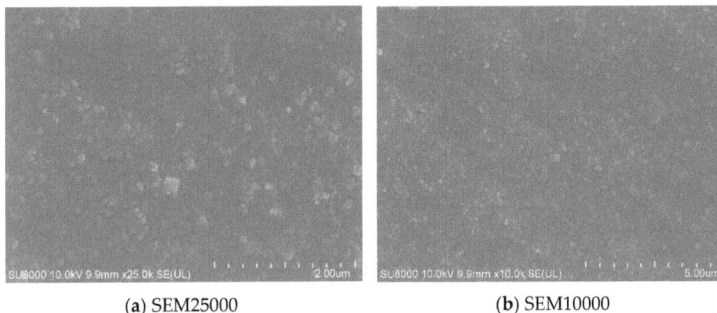

(a) SEM25000   (b) SEM10000

Figure 31. SEM surface morphology of cut-off surface.

Figure 32 is a 5000-fold scanning photo of SEM. It can be seen that the combination of filler and rubber is good, and many irregular gullies and holes are produced after fatigue. This structure is helpful to improve the fatigue life of rubber, because it is a process of generating new surfaces and can consume some energy.

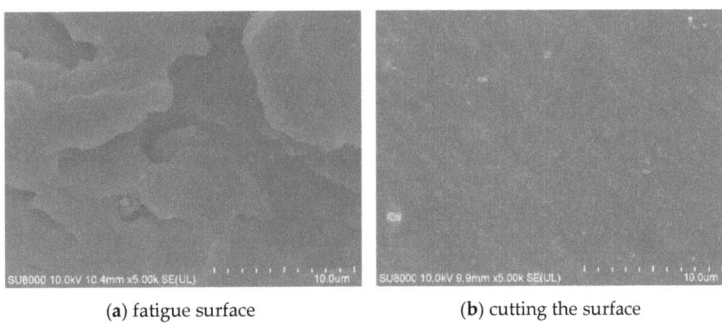

(a) fatigue surface   (b) cutting the surface

Figure 32. Morphology of Rubber Specimen at 5000 times of SEM.

Figure 33a shows a scanning photo of the surface of the rubber specimen cut off with a blade after it has been stretched to 200% strain for six times, and (b) shows a scanning photo of the surface of the specimen that has not been stretched. It can be seen that (b) the surface is smooth and the bonding degree between the filler and the rubber matrix is good, but after the Mullins effect is eliminated, it can be found that some fillers have been separated from the rubber matrix, indicating that the network formed by the filler and the rubber has begun to be destroyed.

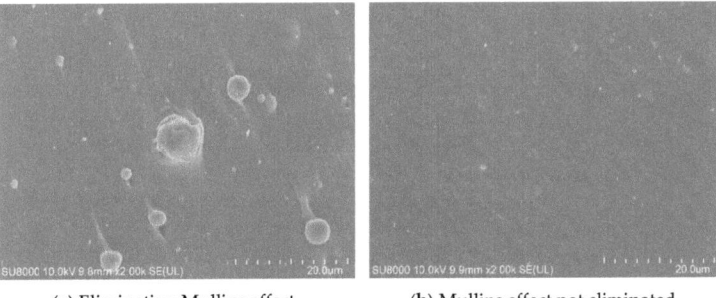

(a) Eliminating Mullins effect   (b) Mullins effect not eliminated

Figure 33. Shape of rubber specimen at SEM2000 times.

4.2. Micromechanism of Rubber Fatigue

From a classical mechanical calculation—the stress problem of a perforated flat plate—it is known that a round hole, i.e., a defect in a flat plate, generates a stress concentration in the place where the defect exists, and when the stress exceeds the critical value, this becomes the initial part of the damage. Similarly, the fatigue damage of rubber is also a concentration of stress at the defective part, which causes the molecular chain to break. The three-dimensional cross-linked network is produced when the rubber is vulcanized, and the molecular chain with low bond energy is destroyed first when the stress is concentrated. The stress condition at the crack tip of the rubber material is an important factor influencing its crack expansion. The shear fatigue test presets the crack to simulate the expansion process of the material after it is cracked. Rubber material is a weak but tough material, and it can be observed that it can have a large deformation with a small force when it is slowly stretched. When the crack tip becomes blunt, then its ability to resist tearing becomes stronger, because at this time the stress is dispersed in several directions, so the reason why rubber can still bear the load when there is damage is the blunting of the crack tip.

Fatigue damage of rubber is actually the result of both physical and chemical damage. When the rubber material is loaded, the chain segments between the cross-linked bonds are oriented, all chain segments are stressed, and the chain segments are Gaussian-distributed, so the stress is also Gaussian-distributed. When the rubber is under a low load and has a small deformation, the conformation of the cross-linked network is sufficient to withstand the current deformation, so it is almost not destroyed, but as the deformation increases, the molecular chain is continuously stretched, and when the maximum deformation is reached, it cannot meet the current deformation. This causes the molecular chain to break and release the strain energy, which is partly used to overcome the new surface energy and partly transferred to the adjacent cross-linked network. Therefore, these adjacent cross-linked networks will also be damaged by the excessive energy. Many fillers are added to rubber products, and when the particle size of these inorganic particles is larger than microns, their interface with the rubber matrix is a defect. When rubber materials are subjected to alternating loads, damage is actually produced in the first cycle, as demonstrated in the previous sections on the Mullins effect. After experiencing the first tension cycle, many microdefects have formed inside the rubber, but the effect of the first few tensile loads on the fatigue life is minimal due to the fact that when the loading is stopped without causing macroscopic visible cracks. These microcracks cause physical–chemical effects, such as infiltration and diffusion, but will be repaired automatically. After undergoing many cycles of loading, more and more molecular chains are damaged, which eventually leads to visible cracks. So one way to improve the fatigue life of rubber is to improve the dispersion of the filler in the rubber.

Fatigue damage often does not occur immediately, but rather over a long period of time since it involves chemical reactions. Chemical reactions generally refer to the erosion of rubber molecular chains by gases, the most important of which is to react with oxygen. When using the universal tensile machine in accordance with national standards for rubber specimens in the tensile test, it was found that the results in the air environment and nitrogen environment are essentially the same, which also shows that chemical reactions need a certain amount of time to occur. A single stretch in oxygen cannot immediately cause it to react with the rubber as the process is not instantaneous. But in different gas environments for fatigue experiments, the results are very different. Sainter studied the fatigue life of styrene–butadiene rubber in air and oxygen environments and found that the fatigue life in air was eight times higher than that in oxygen. In the presence of oxygen, cross-linking and chain breaking occur simultaneously, and the more oxygen content, the more violent the reaction. When cross-linking is predominant, the rubber will harden and become brittle, eventually leading to fracture, and if chain breaking is predominant, then the rubber will become soft and eventually be destroyed because the strength becomes smaller.

5. Conclusions

In this paper, the rubber fatigue test platform was built. The thermal coupling fatigue prediction model of rubber was established based on a large number of experimental data, and the microscopic mechanism of rubber fatigue was analyzed. The main contents include:

(1) By conducting Mullins effect experiments on different formulations of vulcanized rubber, it is found that the Mullins effect can be eliminated after 5–6 stretches.

(2) The effect of different stretching speeds on the stress-strain curves was investigated and it was found that as the stretching speed increased, the stress required to achieve the same strain also increased.

(3) By conducting fatigue experiments under different working conditions, it was found that the life is longer when there is a stress ratio, and the life continues to increase as the stress ratio increases; the effect of the loading method on the crack extension rate is negligible; frequency itself has no effect on fatigue, high frequency makes the rubber rubber temperature rise, thus accelerating the crack extension; the effect of the Mullins effect on the fatigue life of rubber needs to be verified by subsequent system experiments to verify;

(4) By fitting the experimental data, it was found that temperature and tearing energy could fit a power-of-three function, so the influence of temperature was considered in the calculation of tearing energy, and a rubber thermodynamic coupling fatigue prediction model was established; the fatigue life of the specimen at high temperature was calculated using the thermodynamic coupling model, and the error was 2.6% compared with the experimental results, which verified the accuracy of the model;

(5) The surface morphology of rubber before and after fatigue was observed by SEM, and it was concluded that the source of crack extension was the debonding of rubber matrix and agglomerates; from the perspective of molecular mechanism, it was explained that the process of rubber fatigue damage was actually the process of overloading the cross-linked network and thus being destroyed.

Author Contributions: Methodology, X.Q. and H.Y.; Software, X.Q.; Resources, H.Y.; Data curation, Q.J.; Writing—original draft, X.Q.; Writing—review & editing, X.Q. and Q.X. All authors have read and agreed to the published version of the manuscript.

Funding: This research received no external funding.

Institutional Review Board Statement: Not applicable.

Data Availability Statement: Not applicable.

Conflicts of Interest: The authors declare no conflict of interest.

References

1. Tee, Y.L.; Loo, M.S.; Andriyana, A. Recent advances on fatigue of rubber after the literature survey by Mars and Fatemi in 2002 and 2004. *Int. J. Fatigue* **2018**, *110*, 115–129. [CrossRef]
2. Glanowski, T.; Huneau, B.; Marco, Y.; Le Saux, V.; Champy, C.; Charrier, P. Fatigue initiation mechanisms in elastomers: A microtomography-based analysis. *MATEC Web Conf.* **2018**, *165*, 08005. [CrossRef]
3. Huneau, B.; Masquelier, I.; Marco, Y.; Le Saux, V.; Noizet, S.; Schiel, C.; Charrier, P. Fatigue crack initiation in a carbon black–filled natural rubbe. *Rubber Chem. Technol.* **2016**, *89*, 126–141. [CrossRef]
4. Gent, A.N.; Lindley, P.B.; Thomas, A.G. Cut growth and fatigue of rubbers. I. The relationship between cut growth and fatigue. *J. Appl. Polym. Sci.* **1964**, *8*, 455–466. [CrossRef]
5. Mars, W.; Fatemi, A. A literature survey on fatigue analysis approaches for rubber. *Int. J. Fatigue* **2002**, *24*, 949–961. [CrossRef]
6. Cadwell, S.M.; Merrill, R.A.; Sloman, C.M.; Yost, F.L. Dynamic Fatigue Life of Rubber. *Rubber Chem. Technol.* **1940**, *13*, 304–315. [CrossRef]
7. Thomas, A.G. Rupture of rubber. V. Cut growth in natural rubber vulcanizates. *J. Polym. Sci.* **1958**, *31*, 467–480. [CrossRef]
8. Lake, G.J.; Lindley, P.B. Cut growth and fatigue of rubbers. II. Experiments on a noncrystallizing rubber. *J. Appl. Polym. Sci.* **1964**, *8*, 707–721. [CrossRef]
9. Lindley, P.B. Energy for crack growth in model rubber components. *J. Strain Anal.* **1972**, *7*, 132–140. [CrossRef]

10. Rivlin, R.S.; Thomas, A.G. Rupture of rubber. I. Characteristic energy for tearing. *J. Polym. Sci.* **1953**, *10*, 291–318.
11. Mars, W.V.; Fatemi, A. Fatigue crack nucleation and growth in filled natural rubber subjected to multiaxial stress states. *Fatigue Fract. Eng. Mater. Struct.* **2003**, *26*, 779–789. [CrossRef]
12. Lindley, P.B. Relation between hysteresis and the dynamic crack growth resistance of natural rubber. *Int. J. Fatigue* **1973**, *9*, 449–462. [CrossRef]
13. Legorju-Jago, K.; Bathias, C. Fatigue initiation and propagation in natural and synthetic rubbers. *Int. J. Fatigue* **2002**, *24*, 85–92. [CrossRef]
14. Xin, R. Study on fatigue failure mechanism of rubber. *Sci.-Tech. Wind* **2019**, 143.
15. Ruellan, B.; Le Cam, J.B.; Robin, E.; Jeanneau, I.; Canevet, F.; Mortier, F. Influence of the Temperature on Lifetime Reinforcement of a Filled NR. In *Fracture, Fatigue, Failure and Damage Evolution, Volume 6: Proceedings of the 2018 Annual Conference on Experimental and Applied Mechanics*; Springer International Publishing: Cham, Switzerland, 2019; pp. 45–50.
16. Ngolemasango, F.E.; Bennett, M.; Clarke, J. Degradation and life prediction of a natural rubber engine mount compound. *J. Appl. Polym. Sci.* **2008**, *110*, 348–355. [CrossRef]
17. Rey, T.; Chagnon, G.; Le Cam, J.-B.; Favier, D. Influence of the temperature on the mechanical behaviour of filled and unfilled silicone rubbers. *Polym. Test.* **2013**, *32*, 492–501. [CrossRef]
18. Haroonabadi, L.; Dashti, A.; Najipour, M. Investigation of the effect of thermal aging on rapid gas decompression (RGD) resistance of nitrile rubber. *Polym. Test.* **2018**, *67*, 37–45. [CrossRef]
19. Chou, H.-W.; Huang, J.-S.; Lin, S.-T. Effects of thermal aging on fatigue of carbon black–reinforced EPDM rubber. *J. Appl. Polym. Sci.* **2007**, *103*, 1244–1251. [CrossRef]
20. Neuhaus, C.; Lion, A.; Johlitz, M.; Heuler, P.; Barkhoff, M.; Duisen, F. Fatigue behaviour of an elastomer under consideration of ageing effects. *Int. J. Fatigue* **2017**, *104*, 72–80. [CrossRef]
21. Luo, W.; Huang, Y.; Yin, B.; Jiang, X.; Hu, X. Fatigue Life Assessment of Filled Rubber by Hysteresis Induced Self-Heating Temperature. *Polymers* **2020**, *12*, 846. [CrossRef]
22. Ruellan, B.; Le Cam, J.-B.; Jeanneau, I.; Canévet, F.; Mortier, F.; Robin, E. Fatigue of natural rubber under different temperatures. *Int. J. Fatigue* **2019**, *124*, 544–557. [CrossRef]
23. Federico, C.E.; Padmanathan, H.R.; Kotecky, O.; Rommel, R.; Rauchs, G.; Fleming, Y.; Addiego, F.; Westermann, S. Cavitation Micro-mechanisms in Silica-Filled Styrene-Butadiene Rubber Upon Fatigue and Cyclic Tensile Testing. In *Fatigue Crack Growth in Rubber Materials: Experiments and Modelling*; Springer: Berlin/Heidelberg, Germany, 2021; pp. 109–129.
24. Schieppati, J.; Schrittesser, B.; Wondracek, A.; Robin, S.; Holzner, A.; Pinter, G. Temperature impact on the mechanical and fatigue behavior of a non-crystallizing rubber. *Int. J. Fatigue* **2021**, *144*, 106050. [CrossRef]
25. Liu, X.; Shangguan, W.-B.; Zhao, X. Probabilistic fatigue life prediction model of natural rubber components based on the expanded sample data. *Int. J. Fatigue* **2022**, *163*, 107034. [CrossRef]
26. Demiral, M.; Mamedov, A. Fatigue Performance of a Step-Lap Joint under Tensile Load: A Numerical Study. *Polymers* **2023**, *15*, 1949. [CrossRef]
27. Fang, Y.; Zhang, H. High-temperature fatigue life prediction method for rubber bushing of new-energy vehicles based on modified fatigue damage theory. *Mater. Res. Express* **2020**, *7*, 015346.
28. Weng, G.; Huang, G.; Lei, H.; Qu, L.; Zhang, P.; Nie, Y.; Wu, J. Crack initiation of natural rubber under high temperature fatigue loading. *J. Appl. Polym. Sci.* **2012**, *124*, 4274–4280. [CrossRef]
29. Zhang, J.; Xue, F.; Wang, Y.; Zhang, X.; Han, S. Strain energy-based rubber fatigue life prediction under the influence of temperature. *R. Soc. Open Sci.* **2018**, *5*, 180951. [CrossRef]
30. Luo, W.; Li, M.; Huang, Y.; Yin, B.; Hu, X. Effect of temperature on the tear fracture and fatigue life of carbon-black-filled rubber. *Polymers* **2019**, *11*, 768. [CrossRef]
31. Seichter, S.; Archodoulaki, V.M.; Koch, T.; Holzner, A.; Wondracek, A. Investigation of different influences on the fatigue behaviour of industrial rubbers. *Polym. Test.* **2017**, *59*, 99–106. [CrossRef]
32. Grandcoin, J.; Boukamel, A.; Lejeunes, S. A micro-mechanically based continuum damage model for fatigue life prediction of filled rubbers. *Int. J. Solids Struct.* **2014**, *51*, 1274–1286. [CrossRef]
33. Aït-Bachir, M.; Mars, W.; Verron, E. Energy release rate of small cracks in hyperelastic materials. *Int. J. Non-linear Mech.* **2012**, *47*, 22–29. [CrossRef]
34. Julie, D.; Fayolle, B.; Gilormini, P. A review on the Mullins effect. *Eur. Polym. J.* **2009**, *45*, 601–612.
35. Marckmann, G.; Verron, E.; Gornet, L.; Chagnon, G.; Charrier, P.; Fort, P. A theory of network alteration for the Mullins effect. *J. Mech. Phys. Solids* **2002**, *50*, 2011–2028. [CrossRef]
36. Martinez, J.R.S.; Le Cam, J.-B.; Balandraud, X.; Toussaint, E.; Caillard, J. New elements concerning the Mullins effect: A thermomechanical analysis. *Eur. Polym. J.* **2014**, *55*, 98–107. [CrossRef]
37. Li, M. Thermal dissipation and fatigue performance of carbon black filled rubber. Ph.D. Thesis, Xiangtan University, Xiangtan, China, 2019.
38. Ghoreishy, M.H.R.; Sourki, F.A. Development of a new combined numerical/experimental approach for the modeling of the nonlinear hyper-viscoelastic behavior of highly carbon black filled rubber compound. *Polym. Test.* **2018**, *70*, 135–143. [CrossRef]
39. Poisson, J.L.; Méo, S.; Lacroix, F.; Berton, G.; Hosséini, M.; Ranganathan, N. Comparison of Fatigue Criteria under Proportional and Non-Proportional Multiaxial Loading. *Rubber Chem. Technol.* **2018**, *91*, 320–338. [CrossRef]

40. Yong, Z. Preparation of Magnetorheological Elastomer and Study on Fatigue Fracture Mechanism. Doctoral Dissertation, University of Science and Technology, Qingdao, China, 2018.
41. Beatty, J.R. Fatigue of rubber. *Rubber Chem. Technol.* **1964**, *37*, 1341–1364. [CrossRef]
42. Champy, C.; Le Saux, V.; Marco, Y.; Glanowski, T.; Charrier, P.; Hervouet, W. Fatigue of crystallizable rubber: Generation of a Haigh diagram over a wide range of positive load ratios. *Int. J. Fatigue* **2021**, *150*, 106313. [CrossRef]
43. Le Cam, J.B.; Pérocheau, F.; Huneau, B.; Verron, E.; Gornet, L. Cavitation influence on fatigue crack growth in filled natural rubber. *Mec. Ind.* **2006**, *7*, 123–129.
44. Euchler, E.; Schneider, K.; Wiessner, S.; Bernhardt, R.; Heinrich, G.; Roth, S.V.; Tada, T. Cavitation damage in tire rubber materials. In *Constitutive Models for Rubber XI*; CRC Press: Boca Raton, FL, USA, 2019; pp. 87–91.

Disclaimer/Publisher's Note: The statements, opinions and data contained in all publications are solely those of the individual author(s) and contributor(s) and not of MDPI and/or the editor(s). MDPI and/or the editor(s) disclaim responsibility for any injury to people or property resulting from any ideas, methods, instructions or products referred to in the content.

Article

Influence of Delamination Size and Depth on the Compression Fatigue Behaviour of a Stiffened Aerospace Composite Panel

Angela Russo, Rossana Castaldo, Concetta Palumbo and Aniello Riccio *

Department of Engineering, University of Campania Luigi Vanvitelli, Via Roma 29, 81030 Aversa, Italy; angela.russo@unicampania.it (A.R.); rossana.castaldo@studenti.unicampania.it (R.C.); concetta.palumbo@unicampania.it (C.P.)
* Correspondence: aniello.riccio@unicampania.it

Abstract: Delamination in reinforced panels is one of the primary challenges facing the safety and reliability of aerospace structures. This article presents a sensitivity analysis of the fatigue behaviour during the compression of a composite aeronautical stiffened panel experiencing delamination. The main objective is to assess the impact of delamination size and depth on the lifecycle and structural integrity of the panel. Different dimensions and positions of delamination are considered to cover a comprehensive range of damage scenarios. The key feature of this sensitivity analysis is the adoption of a numerical procedure that is mesh- and load-step-independent, ensuring reliable results and providing valuable insight into the criticality of delamination and its impact on the fatigue behaviour during the compression of reinforced aeronautical panels. Sensitivity analyses are essential for enhancing the design process of aerospace structures, thereby contributing to the increased safety and reliability of structural components. In this regard, the use of robust and effective numerical procedures is of crucial significance. This may be seen as the real added value of this paper.

Keywords: stiffened panel; circular delamination; structural design; finite element analysis

Citation: Russo, A.; Castaldo, R.; Palumbo, C.; Riccio, A. Influence of Delamination Size and Depth on the Compression Fatigue Behaviour of a Stiffened Aerospace Composite Panel. *Polymers* **2023**, *15*, 4559. https://doi.org/10.3390/polym15234559

Academic Editors: Alexey V. Lyulin and Valeriy V. Ginzburg

Received: 10 October 2023
Revised: 20 November 2023
Accepted: 22 November 2023
Published: 28 November 2023

Copyright: © 2023 by the authors. Licensee MDPI, Basel, Switzerland. This article is an open access article distributed under the terms and conditions of the Creative Commons Attribution (CC BY) license (https://creativecommons.org/licenses/by/4.0/).

1. Introduction

Fracture has always been a significant issue for the aerospace industry during the fabrication of aerospace structures, and, over the years, it has become more severe due to the extensive use of innovative materials, such as composite materials [1,2], which are very prone to interlaminar fractures [3]. To take full advantage of the benefits offered by these materials, they are used to fabricate stiffened panels that are used in the aircraft fuselages, wings, and horizontal/vertical tail planes. However, the latter, when in service, are exposed to impact phenomena, which can cause extensive damage, sometimes not visible on first visual inspection, leading to catastrophic events [4–6]. Moreover, manufacturing flaws and maintenance processes may cause the onset of cracks that can easily propagate under service loads [7,8]. The phenomenon of fatigue amplifies these issues, significantly reducing the strength properties of stiffened composite aircraft components through material degradation [9,10].

Many studies have been carried out in the literature to consider the delamination phenomenon due to a variety of causes. The study in [11] introduced a mesh-independent computational method, utilising the extended finite element methodology, to predict transverse matrix crack and delamination evolution. The approach demonstrated good agreement with conventional numerical methods and experiments. In [12], a review on how delamination contributes to the failure of fibre-reinforced composites is presented. The study in [13] explored hydrothermal aging in a carbon-fibre-reinforced laminate and its epoxy matrix in bulk conditions. Minimal changes in bulk resin fracture toughness but variable trends in composite interlaminar fracture toughness were observed. In [14], composite laminates impacted at low velocities were studied. Drop weight tests and ultrasonic C-scan were employed to assess delamination evolution. Numerical modelling, including the cohesive contact method, has been used to simulate delamination. The FEM results have been proven to be aligned with experiments. Liu et al. [15] conducted a 2D and 3D parametric finite element analysis of composite flat laminates with two types of through-the-width delamination. The effect of multiple delaminations on postbuckling properties has been studied by using the virtual crack closure technique. The study in [16] employed Abaqus, enhancing fatigue analysis by integrating R-curve effects into the Paris law through an empirical method. Implemented via a user-defined subroutine, the novel method accurately predicts fatigue life within 5% of the test results. The work in [17] explored disbonding in adhesively bonded panels through Lamb waves during fatigue tests, emphasising frequency, mode, sensor placement, and parameter selection. Piezoelectric transducers act as actuators and sensors, correlating crack propagation rate with A0 mode velocity. Šedková et al. [18] investigated debonding and delamination assessment by means of Lamb waves in adhesively bonded composite joints.

The study of damage propagation in aircraft structures must begin at the preliminary design stage. Indeed, the use of new design concepts and innovative materials, whose damage behaviour is not well understood, often leads to the over-dimensioning of aircraft structures. Furthermore, in a damage-tolerant design philosophy, the loads and stresses that the aircraft is expected to withstand in operation must be considered in detail, and the effects of the damages on the structural integrity of the component must be assessed. The work in [19] delved into the numerical methods employed to forecast impact damage in advanced composite structures, emphasising their significance for industry approval in designing and certifying composite aircraft. The article in [20] addresses the low-velocity impact (LVI) challenge in the composites industry, highlighting how factors are crucial for improving the impact resistance and damage tolerance of fibre-reinforced composites (FRCs). The review identifies gaps in the literature and suggests future research directions, while also discussing various damage modelling strategies to predict the impact resistance and damage tolerance of FRCs. Jones [21] addressed challenges related to aging aircraft and infrastructure, emphasising the need for tools to predict crack growth from small material discontinuities. They discuss the differences in the analysis tools used for design and sustainment, modelling crack growth, determining short crack data, predicting growth using existing equations, and accounting for variations in crack growth histories. In [22], the damage tolerance of a stitched carbon/epoxy laminate was studied. Investigating the impact of stitching on a carbon fibre epoxy laminate revealed significantly improved damage tolerance compared with that of brittle epoxy.

Appropriate experimental testing campaigns should be conducted to identify structural weaknesses. However, the production of aircraft structures, even on a small scale, to identify the best structural configurations can be very expensive as well as time-consuming. Aircraft designers and engineers use a variety of tools and techniques to analyse and predict the behaviour of materials and structures under operating loads, including finite element analyses [23–28].

Sensitivity analyses are essential in structural design and projects to identify the critical geometrical parameters that have the most significant impact on the structural design or project outcome [15,29–31]. By adapting these variables (within a reasonable range), engineers can understand which factors play a crucial role in the performance and damage behaviour of a structure. Moreover, sensitivity analysis allows engineers to evaluate the robustness of the design against variations in input parameters. In general, sensitivity analyses provide valuable insights into the behaviour and performance of structural configurations. This helps designers make proper decisions, optimise designs, manage risks, and ensure the structural integrity and efficiency of the final product.

The main added value of this article is the use of the FT-SMXB methodology, due to its inherent characteristics, to study, specifically, the influence of the size and depth of typical impact damage on composite panels. The standard virtual crack closure technique (VCCT) would probably influence the results due to its dependence on the mesh and time step of the finite element analysis. On the contrary, one of the advantages of the cohesive zone model (CZM) methodology is its lower sensitivity to mesh size. However, defining the correct cohesive stiffness is still a challenge [32], and a rigorous calibration of the cohesive parameters is needed to avoid mesh sensitivity [33]. For the finite element method, the dependence of the results on the size of the elements selected for the numerical model discretisation and the timestep used for the load application during the analyses play a key role in the obtained results. The FT-SMXB numerical approach has been proven to overcome these FEM limitations. This procedure has been validated both at the coupon level, such as in the double cantilever beam test, end-notched flexure test and single leg bending shear test, and for more complex structures, such as stiffened panels characterised by skin-stringer debonding [34,35]. The use of this numerical procedure is the real added value of this work, which includes a sensitivity analysis of a typical aeronautical reinforced panel subjected to typical impact damage of varying size and depth in the skin panel bay. This manuscript can be considered an evolution of previous work [36], where the static failure of an aeronautical stiffened composite panel under compressive loading conditions was studied. The results showed a strong dependence of the panel compressive behaviour on the geometric parameters, including the depth and size of the circular delamination. Here, the fatigue behaviour of the panel was assessed by varying the radius and the position in the thickness of a circular delamination. The influence of such variables on the delamination evolution and the stiffness degradation was assessed. The delamination between two interfaces was considered. The FT-SMXB procedure does not consider the delamination migration phenomenon.

2. Numerical Model Description

A typical aeronautical stiffened panel was studied under compressive fatigue loading conditions. The panel was 300 mm in length and 400 in width. Two T-shape stringers were tied on the skin, which has a 60 mm foot and 30 mm web. The stacking sequences of the different panel components, which are listed in Table 1, were chosen according to typical manufacturing process needs. A 0.165 mm ply of a carbon fibre/epoxy resin material system was used to model the thicknesses of the panel parts. Circular delamination damage in the panel bay was placed, representing a typical impact damage. It is known that the shape of delamination after impact in composite structures can change depending on several factors, including the nature and extent of the impact, the material properties, and the geometry of the structure. While circular delamination is commonly used in theoretical and analytical models for simplicity, real delamination resulting from impacts can have different shapes. However, in some cases, the delamination may have a circular or nearly circular shape, especially if the impact is relatively symmetrical and occurs at a specific point.

Table 1. Stacking sequences.

Component	Number of Plies	Layup	Thickness
Skin panel	16	$[45,90,0,-45]_{2s}$	2.64 mm
Foot stringer	24	$[(0,90,90,0)_s(45,90,0,-45)_{2s}]$	3.96 mm
Web stringer	8	$[0,90,90,0]_s$	1.32 mm

The elastic and interlaminar properties of the material were evaluated through an experimental characterisation, based on the ASTM standards in [37–42], and are reported in Table 2. Static tensile, compressive, and shear tests were performed on samples manufactured according to the standards. The tests summary is shown in Figure 1.

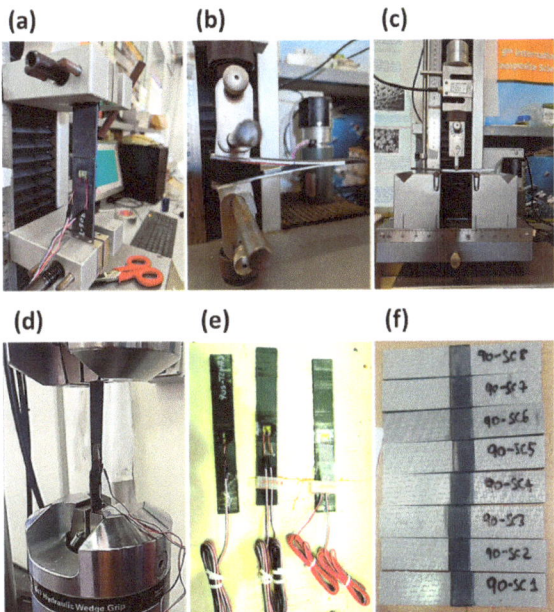

Figure 1. Experimental characterisation of the material: (**a**) matrix tensile test; (**b**) double cantilever beam test; (**c**) end-notched flexure test; (**d**) fibre tensile test; (**e**) matrix tensile specimens; (**f**) fibre compressive specimens.

Table 2. Material properties.

Property	Value	Description
E_{11}	122,000 MPa	Young's modulus in the fibre's direction
$E_{22} = E_{33}$	6265 MPa	Young's modulus in the transverse direction
$G_{12} = G_{13}$	4649 MPa	Shear modulus in the 1–2 and 1–3 planes
G_{23}	4649 MPa	Shear modulus in the 2–3 plane
$\nu_{12} = \nu_{13}$	0.3008	Poisson's ratio in the 1–2 and 1–3 planes
ν_{23}	0.02	Poisson's ratio in the 2–3 plane
G_{Ic}	180 J/m²	Mode I critical energy release rate
$G_{IIc} = G_{IIIc}$	1900 J/m²	Mode II and Mode III critical strain energy release rate
c1	0.7188	Mode I Paris constant
n1	8	Mode I Paris exponent
c2	6.5938	Mode II Paris constant
n2	6	Mode II Paris exponent

The panel was modelled through the Ansys Parametric Design Language (APDL) to easily modify the value of the variable parameters (radius and depth). The global–local approach was chosen as modelling strategy in order to reduce the computational cost and the analyses' duration. Indeed, a coarse mesh was considered for the global part, while a more refined discretisation was performed in the local zone, which is the region characterised by the delamination damage and is involved in the propagation. Elements of 12.5 mm were used for the global model, while elements of approximately 2 mm and 5 mm were considered in the local region. According to the fail release approach, typically used in conjunction with the virtual crack closure technique (VCCT), the local zone was built with two overlapping identical solids, as shown in Figure 2a. Identical discretisation was carried out, and the nodes, which are characterised by the same spatial coordinates, were connected with constraints with "birth and death" capabilities, except in the delaminated area, where nodes are released, and only interactions were considered to avoid penetration. The FEM model is displayed in Figure 2b. A schematic of the panel is reported in Figure 2c.

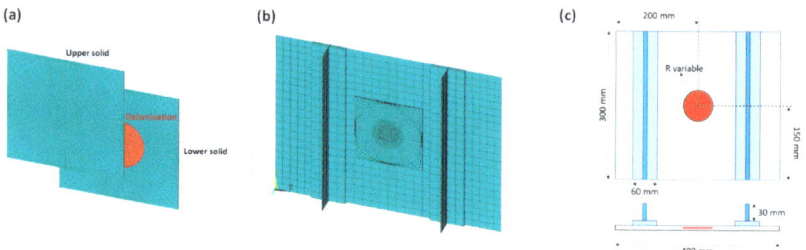

Figure 2. (a) Fail release approach; (b) FEM model of the panel; (c) schematic of the panel.

The translational and rotational degrees of freedom on one side of the panel were clamped (interlocking constraint); on the other side, the terminal nodes were connected through rigid connections to a reference point (pilot node) on which the compressive force was applied. The scheme of the boundary conditions is shown in Figure 3.

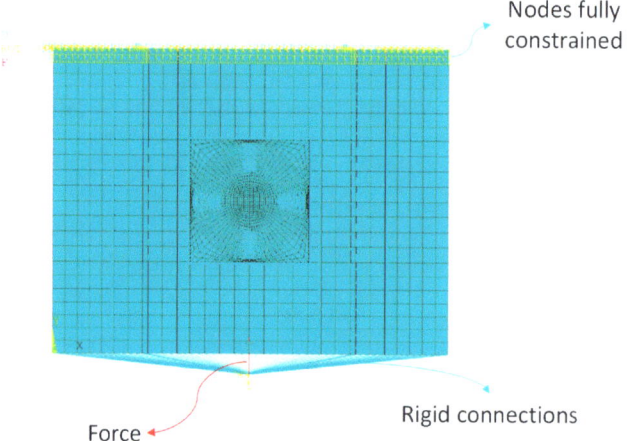

Figure 3. Boundary conditions of the fatigue analyses.

2.1. FT-SMXB Approach

Fracture mechanics relations are the basis of the FT-SMXB numerical method. In particular, the modified virtual crack closure technique (VCCT) equations were implemented to calculate the energy release rate (ERR) values on the nodes of the delamination front. Such values were compared with the critical fracture toughness values to assess the propagation of delamination, as described by Equation (1).

$$\frac{G_I}{G_{Ic}} + \frac{G_{II}}{G_{IIc}} + \frac{G_{III}}{G_{IIIc}} = E_d \geq 1 \quad (1)$$

The ERR G_j with $j = I, II, III$ depends on the increase in delaminated area ΔA, as reported in Equation (2), where F_j is the force at the crack tip, and u_j is the opening displacement (see Figure 4).

$$G_j = \frac{F_j u_j}{2\Delta A} \text{ with } j = I, II, III \quad (2)$$

According to the VCCT, G_j is the amount of energy needed to close an area ΔA of the fracture surface, while G_{jc} is the critical value that the energy must reach to open the crack. However, considering the FEM approach, the delamination growth is influenced by the chosen element size. Therefore, even if, at a specific load step, the delaminated area is smaller than that of an entire element (for the computed energy values), it is overestimated.

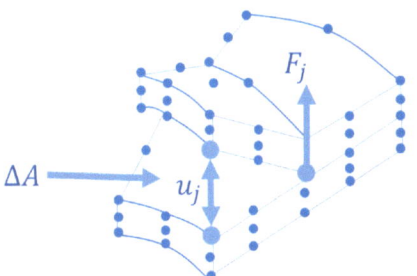

Figure 4. Crack tip.

In the FT-SMXB approach, the VCCT limits in terms of mesh and dependence on load step size are avoided through the implementation of an iterative procedure, named SMART TIME, that can iteratively change the load step size until the calculated energy corresponds exactly to the amount needed to propagate an area equal to that of one or more elements. In detail, the equivalence in Equation (3) must be satisfied. The subscript i refers to the nodes of the delamination front.

$$\sum_{i=1}^{N} A_i = \sum_{i=1}^{N} \left(\sum_{j=1}^{3} \frac{F_{ji}^{(t+1)} u_{ji}^{(t+1)}}{2G_{jc}} \right) \cong \sum_{i=1}^{N} A_i^{Element} \qquad (3)$$

Cyclic load can be applied on the structure via a succession of static analyses, which can be considered as fatigue cycles. The load applied on the structure in each static analysis is calculated as a percentage of the maximum load that the structure can support when subjected to static loading conditions. Basically, a first static analysis is performed, and the ERR values on the delamination front nodes are calculated and used as input in the Paris law criterion, defined by Equation (4). N represents the number of cycles, while C and n are the Paris law constants evaluated experimentally.

$$\frac{\Delta A_i}{\Delta N_i} = C\,f(G)^n \qquad (4)$$

The number of cycles needed to propagate delamination damage is calculated by reversing the Paris law, as described in Equation (5). Finally, the node with the higher criterion value, which is the node where the energy values are closest to the critical values, is selected and released.

$$\Delta N_i = \frac{\Delta A_i^t}{C f(G)^n} \qquad (5)$$

Starting from the last converged solution, subsequent analyses are performed until the number of cycles to failure is reached, or until the user-defined limits are encountered. Damage accumulation is considered. A damage variable is considered in each node of the delamination front to account for damage accumulation, and the number of cycles starts from that reached in the previous analysis.

2.2. Preliminary Static Analysis

Three different radii of circular delamination were studied: 20 mm (R20), 30 mm (R30), and 40 mm (R40). Each of them was positioned at three depths along the thickness (see

Table 3). This placed the delamination between plies of different orientations. Specifically, under two laminae (2PLY), the delamination was between 0 and 90 degrees; under three laminae (3PLY), the delamination was between 90 and 45 degrees; and under four laminae (4PLY), it was between 45 and −45 degrees.

Table 3. Panel configurations.

Id Configuration	Delamination Radius	Delamination Depth
R20-2PLY	20 mm	0.33 mm
R20-3PLY	20 mm	0.495 mm
R20-4PLY	20 mm	0.66 mm
R30-2PLY	30 mm	0.33 mm
R30-3PLY	30 mm	0.495 mm
R30-4PLY	30 mm	0.66 mm
R40-2PLY	40 mm	0.33 mm
R40-3PLY	40 mm	0.495 mm
R40-4PLY	40 mm	0.66 mm

The propagation initiation load and the overall buckling behaviour of the panel under static loading conditions were needed to proceed with fatigue analyses. Hence, preliminary analyses were conducted considering the panel with one edge fixed (all degrees of freedom constrained) and the other subjected to compressive displacement (free degrees of freedom in the compression direction). Figure 5 shows the boundary conditions.

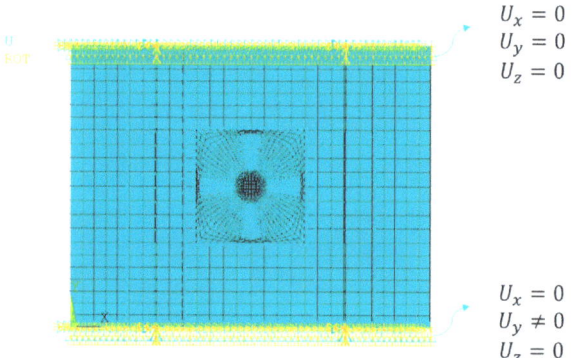

$U_x = 0$
$U_y = 0$
$U_z = 0$

$U_x = 0$
$U_y \neq 0$
$U_z = 0$

Figure 5. Boundary conditions of the static analyses.

Figure 5 shows the load as a function of applied displacement curves and the trend of delaminated area with the load for each considered configuration. The charts in Figure 6 indicate that increasing the depth of delamination, with the same radius, reduces the load supported by the structure, and the propagation of delamination becomes unstable and rapid, leading to sudden delamination evolution. The ultimate displacement applied to the structure, visible in Figure 6, does not represent its point of failure but only the displacement applied in the last step of the analysis. The end of the simulation is due to the achievement of the maximum extent of delamination, which is limited to the circular region around the initial defect.

Fixing the position along the thickness of the delamination damage, as the radius increases, there are no significant variations in the delaminated area trend with increasing load; as demonstrated in Figure 7, only a slight increase in the speed of propagation can be noticed for larger radii.

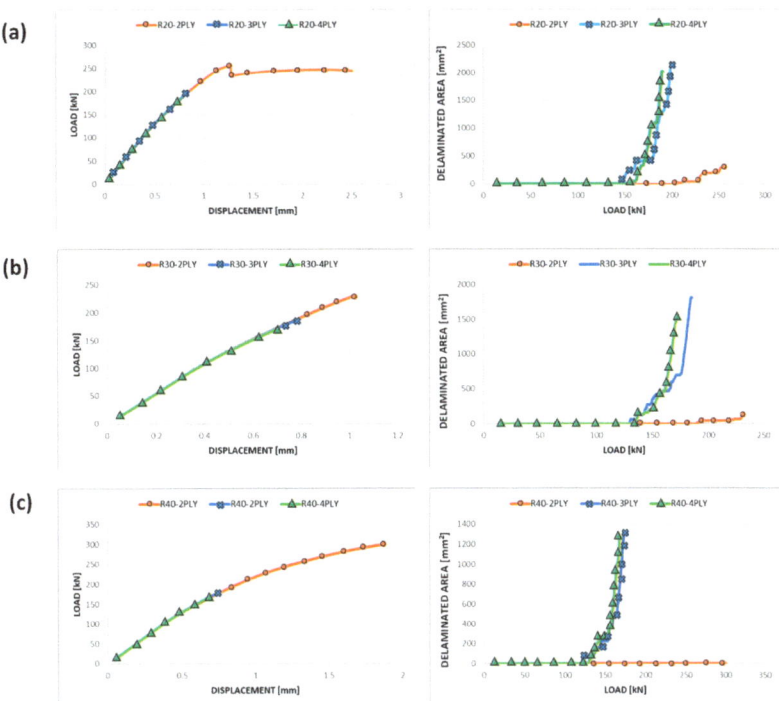

Figure 6. Load vs. applied displacement and delaminated area vs. load curves—static analyses: (**a**) R20 panel; (**b**) R30 panel; (**c**) R40 panel.

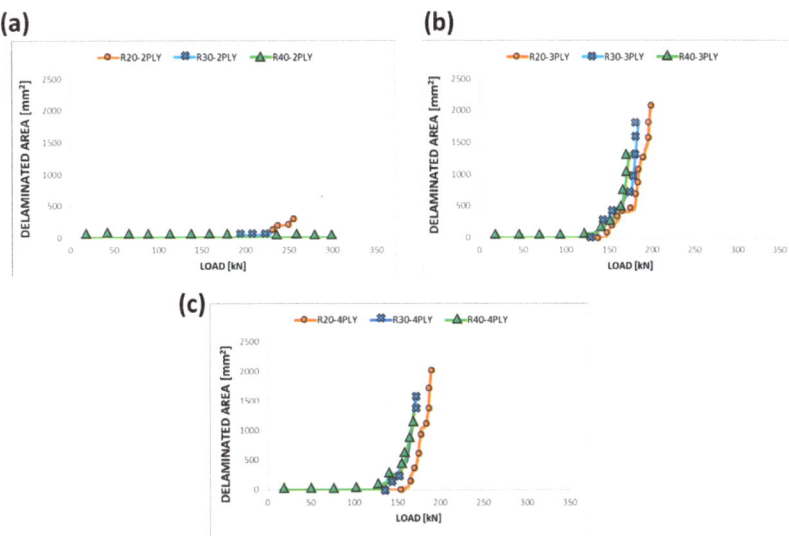

Figure 7. Delaminated area vs. load curves at fixed delamination depth—static analyses: (**a**) 2-ply depth of delamination; (**b**) 3-ply depth of delamination; (**c**) 4-ply depth of delamination.

It is worth specifying that the analyses automatically terminate when the delamination (and thus, the nodes released using the fail release approach) reach the user-defined limit. In this specific case, the limit was set at the outmost nodes of the circular propagation crown considered, as shown in Figure 8.

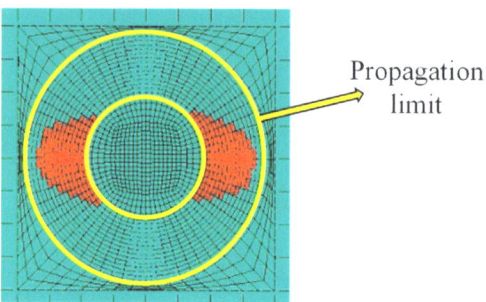

Figure 8. Delamination propagation limit.

Based on the obtained results, considering the load at which delamination starts propagating, fatigue analyses were conducted, according to Table 4.

Table 4. Fatigue analyses.

Id Configuration	Static Delamination Onset Load	Fatigue Loads	Onset Cycle	Failure Cycle
R20-3PLY	147 kN	80%	113,593	>1 × 10^6
		90%	939	16,459
R20-4PLY	158 kN	80%	>1 × 10^6	-
		90%	>1 × 10^6	
R30-3PLY	130 kN	80%	50,507	672,364
		90%	461	44,866
R30-4PLY	135 kN	80%	>1 × 10^6	-
		90%	37,493	200,678
R40-3PLY	121 kN	80%	15,049	>1 × 10^6
		90%	792	91,584
R40-4PLY	130 kN	80%	170,487	>1 × 10^6
		90%	1562	20,665

3. Results and Discussion

The summary of the performed fatigue analysis is reported in Table 4. The configuration R20/30/40 2P was not considered for fatigue assessment because the results in Figure 7 show that a slight propagation of the damage is experienced by the panel, which undergoes global buckling by preventing the growth in the delamination.

The fatigue damage initiation life refers to the duration or number of loading cycles that a material or structure can undergo before the initiation of fatigue damage. Evaluating fatigue damage initiation life involves analytical methods, such as cycle counting methods and the damage accumulation model, which are taken into account in FT-SMXB. The delamination initiation cycle was determined through the Paris law relation, as defined in Equation (5). Figure 9 shows a graph of the fatigue crack initiation life values for the various configurations analysed at 90% and 80% of the static delamination initiation load.

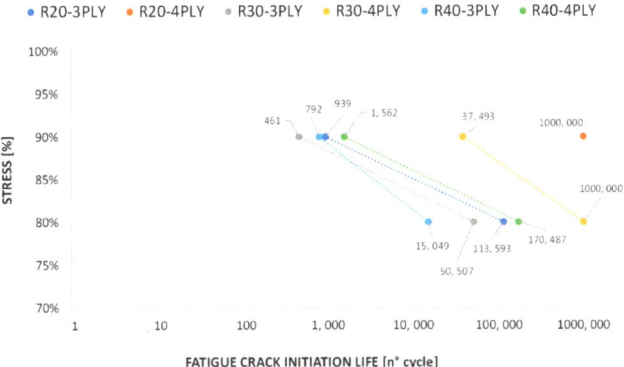

Figure 9. Fatigue crack initiation life curves.

Figure 10 displays the delamination evolution over as a function of the number of cycles for three different radii of circular delamination under three plies. According to the curves' trend, the larger delamination size (R40-3PLY) resulted in better fatigue behaviour of the composite panel. Indeed, a smoother growth of the delaminated area can be seen, achieving a higher number of cycles, although with a lower value of total delaminated area. In fact, in some specific cases, a larger delamination in a composite panel may create a more gradual stress transition in the surrounding area, which means that there are lower stress gradients in the material surrounding the delamination, thus reducing stress concentrations and the probability of further cracking and propagation.

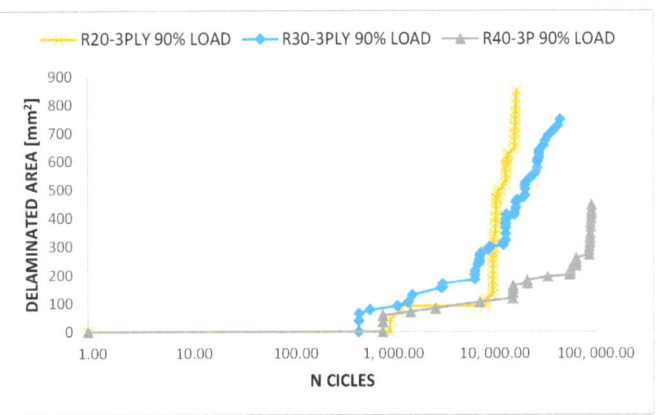

Figure 10. Delaminated area vs. number of cycles (log-scale)—fatigue analyses.

Increasing the depth of delamination for the same radius can have complex effects on the fatigue behaviour of a composite panel. According to Figure 11, increasing the depth of the delamination can increase the onset cycle, which is the number of cycles required for the crack to initiate. The relationship between delamination depth and fatigue behaviour is typically not linear. Primarily, deeper delamination could delay crack initiation due to the higher load required to propagate the crack deeper into the material [43,44]. However, as the delamination extends, it becomes a more significant structural defect, leading to higher stress concentrations and potentially accelerated crack growth.

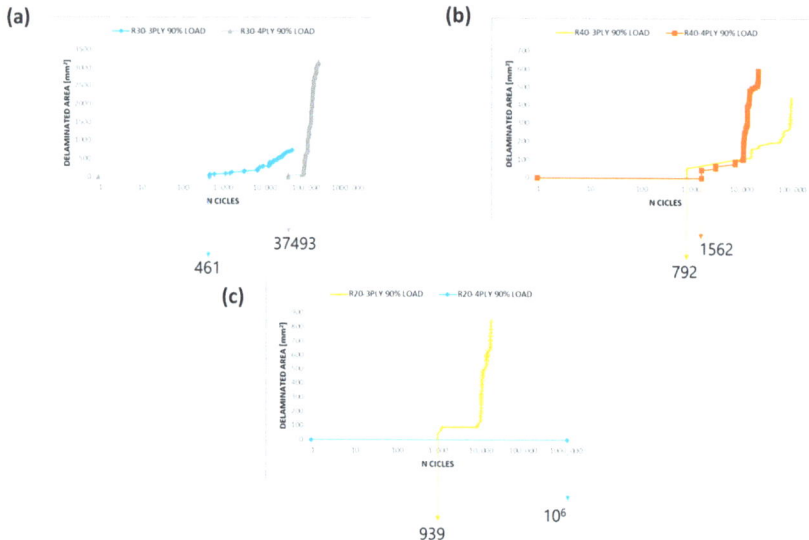

Figure 11. Delaminated area vs. number of cycles (log-scale)—fatigue analyses: (**a**) R30 configuration; (**b**) R40 configuration; (**c**) R20 configuration.

According to Figure 12, the stiffness (defined as load/displacement and measured on the nodes of load application) of all the analysed configurations decreased by less than 5% because, in the considered cases, the delamination extension is limited and hence does not affect the overall stiffness of the structure. The investigated structures maintain their overall integrity, with relatively small changes in stiffness within an acceptable range. Even if the stiffness remains almost constant, different levels of damage size and depth can influence the stress intensity in the damaged region. This can affect the number of loading cycles required for crack initiation and the rate of crack growth during propagation.

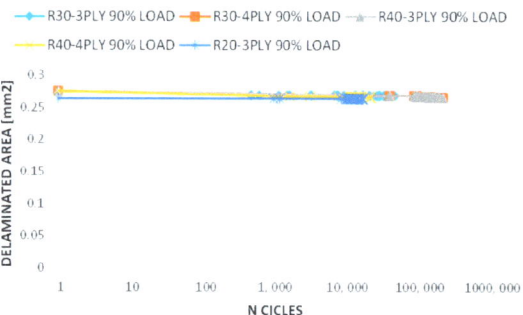

Figure 12. Stiffness vs. number of cycles (log-scale)—fatigue analyses.

The evolution of delamination for different numbers of cycles is shown in Figure 13 for one of the analysed configurations (R30 4PLY), as an example. The red region indicates the elements that correspond to the nodes released by the VCCT, which reached complete failure according to the Paris law calculations. Therefore, the red portion represents the growth achieved in the delaminated area.

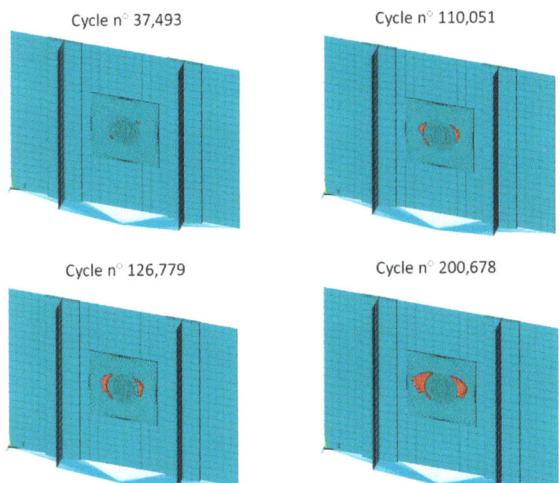

Figure 13. Evolution of the damage for different numbers of cycles.

4. Final Remarks

In this work, the mesh- and load-step-independent FT-SMXB numerical methodology was employed to assess the compressive fatigue behaviour of a composite material, an aeronautical stiffened panel, affected by typical impact damage in the bay region. Two percentages of the static damage initiation load were considered (80% and 90%). The results revealed that for greater radii of the circular delamination, at a fixed damage depth, delaminated areas experience smooth growth, achieving a higher number of fatigue cycles with a lesser damage extent. Furthermore, increasing the depth of the damage, at a fixed radius, increases the delamination onset cycle but leads to faster and unstable delamination evolution. The overall stiffness of the panel remains almost constant with the fatigue cycles for all the considered radii and depths of delamination.

It is important to note that the specific behaviour of composites and the effect of delamination depend on various factors, including the type of composite material, its layup configuration, the loading conditions, and the size and shape of the delamination. However, the use of effective and robust computational methodologies can provide help in designing damage-tolerant composite structures for various applications and loads, thus reducing the waste of money and time derived from manufacturing and experimental tests. The primary contribution of this article lies in the application of the FT-SMXB methodology, which, owing to its intrinsic characteristics, proves exceptionally valuable in examining the impact of size and depth variations in a typical damage scenario on composite panels. Unlike the standard VCCT and CZM methodologies, which are likely to impact the results due to their reliance on the mesh and time step parameters in finite element analysis, the FT-SMXB methodology stands out for its independence from these factors.

Author Contributions: Conceptualization, A.R. (Angela Russo), R.C., C.P. and A.R. (Aniello Riccio); methodology, A.R. (Angela Russo), R.C., C.P. and A.R. (Aniello Riccio); formal analysis, A.R. (Angela Russo), R.C., C.P. and A.R. (Aniello Riccio); writing, A.R. (Angela Russo), R.C., C.P. and A.R. (Aniello Riccio); writing—review and editing, A.R. (Angela Russo), R.C., C.P. and A.R. (Aniello Riccio); supervision, A.R. (Angela Russo), R.C., C.P. and A.R. (Aniello Riccio); funding acquisition, A.R. (Aniello Riccio). All authors have read and agreed to the published version of the manuscript.

Funding: This study based upon research supported by the Office of Naval Research under Award Number (N62909-20-1-2042) and supervised by Anisur Rahman (Office of Naval Research) and William Nickerson (Office of Naval Research).

Data Availability Statement: Data is contained within the article.

Conflicts of Interest: The authors declare no conflict of interest.

References

1. Baker, A.A. *Composite Materials for Aircraft Structures*; AIAA: Reston, VA, USA, 2004.
2. Irving, P.E.A.C.S. *Polymer Composites in the Aerospace Industry*; Woodhead Publishing: Sawston, UK, 2019.
3. Rikards, R. *Interlaminar Fracture Behaviour of Laminated Composites*; Computers & Structures: Walnut Creek, CA, USA, 2000; pp. 11–18.
4. Christoforou, A.P.A.A.S.Y. Scaling of low-velocity impact response in composite structures. *Compos. Struct.* **2009**, *91*, 358–365. [CrossRef]
5. Caputo, F.E.A. Numerical study for the structural analysis of composite laminates subjected to low velocity impact. *Compos. Part B Eng.* **2014**, *67*, 296–302. [CrossRef]
6. Maio, L.E.A. Simulation of low velocity impact on composite laminates with progressive failure analysis. *Compos. Struct.* **2013**, *103*, 75–85. [CrossRef]
7. Suriani, M.J.; Rapi, H.Z.; Ilyas, R.A.; Petrů, M.; Sapuan, S.M. Delamination and manufacturing defects in natural fiber-reinforced hybrid composite: A review. *Polymers* **2021**, *13*, 1323. [CrossRef] [PubMed]
8. Zhang, J.; Fox, B.L. Manufacturing influence on the delamination fracture behavior of the T800H/3900-2 carbon fiber reinforced polymer composites. *Mater. Manuf. Process.* **2007**, *22*, 768–772. [CrossRef]
9. Yao, L.; Alderliesten, R.C.; Zhao, M.; Benedictus, R. Discussion on the use of the strain energy release rate for fatigue delamination characterization. *Compos. Part A Appl. Sci. Manuf.* **2014**, *66*, 65–72. [CrossRef]
10. Tanulia, V.; Wang, J.; Pearce, G.M.; Baker, A.; David, M.; Prusty, B.G. A procedure to assess disbond growth and determine fatigue life of bonded joints and patch repairs for primary airframe structures. *Int. J. Fatigue* **2020**, *137*, 105664. [CrossRef]
11. Iarve, E.V.; Gurvich, M.R.; Mollenhauer, D.H.; Rose, C.A.; Dávila, C.G. Mesh-independent matrix cracking and delamination modeling in laminated composites. *Int. J. Numer. Methods Eng.* **2011**, *88*, 749–773. [CrossRef]
12. Wisnom, M.R. The role of delamination in failure of fibre-reinforced composites. *Phil. Trans. R. Soc. A* **2012**, *370*, 1850–1870. [CrossRef]
13. Alessi, S.; Pitarresi, G.; Spadaro, G. Effect of hydrothermal ageing on the thermal and delamination fracture behaviour of CFRP composites. *Compos. Part B Eng.* **2014**, *67*, 145–153. [CrossRef]
14. Long, S.; Yao, X.; Zhang, X. Delamination prediction in composite laminates under low-velocity impact. *Compos. Struct.* **2015**, *132*, 290–298. [CrossRef]
15. Liu, P.F.; Hou, S.J.; Chu, J.K.; Hu, X.Y.; Zhou, C.L.; Liu, Y.L.; Zheng, J.Y.; Zhao, A.; Yan, L. Finite element analysis of postbuckling and delamination of composite laminates using virtual crack closure technique. *Compos. Struct.* **2011**, *93*, 1549–1560. [CrossRef]
16. Pennington, A.; Goyal, V.K. Durability of Buckled Composites Using Virtual Crack Closure Technique Fatigue R-Curve Implementation. *AIAA J.* **2023**, *61*, 3644–3663. [CrossRef]
17. Michalcová, L.; Rechcígel, L.; Blský, P.; Kuchařský, P. Fatigue disbonding analysis of wide composite panels by means of Lamb waves. *Proc. SPIE Int. Soc. Opt. Eng.* **2018**, *10599*, 105991E.
18. Šedková, L.; Vlk, V.; Šedek, J. Delamination/disbond propagation analysis in adhesively bonded composite joints using guided waves. *J. Struct. Integr. Maint.* **2022**, *7*, 25–33. [CrossRef]
19. Johnson, A.F.; Toso-Pentecôte, N.; Schueler, D. Numerical modelling of impact and damage tolerance in aerospace composite structures. *Numer. Model. Fail. Adv. Compos. Mater.* **2015**, 479–506. [CrossRef]
20. Shah, S.Z.H.; Karuppanan, S.; Megat-Yusoff, P.S.M.; Sajid, Z. Impact resistance and damage tolerance of fiber reinforced composites: A review. *Compos. Struct.* **2019**, *217*, 100–121. [CrossRef]
21. Jones, R. Fatigue crack growth and damage tolerance. *Fatigue Fract. Eng. Mater. Struct.* **2014**, *37*, 463–483. [CrossRef]
22. Larsson, F. Damage tolerance of a stitched carbon/epoxy laminate. *Compos. Part A Appl. Sci. Manuf.* **1997**, *28*, 923–934. [CrossRef]
23. Müzel, S.D.; Bonhin, E.P.; Guimarães, N.M.; Guidi, E.S. Application of the finite element method in the analysis of composite materials: A review. *Polymers* **2020**, *12*, 818. [CrossRef]
24. Nguyen, M.Q.; Elder, D.J.; Bayandor, J.; Thomson, R.S.; Scott, M.L. A review of explicit finite element software for composite impact analysis. *J. Compos. Mater.* **2005**, *39*, 375–386. [CrossRef]
25. Sayyad, A.S.; Ghugal, Y.M. Bending, buckling and free vibration of laminated composite and sandwich beams: A critical review of literature. *Compos. Struct.* **2017**, *171*, 486–504. [CrossRef]
26. Almeida, F.S.; Awruch, A.M. Design optimization of composite laminated structures using genetic algorithms and finite element analysis. *Compos. Struct.* **2009**, *88*, 443–454. [CrossRef]
27. Muc, A.; Gurba, W. Genetic algorithms and finite element analysis in optimization of composite structures. *Compos. Struct.* **2001**, *54*, 275–281. [CrossRef]
28. Her, S.-C.; Liang, Y.-C. The finite element analysis of composite laminates and shell structures subjected to low velocity impact. *Compos. Struct.* **2004**, *66*, 277–285. [CrossRef]
29. Guynn, E.G.; Ochoa, O.O.; Bradley, W.L. A Parametric Study of Variables That Affect Fiber Microbuckling Initiation in Composite Laminates: Part 1-Analyses. *J. Compos. Mater.* **1992**, *26*, 1594–1616. [CrossRef]

30. Zona, A.; Barbato, M.; Conte, J.P. Finite element response sensitivity analysis of steel-concrete composite beams with deformable shear connection. *J. Eng. Mech.* **2005**, *131*, 1126–1139. [CrossRef]
31. Pegorin, F.; Pingkarawat, K.; Daynes, S.; Mouritz, A.P. Influence of z-pin length on the delamination fracture toughness and fatigue resistance of pinned composites. *Compos. Part B Eng.* **2015**, *78*, 298–307. [CrossRef]
32. Lu, X.; Ridha, M.; Chen, B.Y.; Tan, V.B.C.; Tay, T.E. On cohesive element parameters and delamination modelling. *Eng. Fract. Mech.* **2019**, *206*, 278–296. [CrossRef]
33. Blal, N.; Daridon, L.; Monerie, Y.; Pagano, S. *On Mesh Size to Cohesive Zone Parameters Relationships*; ACOMEN: Liège, Belgium, 2011; p. Cd–Rom.
34. Russo, A.; Riccio, A.; Palumbo, C.; Sellitto, A. Fatigue driven delamination in composite structures: Definition and assessment of a novel fracture mechanics based computational tool. *Int. J. Fatigue* **2023**, *166*, 107257. [CrossRef]
35. Russo, A.; Riccio, A.; Sellitto, A. A robust cumulative damage approach for the simulation of delamination under cyclic loading conditions. *Compos. Struct.* **2022**, *281*, 114998. [CrossRef]
36. Riccio, A.; Castaldo, R.; Palumbo, C.; Russo, A. Delamination Effect on the Buckling Behaviour of Carbon–Epoxy Composite Typical Aeronautical Panels. *Appl. Sci.* **2023**, *13*, 4358. [CrossRef]
37. *ASTM D3039/D3039M-00*; Standard Test Method for Tensile Properties of Polymer Matrix Composite Materials. ASTM International: West Conshohocken, PA, USA, 2000.
38. *ASTM D3410*; Standard Test Method for Compressive Properties of Unidirectional or Cross-ply Fiber-Resin Composites. American Society for Testing and Materials: West Conshohocken, PA, USA, 1987; pp. 132–139.
39. *ASTM D3518/D3518M-94*; Standard Test Method for in-Plane Shear Response of Polymer Matrix Composite Materials by Tensile Test of a ±45° Laminate. ASTM International: West Conshohocken, PA, USA, 2001.
40. *ASTM D5528*; Standard Test Method for Mode I Interlaminar Fracture Toughness of Unidirectional Fiber Reinforced Polymer Matrix Composites. ASTM International: West Conshohocken, PA, USA, 1994.
41. *ASTM Standard D6115*; Mode I Fatigue Delamination Growth Onset of Unidirectional Fiber-Reinforced Polymer Matrix Composites. American Society for Testing and Materials: West Conshohocken, PA, USA, 2011.
42. *ASTM D 7905/D7905M-14*; ASTM Int. Standard Test Method for Determination of the Mode II Interlaminar Fracture Toughness of Unidirectional Fiber-Reinforced Polymer Matrix Composites. ASTM International: West Conshohocken, PA, USA, 2014.
43. Li, Y.; Wang, B.; Zhou, L. Study on the effect of delamination defects on the mechanical properties of CFRP composites. *Eng. Fail. Anal.* **2023**, *153*, 107576. [CrossRef]
44. Mekonnen, A.A.; Woo, K.; Kang, M.; Kim, I.-G. Effects of Size and Location of Initial Delamination on Post-buckling and Delamination Propagation Behavior of Laminated Composites. *Int. J. Aeronaut Space Sci.* **2020**, *21*, 80–94. [CrossRef]

Disclaimer/Publisher's Note: The statements, opinions and data contained in all publications are solely those of the individual author(s) and contributor(s) and not of MDPI and/or the editor(s). MDPI and/or the editor(s) disclaim responsibility for any injury to people or property resulting from any ideas, methods, instructions or products referred to in the content.

Article

Surface-Initiated Polymerization with an Initiator Gradient: A Monte Carlo Simulation

Zhining Huang [1], Caixia Gu [1], Jiahao Li [1], Peng Xiang [1], Yanda Liao [2], Bang-Ping Jiang [1], Shichen Ji [1,3,*] and Xing-Can Shen [1,*]

[1] State Key Laboratory for Chemistry and Molecular Engineering of Medicinal Resources, Key Laboratory for Chemistry and Molecular Engineering of Medicinal Resources, Ministry of Education of China, Collaborative Innovation Center for Guangxi Ethnic Medicine, School of Chemistry and Pharmaceutical Sciences, Guangxi Normal University, Guilin 541004, China; huang_z_n@stu.gxnu.edu.cn (Z.H.); jiangbangping@mailbox.gxnu.edu.cn (B.-P.J.)

[2] School of Computer Science and Engineering & School of Software, Guangxi Normal University, Guilin 541004, China; lyd@gxnu.edu.cn

[3] State Key Laboratory of Molecular Engineering of Polymers, Fudan University, Shanghai 200433, China

* Correspondence: shichen.ji@mailbox.gxnu.edu.cn (S.J.); xcshen@mailbox.gxnu.edu.cn (X.-C.S.)

Abstract: Due to the difficulty of accurately characterizing properties such as the molecular weight (M_n) and grafting density (σ) of gradient brushes (GBs), these properties are traditionally assumed to be uniform in space to simplify analysis. Applying a stochastic reaction model (SRM) developed for heterogeneous polymerizations, we explored surface-initiated polymerizations (SIPs) with initiator gradients in lattice Monte Carlo simulations to examine this assumption. An initial exploration of SIPs with 'homogeneously' distributed initiators revealed that increasing σ slows down the polymerization process, resulting in polymers with lower molecular weight and larger dispersity (Đ) for a given reaction time. In SIPs with an initiator gradient, we observed that the properties of the polymers are position-dependent, with lower M_n and larger Đ in regions of higher σ, indicating the non-uniform properties of polymers in GBs. The results reveal a significant deviation in the scaling behavior of brush height with σ compared to experimental data and theoretical predictions, and this deviation is attributed to the non-uniform M_n and Đ.

Keywords: gradient brush; surface-initiated polymerization; stochastic reaction model; Monte Carlo simulation

Citation: Huang, Z.; Gu, C.; Li, J.; Xiang, P.; Liao, Y.; Jiang, B.-P.; Ji, S.; Shen, X.-C. Surface-Initiated Polymerization with an Initiator Gradient: A Monte Carlo Simulation. *Polymers* **2024**, *16*, 1203. https://doi.org/10.3390/polym16091203

Academic Editors: Alexey V. Lyulin and Valeriy V. Ginzburg

Received: 21 March 2024
Revised: 22 April 2024
Accepted: 23 April 2024
Published: 25 April 2024

Copyright: © 2024 by the authors. Licensee MDPI, Basel, Switzerland. This article is an open access article distributed under the terms and conditions of the Creative Commons Attribution (CC BY) license (https:// creativecommons.org/licenses/by/ 4.0/).

1. Introduction

Gradient brushes (GBs) are polymer brushes wherein properties, such as molecular weight, grafting density, or chemical composition, gradually vary in one or more directions along the substrate. GBs are powerful tools for high-throughput and low-cost investigations in the areas of physics, chemistry, biomaterials science, and biology [1–8]. In a single sample, a given surface parameter across a wide range can be systematically explored, avoiding the need for lengthy repetitive procedures and enhancing the efficiency of research and development [8]. Additionally, GBs are widely used to study interfacial phenomena like the directional transport of liquids, cell adhesion, and migration [9,10].

Surface-initiated polymerization (SIP) is a promising approach to synthesizing GBs with higher grafting density. There are two major forms of SIP classified by initiator distribution [3,11]: one with a homogeneous distribution of initiators and another with a gradient distribution. In the former case, GBs are obtained by controlling the spatial polymerization time, for example, using a movable mask or reaction solution, or by adjusting the spatial polymerization rate through methods such as varying the intensity of transmitted light with a filter in photopolymerization [12–16]. In the latter case, initiators with a gradient density are firstly anchored to the surface using methods such as the silane diffusion method, nanolithography methods, or methods involving gradients of temperature or electrochemical potential [15–21]. Subsequently, SIP is carried out to yield GBs. Generally, the former is a simpler and more feasible method, while the latter is suitable for small-sized patterns and arbitrary structures [3].

Despite significant progress in the preparation of GBs, characterizing crucial properties such as grafting density (σ), molecular weight (M_n), and molecular weight distribution remains a challenging task [13,22,23]. The characterization of σ and M_n is interrelated since

$$\sigma = h\rho N_A / M_n, \qquad (1)$$

where h is the height of a brush in the dry state, ρ is the density of a polymer, and N_A is Avogadro's constant. Gel permeation chromatography (GPC) is the most common method for directly determining the molecular weight of grafted polymers, involving the degrafting of polymer chains from a substrate. However, the GPC method requires a sufficient amount of a sample, posing a challenge for SIP, especially regarding polymerization on a flat substrate [2]. On the other hand, the accuracy of indirect measurement, achieved by incorporating sacrificial initiators for simultaneous bulk- and surface-initiated polymerizations and characterizing the resulting polymers in solution, has been a topic of debate [22,23]. Notably, neither direct nor indirect GPC methods offer insights into the spatial distribution of these properties in GBs.

The lack of information on molecular weight and grafting density has significantly hindered efforts toward comprehensively understanding gradient polymer brushes and applying them. An early study examined the scaling behavior between brush height and grafting density for a gradient polymer brush [17]. In this study, a polyacrylamide (PAAm) brush with a grafting density gradient was obtained via atom-transfer radical polymerization (ATRP) with an initiator gradient generated via the silane diffusion method [17,24]. Due to the absence of information on molecular weight, two assumptions were made to determine the grafting density in space: (1) the molecular weight of polymers along the substrate in GBs is uniform, and (2) there is similarity in the molecular weight between polymers in the GBs and those obtained in solution polymerization under the same conditions. Upon determining the dry brush height using variable-angle spectroscopic ellipsometry, the spatial distribution of grafting density was obtained using Equation (1). Subsequently, we examined the scaling relationship between the wet height of the brush and grafting density, revealing only a slight deviation from the theoretical prediction [17].

Although researchers have recognized the limitations of the uniform molecular weight assumption [17], it remains prevalent in experimental studies due to its simplicity. This raises a question: what is the significance of the effect of this assumption?

To answer this question, the polymerization mechanism should be examined, as it strongly influences properties such as molecular weight and dispersity. Computer simulations have played an important role in revealing the mechanisms of SIPs. A pioneering study was performed by Genzer using a Monte Carlo (MC) simulation [25], and it inspired studies using different simulation methods [26–35]. Typically, SIPs result in polymers with larger dispersity and smaller molecular weight compared to those generated via bulk-initiated polymerizations (BIPs). This trend holds even in simultaneous bulk- and surface-initiated polymerization [26,28,35]. Moreover, grafting density is a key parameter in SIP, impacting both the kinetics of the reaction and the properties of the polymers, such as molecular weight and dispersity. The main reason is that SIP is a heterogeneous polymerization, as the homogeneous distribution of free monomers is altered by the newly formed polymer brush [34], while BIP is a homogeneous polymerization.

According to the existing simulations of SIPs, it is natural to expect that the molecular weight should be non-uniform in an SIP with an initiator gradient. However, the significance of this difference in molecular weight and its potential impact on the scaling behavior observed in Ref. [17] remain uncertain because no simulations, to the best of our knowledge, have directly addressed this issue.

To address this gap, we conducted a lattice Monte Carlo simulation to examine SIP with initiator gradients (referred to as gradient polymerization in the remainder of this paper), using a stochastic reaction model (SRM) developed for heterogeneous polymerizations [34–36]. We explored two systems: one with a series of homogeneous SIPs with varying grafting densities, wherein the initiators were homogeneously distributed, and the other with SIPs with an initiator gradient. In both systems, the properties of the polymers are significantly affected by grafting density. Notably, the scaling relationship between the brush height and grafting density in GBs, as obtained in the simulation, diverges from the experimental results, highlighting the need for a more in-depth exploration of gradient polymerization. The lattice MC model and SRM algorithms are introduced in Section 2, while the results of living polymerizations with homogeneous and gradient initiators are shown in Section 3. Brief conclusions are provided in Section 4.

2. Models and Simulation Methods

2.1. Lattice Monte Carlo Simulation

In this study, we employed the Larson-type bond fluctuation model [37,38], which was previously utilized in our investigations of SIP and the flow behavior of polymer brushes [34–36,39]. Briefly, the simulation was carried out in a simple cubic lattice with a volume, V, equal to $L_x \times L_y \times L_z$. Each lattice site can be occupied by a monomer or initiator only once, and the bond length was set to 1 or $\sqrt{2}$. During relaxation, a monomer is randomly selected to be exchanged with one of its 18 nearest or next-nearest neighbor sites. This exchange will be accepted under the conditions that the neighbor site is vacant and that the exchange would not break the chain and cause bond intersection (possible bond intersections are shown in Figure S1). The excluded volume effect and entanglement were well considered in this model. The simulation time was measured in units of Monte Carlo steps (MCs), defined as all monomers attempting to move once, on average.

Two impenetrable walls were set in the $y = 1$ and L_y planes, respectively, while periodic boundary conditions were applied in both the x and z directions. In this simulation, the x direction represents the initiator gradient direction, the y direction indicates the chain growth direction, and the z direction corresponds to the equivalent direction. In the beginning, all the free monomers were randomly distributed in the system. The immobilized initiators were randomly positioned on the $y = 1$ plane during the investigation of homogeneous SIP. It should be noted that the term 'homogeneous' does not imply a perfectly 'regular' distribution of initiators, as shown in Figure S2. The properties of these two systems show subtle differences [30]. Instead, "homogeneous" is used in comparison to the gradient polymerization.

While studying gradient polymerization, the $y = 1$ plane was divided into multiple stripes in the x direction (initiator gradient direction), as illustrated in Scheme 1. Each stripe has a width w, resulting in a total number of L_x/w stripes. In the left part of the simulation box, the grafting density of the leftmost stripe is σ_{min} and gradually increases with the value of $\Delta\sigma$ (the difference in density between successive stripes) with an increasing number of stripe locations until it reaches the maximum grafting density σ_{max}. The grafting density in the right part mirrored that of the left part, i.e., the grafting density decreased from the middle to the rightmost stripe as the location of stripes shifted forward further. Within each stripe, the number of initiators can be calculated as $\sigma(x) \times w \times L_z$, with $\sigma(x)$ denoting the grafting density of a stripe, and these initiators exhibited a random distribution within the stripes.

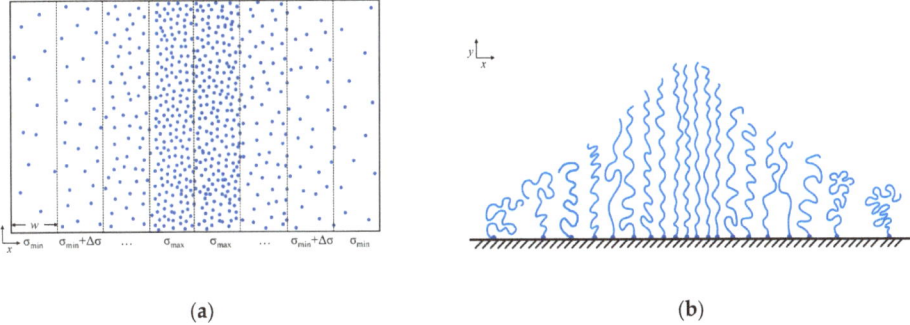

Scheme 1. An illustration of the gradient distribution of initiators in the $y = 1$ plane (**a**) and the projection of gradient polymer brush in the xy direction (**b**). In this simulation, the x direction represents the initiator gradient direction, the y direction indicates the chain growth direction, and the z direction indicates the equivalent direction.

2.2. Implementation of Polymerization

In this study, a living polymerization was considered, which occurred at intervals of every τ MCs during the relaxation process. Here, τ is defined as the characteristic delay time, or reaction interval time [27,34–36,40]. By decreasing or increasing the value of τ, the reaction can be adjusted to make it diffusion-limited or reaction-limited.

The stochastic reaction model (SRM) proposed by our group was applied to model the polymerization [34–36]. Firstly, an initiator or active center was randomly selected. Then, the number of free monomers m within the radius of $\sqrt{2}$ was determined. The initiator or active center tries to react with a random monomer among these m free monomers with a given probability P_r, which is determined by the local number of free monomers and calculated as mP_0 (where P_0 is a constant representing the elementary reaction probability between one active center and one free monomer). If the reaction is accepted, the free monomer transforms into an active center for future reactions. Since P_r is determined by the local reaction environment, each active center reacts with its own probability. Thus, our SRM model fully accounts for the heterogeneous reaction microenvironment, which is the inherent character of SIP [34–36]. As demonstrated by living bulk-initiated polymerization, the polymerization kinetics obtained by this SRM were found to be very consistent with the theoretical predictions [34,36].

3. Results and Discussion

3.1. Homogeneous Surface-Initiated Polymerization

We first investigated a series of homogeneous SIPs with varying grafting densities and compared the properties at the same polymerization times. The parameters were fixed and set as follows: the dimensions of the simulation box were $L_x \times L_y \times L_z = 50 \times 77 \times 50$, the initial concentration of free monomers was $[M]_0 = 0.4$ monomers per lattice, the reaction interval time was $\tau = 10$ MCs, the simulation time was 2×10^6 MCs, and the elementary reaction probability was $P_0 = 0.001$. The results were averaged over 60 independent runs.

Figure 1a shows the number-average molecular weight M_n during polymerization with a given grafting density σ. It is evident that σ significantly influences the polymerization kinetics, with M_n increasing more rapidly at lower σ compared to higher values. Figure 1b shows M_n as a function of σ with a given polymerization time t. When σ is very low, M_n is nearly constant. However, when σ is high, M_n exhibits a monotonic decrease with an increasing σ. The decrease in M_n with increasing σ is related to the polymerization time. For example, when $t = 400{,}000$ MCs, the values of M_n are 21.7 and 13.5 at $\sigma = 0.1$ and 0.4, respectively. The latter ($M_n = 13.5$) is only about 62% of the former, and by $t = 800{,}000$ MCs, this ratio further decreases to 54%. Besides M_n, the dispersity ($Đ = M_w/M_n$) and molecular weight distribution (MWD) are also influenced by σ. The dispersity increases with increasing σ (Figure 1c), and MWD becomes broader and more asymmetric (Figure 1d).

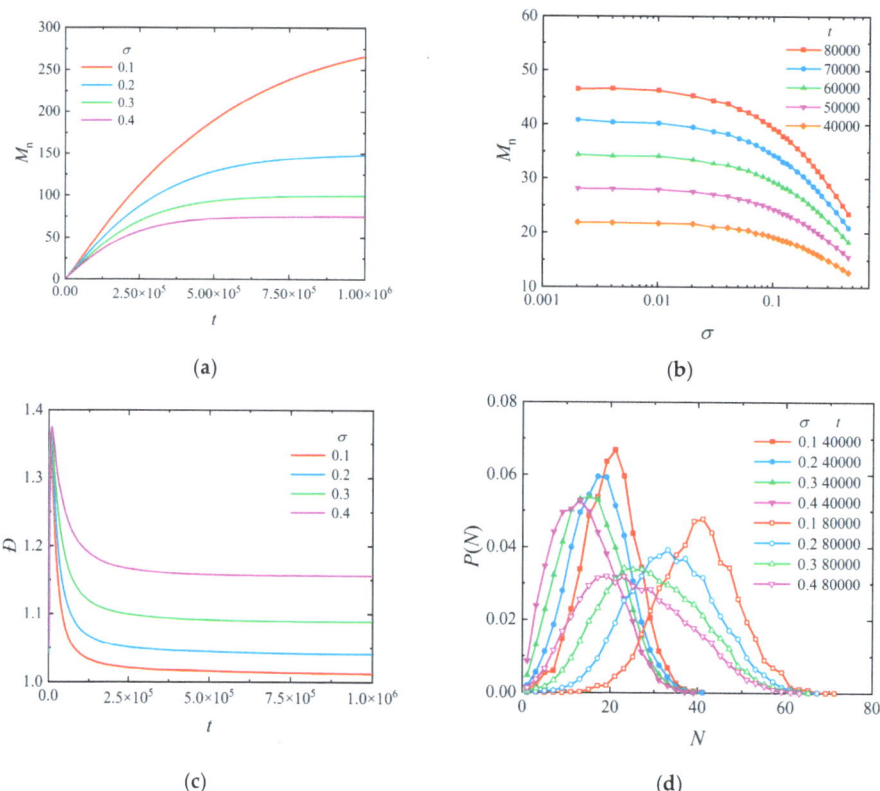

Figure 1. Influence of grafting density σ on surface-initiated polymerization with a homogeneous distribution of initiators. (**a**) Number-average molecular weight M_n as a function of polymerization time t with a given σ. The molecular weight with $\sigma = 0.1$ was saturated at the end of the simulation (2×10^6 MCs). (**b**) M_n as a function of σ with a given t. (**c**) Dispersity ($Đ$) as a function of t. (**d**) The molecular weight distribution $P(N)$ with a given σ.

The results suggest that in SIP, systems with different values of σ exhibit variations in polymer properties at the same polymerization time, preliminarily indicating that the molecular weight in GBs might be non-uniform due to the initiator gradient. It should be pointed out that molecular weight is independent of the concentration of initiators in living BIPs with low monomer conversion, and this can be proved as follows. The monomer conversion C of BIP can be written as

$$C = 1 - \frac{[M]_t}{[M]_0} = 1 - \exp(-(m_{\max} - 1)[I]_0 P_0 t/\tau), \tag{2}$$

where $[M]_t$ is the concentration of free monomers at time t, $[I]_0$ is the concentration of the initiator, and $m_{\max}$ is the maximum number of free monomers around an active center, equal to 18 in this simulation. At low conversion, M_n linearly increases with t as

$$M_n = \frac{[M]_0 C}{[I]_0} \approx [M]_0 (m_{\max} - 1) P_0 t/\tau. \tag{3}$$

Thus, in BIP, molecular weight is only determined by reaction time, and it is independent of the number of initiators.

Why does this assumption hold for BIP but not SIP? The reason is that Equation (1) was deduced from a homogeneous polymerization system, such as BIP. However, SIP is a heterogeneous system, especially when σ is high. When σ is low, the active centers are far apart and react with free monomers like isolated active centers, resembling a homogeneous polymerization system. Conversely, when σ is large, brush-like polymers are obtained, and the system is no longer homogeneous as free monomers are expelled by the nascent polymers from the surface. The active centers near the surface react in an environment with a lower concentration of free monomers compared with those far from the surface. Such a heterogeneous reaction environment is the key feature of SIP, and the heterogeneity of the reaction environment increases with σ [34,35].

3.2. Surface-Initiated Polymerization with Initiator Gradient

We further investigated the gradient polymerization with a simulation box for which $L_x \times L_y \times L_z = 288 \times 72 \times 100$. The initial concentration of free monomers was $[M]_0 = 0.4$ monomers per lattice. As shown in Scheme 1, the grafting plane was divided into stripes with a width $w = 4$. In this study, the maximum grafting density of a stripe is $\sigma_{max} = 0.42$, and the minimum is $\sigma_{min} = 0.07$. We did not explore lower grafting densities as our primary interest lay in the scaling behavior of polymer brushes within regions with high grafting density. The difference in grafting density between the adjacent stripes is $\Delta\sigma = 0.01$. The corresponding steepness of the gradient is $\delta = \Delta\sigma/w = 0.0025$. As discussed later, such a steepness is low enough to examine the gradient polymerization process. Polymerization stops when the number-average molecular weight of the brush reaches 50. Such a molecular weight is large enough to ensure the system stays in the brush region; meanwhile, it can avoid the situation wherein some very long chains might approach the $y = L_y$ plane.

The density contour map (Figure 2a) clearly confirms the formation of a gradient brush. In the vertical direction, the polymer density decreases with an increasing distance from the surface. Horizontally, there is a gradient increase in the density from the low-grafting region ($x = 1$) to the high-grafting region ($x = 144$), which then decreases upon further shifting the x position forward. Meanwhile, the density contour map of free monomers exhibits the opposite trend (Figure 2b).

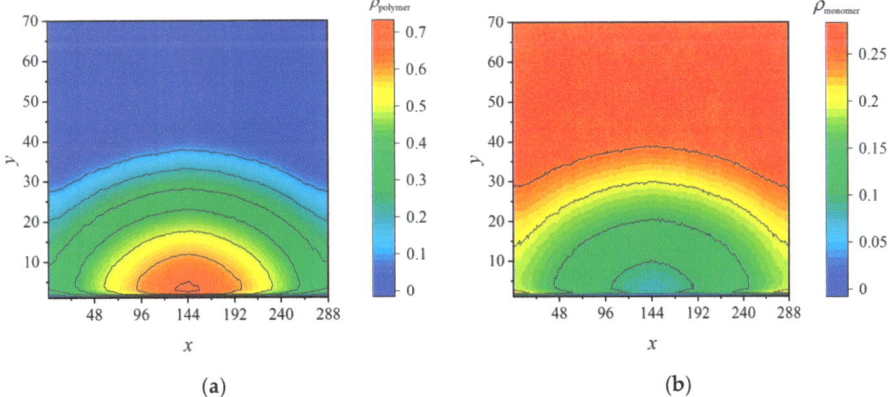

Figure 2. The density contour maps of polymer brush (**a**) and free monomers (**b**) in the gradient polymerization. The number-average molecular weight in the system is $M_n = 50$.

The molecular weights of the polymers at each stripe were examined (Figure 3a), with the results clearly proving that there is a non-uniform molecular weight in GBs. M_n decreases from the low-grafting regions (outsides) to the high-grafting regions (middle).

Although the overall M_n of the system is 50, it is only 42.4 in the middle, contrasting with the higher value of 65.5 on the outside areas. Such a notable difference in M_n should not be overlooked. We further compared the molecular weight and dispersity in gradient polymerization with those in SIP at the same reaction time (Figure 3b). The variation in M_n between different grafting density positions in gradient polymerization is smaller than that in SIP. This might be attributed to the competition among different polymerization regions in gradient polymerization. Higher-grafting-density regions tend to consume more monomers during the reaction. The dispersity of polymers in gradient polymerization increases with grafting density, which is similar to the trend observed in SIP (Figure 3b).

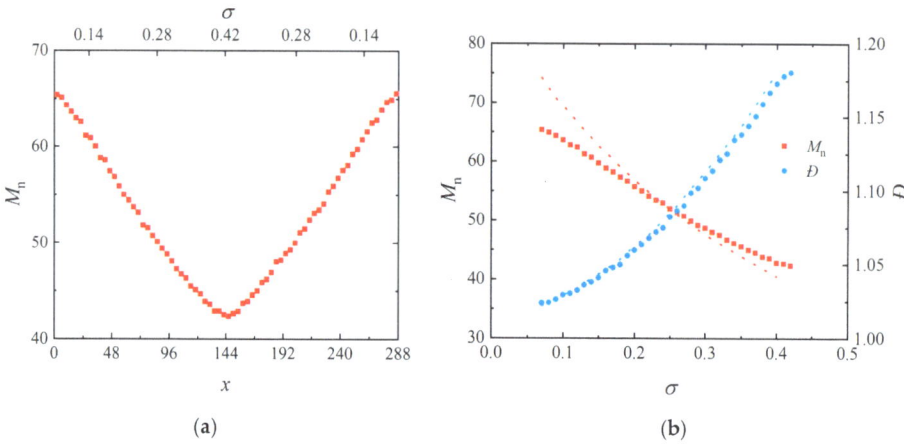

Figure 3. (a) M_n as a function of positions and corresponding grafting densities in gradient polymerization when the overall M_n of the system is 50. (b) M_n and $Đ$ as a function of σ in both gradient polymerization (solid symbols) and SIP (dotted lines).

In experiments, the height of a brush in the dry state is widely used to estimate grafting density in accordance with Equation (1). In this study, we have supposed that the brush shown in Figure 2a vertically collapses onto the surface; thus, the dry height at position x, denoted as $h(x)$, can be calculated as follows:

$$h(x) = \sum_{y=1}^{L_y} \rho(x,y). \tag{4}$$

Figure 4a suggests that a gradient brush is obtained but that the dry height $h(x)$ does not linearly increase with the grafting density $\sigma(x)$, as depicted by Equation (1). This deviation from a linear relationship is evidently caused by the variations in M_n at different positions, as shown in Figure 3a. After normalizing the height with respect to the corresponding molecular weight, a linear relationship between $h(x)$ and $\sigma(x)$ can be restored.

The height in solution H is a key property of a polymer brush and is calculated as follows:

$$H(x) = \sum_{y=1}^{L_y} y\rho(x,y) / \sum_{y=1}^{L_y} \rho(x,y) \tag{5}$$

Figure 4b suggests that H increased from the outside areas to the middle, indicating the formation of a gradient brush. We are more interested in the relationship between $H(x)$ and $\sigma(x)$. In the low-grafting-density region, H increases only slightly with σ (Figure 4c). Subsequently, a scaling relationship between them with a scaling exponent of 0.15 can be observed. In the even-higher-grafting-density region, the increase in thickness slows down again. We speculate that the absence of pronounced scaling behavior on both sides is due to the unidirectional extension of the chains. In the region with the highest grafting density, the chains primarily extend towards the low-density region due to the significant compression between chains. The opposite behavior is observed in the low-density region. This explanation is supported by Figure S3. In the high (low)-grafting-density region, the number of monomers consumed by corresponding initiators during polymerization is greater (less) than the actual number of monomers in that region.

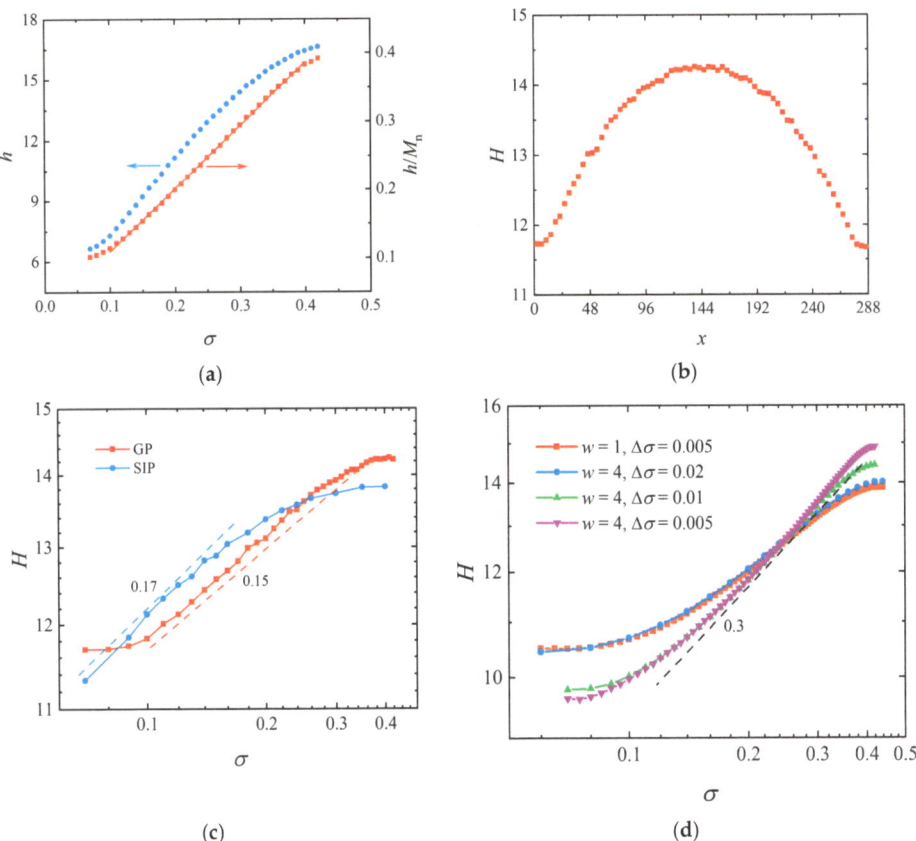

Figure 4. (**a**) The height of gradient brush in dry state h (circles) and the normalized height h/M_n (squares) as a function of grafting density. (**b**) The height of gradient brush in solution as a function of position x. (**c**) The height of gradient polymer brush (square) as a function of grafting density. For comparison, the heights of series polymer brushes formed by surface-initiated polymerization (SIP) at the same reaction time as the gradient polymerization are represented (circles). (**d**) The effect of steepness on the height of gradient brushes induced by varying the width of the stripes (w) and the difference in grafting density between the adjacent stripes ($\Delta\sigma$). It should be pointed that the gradient brushes shown in (**d**) are formed by monodisperse polymers for which $M_n = 50$. The slope of the dashed line is 0.3.

If we focus on the scaling behavior in the middle region, the obtained scaling exponent (0.15) is much smaller than both the theoretical value (1/3) [41,42] and the experimental value (0.37–0.4) obtained in Ref. [17]. The theoretical scaling exponent has been extensively validated through simulations [39,43–46], although some simulations suggest that it may not be a constant but instead slightly vary with the grafting density [47]. This raises the following question: how can we interpret such a smaller scaling exponent obtained in this simulation?

First, the theoretical scaling exponent of one-third was validated for brushes of monodisperse polymers in good solvents. As shown in Figure 3a, in gradient polymerization, M_n decreases with an increasing grafting density in the x direction, and the difference in M_n is significant. Moreover, the polymers at each stripe are polydisperse, and the dispersity increases with the increase in grafting density. Studies have shown that the dispersity of polymers also influences the height of polydisperse polymer brushes [34,48,49]. Thus, in a gradient polymerization system, if we only examine the relationship between height and grafting density, the scaling exponent does not need to adhere to the one-third scaling behavior. We are particularly interested in the experimental value [17], as the non-uniform molecular weight and the dispersity should also be present in the experiment as in this simulation. The fact that the experimental value is close to the theoretical prediction might be related to the method used to calculate the grafting density or instead be a coincidental approximation.

The observed small exponent might relate to the steepness of the initiator gradient. In GBs, polymers experience unbalanced lateral compression in the gradient direction, causing chains in high-grafting-density regions to extend towards regions with lower grafting density. In the experiment, the steepness is negligible as the lateral size of the substrate is 10^6 times larger than the height of the polymer brush [17], while in simulations, the role of steepness should be considered due to the finite simulation size. We can expect that the larger the steepness, the smaller the exponent.

To address this, we studied GBs with different levels of steepness (Figure 4d). It should be pointed out that these GBs consist of polymers with a fixed value of $M_n = 50$, avoiding the influence of non-uniform molecular weight and dispersity observed in gradient polymerization. Figure 4d shows the relationship between brush height and grafting density. When the steepness decreases from 0.005 to 0.0025, the curve exhibits a steeper incline, indicating the influence of steepness. A further reduction in steepness to 0.00125 causes only a minor change in the curve. In this case, the scaling exponent in the middle region is about 0.3, close to the theoretical value. Additionally, we varied the stripe width from $w = 4$ to $w = 1$ while fixing the steepness (Figure 4d), revealing almost identical results. Thus, we believe that the applied steepness is small enough, and this factor is not the main cause of the observed small scaling exponent.

Although the steepness cannot be infinitely small in simulations, we can examine a series of homogeneous SIPs with different initiator densities, the reaction time of which is the same as that of gradient polymerization. This special case allows us to approximate the behavior of gradient brushes with zero steepness. Figure 4c demonstrates that the heights of brushes obtained through SIP display a certain scaling relationship, with a scaling exponent (0.17) slightly larger than that of gradient polymerization (0.15). Thus, we can conclude that the steepness is not the primary cause of the observed small scaling exponent in our simulation.

4. Conclusions

The application of gradient brushes requires critical information regarding properties such as molecular weight and grafting density, which are difficult to characterize experimentally. The assumption that polymers in gradient brushes have uniform properties was commonly adopted in previous experiments, and its validity was questioned but not directly examined. In this study, we employed a stochastic reaction model to investigate surface-initiated polymerization with initiator gradients and analyzed the properties of the resulting gradient brushes.

We first examined surface-initiated polymerizations with homogeneously distributed initiators. The results indicated that, at a given reaction time, polymers with lower molecular weight and higher dispersity were obtained when there was increasing grafting density. This trend can be attributed to the heterogeneous reaction environment inherent in SIP.

Similarly, in SIPs with an initiator gradient, the corresponding polymers exhibited position-dependent properties, with lower molecular weights and higher dispersity at positions with higher grafting density. The difference in molecular weight in gradient brushes, reaching up to 154% (65.5/42.4) in this study, strongly supports the notion that the properties of polymers in gradient brushes are non-uniform.

Subsequent investigation into the height of gradient brushes in solution revealed a small scaling exponent (0.15) in scaling behavior with respect to grafting density, notably deviating from the expected scaling exponent of 1/3. We attributed this discrepancy to the variations in molecular weight and dispersity across space while also excluding the influence of the steepness of initiator gradient. It is noteworthy that a 1/3 scaling exponent is conventionally applied to monodispersed polymer brushes.

We are intrigued by the proximity of the experimental scaling exponent to the theoretical value since the non-uniform properties of polymers should also be present in an experiment. However, we failed to find any other experimental studies of scaling behavior with respect to gradient brushes, and we are uncertain whether this behavior is just a coincidence or if there are underlying mechanisms. It is important to recognize the differences between the simulation and experiments. In the simulation, a living polymerization was modeled, and all initiators reacted, while in the experiment, ATRP was applied [17], and the initiator efficiency was typically low [23].

In summary, this study provides direct evidence of the significant non-uniform properties of polymers in surface-initiated polymerizations with initiator gradients. We hope experimental studies are conducted in the future to better clarify the experimental results in Ref. [17]. Additionally, it would be interesting to investigate surface-initiated polymerizations with other gradients, such as reaction time [13].

Supplementary Materials: The following supporting information can be downloaded at https://www.mdpi.com/article/10.3390/polym16091203/s1. Figure S1: Illustration of two possible exchanges between a monomer and a vacancy, which involve bond intersections and are forbidden in this simulation; Figure S2: Illustration of initiators with a homogeneous distribution (left) and a regular distribution (right). Here, the homogeneous distribution means that all initiators are randomly placed on the substrate. While the regular distribution means that initiators are arranged in a certain lattice patter; Figure S3: In surface-initiated polymerization with initiator gradient, at a given stripe, the number of free monomers consumed by corresponding initiators (red circles), and the number of actual beads of polymers above the stripe (black squares). The former equals the number of chains Nchain(x) multiplied by the corresponding number-average molecular weight $M_n(x)$. While the latter equals the area ($w \times L_z$) multiplied by the corresponding dry height $h(x)$. A smaller number of polymer beads are found in the high grafting regions since the chains tend to the extend to lower grafting regions due to the unbalance lateral compression.

Author Contributions: S.J., B.-P.J. and X.-C.S. designed the research and wrote the paper. Z.H., C.G., J.L., P.X. and Y.L. carried out the simulations and analyzed the data. C.G. drew the schematic diagrams. S.J. interpreted the results. All authors have read and agreed to the published version of the manuscript.

Funding: This work was supported by the "Overseas 100 Talents Program" of Guangxi Higher Education, the Natural Science Foundation of China (No. 22263002).

Institutional Review Board Statement: Not applicable.

Data Availability Statement: Data are contained within the article and Supplementary Materials.

Acknowledgments: We thank the reviewer for pointing out the inappropriate expressions used in the manuscript.

Conflicts of Interest: The authors declare no conflicts of interest.

References

1. Lin, X.; He, Q.; Li, J. Complex polymer brush gradients based on nanolithography and surface-initiated polymerization. *Chem. Soc. Rev.* **2012**, *41*, 3584–3593. [CrossRef]
2. Barbey, R.; Lavanant, L.; Paripovic, D.; Schuwer, N.; Sugnaux, C.; Tugulu, S.; Klok, H.-A. Polymer brushes via surface-initiated controlled radical polymerization: Synthesis, characterization, properties, and applications. *Chem. Rev.* **2009**, *109*, 5437–5527. [CrossRef]
3. Zhou, X.; Liu, X.; Xie, Z.; Zheng, Z. 3D-patterned polymer brush surfaces. *Nanoscale* **2011**, *3*, 4929–4939. [CrossRef]
4. Genzer, J. Surface-bound gradients for studies of soft materials behavior. *Annu. Rev. Mater. Res.* **2012**, *42*, 435–468. [CrossRef]
5. Bhat, R.R.; Tomlinson, M.R.; Wu, T.; Genzer, J. Surface-grafted polymer gradients: Formation, characterization, and applications. *Adv. Polym. Sci.* **2006**, *198*, 51–124.
6. Chen, W.-L.; Cordero, R.; Tran, H.; Ober, C.K. 50th anniversary perspective: Polymer brushes: Novel surfaces for future materials. *Macromolecules* **2017**, *50*, 4089–4113. [CrossRef]
7. Murad Bhayo, A.; Yang, Y.; He, X. Polymer brushes: Synthesis, characterization, properties and applications. *Prog. Mater. Sci.* **2022**, *130*, 101000. [CrossRef]
8. Morgenthaler, S.; Zink, C.; Spencer, N.D. Surface-chemical and -morphological gradients. *Soft Matter* **2008**, *4*, 419–434. [CrossRef]
9. Afzali, Z.; Matsushita, T.; Kogure, A.; Masuda, T.; Azuma, T.; Kushiro, K.; Kasama, T.; Miyake, R.; Takai, M. Cell adhesion and migration on thickness gradient bilayer polymer brush surfaces: Effects of properties of polymeric materials of the underlayer. *ACS Appl. Mater. Interfaces* **2022**, *14*, 2605–2617. [CrossRef]
10. Kajouri, R.; Theodorakis, P.E.; Deuar, P.; Bennacer, R.; Židek, J.; Egorov, S.A.; Milchev, A. Unidirectional droplet propulsion onto gradient brushes without external energy supply. *Langmuir* **2023**, *39*, 2818–2828. [CrossRef]
11. Luzinov, I.; Minko, S.; Tsukruk, V.V. Responsive brush layers: From tailored gradients to reversibly assembled nanoparticles. *Soft Matter* **2008**, *4*, 714–725. [CrossRef] [PubMed]
12. Tomlinson, M.R.; Efimenko, K.; Genzer, J. Study of kinetics and macroinitiator efficiency in surface-initiated atom-transfer radical polymerization. *Macromolecules* **2006**, *39*, 9049–9056. [CrossRef]
13. Harris, B.P.; Metters, A.T. Generation and characterization of photopolymerized polymer brush gradients. *Macromolecules* **2006**, *39*, 2764–2772. [CrossRef]
14. Krabbenborg, S.O.; Huskens, J. Electrochemically generated gradients. *Angew. Chem. Int. Ed.* **2014**, *53*, 9152–9167. [CrossRef] [PubMed]
15. Poelma, J.E.; Fors, B.P.; Meyers, G.F.; Kramer, J.W.; Hawker, C.J. Fabrication of complex three-dimensional polymer brush nanostructures through light-mediated living radical polymerization. *Angew. Chem.* **2013**, *125*, 6982–6986. [CrossRef]
16. Zhang, C.; Wang, L.; Jia, D.; Yan, J.; Li, H. Microfluidically mediated atom-transfer radical polymerization. *Chem. Commun.* **2019**, *55*, 7554–7557. [CrossRef] [PubMed]
17. Wu, T.; Efimenko, K.; Genzer, J. Combinatorial study of the mushroom-to-brush crossover in surface anchored polyacrylamide. *J. Am. Chem. Soc.* **2002**, *124*, 9394–9395. [CrossRef]
18. Liu, Y.; Klep, V.; Zdyrko, B.; Luzinov, I. Synthesis of high-density grafted polymer layers with thickness and grafting density gradients. *Langmuir* **2005**, *21*, 11806–11813. [CrossRef]
19. Wang, X.; Tu, H.; Braun, P.V.; Bohn, P.W. Length scale heterogeneity in lateral gradients of poly (N-isopropylacrylamide) polymer brushes prepared by surface-initiated atom transfer radical polymerization coupled with in-plane electrochemical potential gradients. *Langmuir* **2006**, *22*, 817–823. [CrossRef]
20. Coad, B.R.; Bilgic, T.; Klok, H.-A. Polymer brush gradients grafted from plasma-polymerized surfaces. *Langmuir* **2014**, *30*, 8357–8365. [CrossRef]
21. Sheng, W.; Li, W.; Xu, S.; Du, Y.; Jordan, R. Oxygen-tolerant photografting for surface structuring from microliter volumes. *ACS Macro Lett.* **2023**, *12*, 1100–1105. [CrossRef]
22. Pester, C.W.; Klok, H.-A.; Benetti, E.M. Opportunities, challenges, and pitfalls in making, characterizing, and understanding polymer brushes. *Macromolecules* **2023**, *56*, 9915–9938. [CrossRef]
23. Zoppe, J.O.; Ataman, N.C.; Mocny, P.; Wang, J.; Moraes, J.; Klok, H.-A. Surface-initiated controlled radical polymerization: State-of-the-art, opportunities, and challenges in surface and interface engineering with polymer brushes. *Chem. Rev.* **2017**, *117*, 1105–1318. [CrossRef]
24. Wu, T.; Efimenko, K.; Vlček, P.; Šubr, V.; Genzer, J. Formation and properties of anchored polymers with a gradual variation of grafting densities on flat substrates. *Macromolecules* **2003**, *36*, 2448–2453. [CrossRef]
25. Genzer, J. In silico polymerization: Computer simulation of controlled radical polymerization in bulk and on flat surfaces. *Macromolecules* **2006**, *39*, 7157–7169. [CrossRef]
26. Turgman-Cohen, S.; Genzer, J. Computer simulation of concurrent bulk-and surface-initiated living polymerization. *Macromolecules* **2012**, *45*, 2128–2137. [CrossRef]
27. Liu, H.; Li, M.; Lu, Z.-Y.; Zhang, Z.-G.; Sun, C.-C. Influence of surface-initiated polymerization rate and initiator density on the properties of polymer brushes. *Macromolecules* **2009**, *42*, 2863–2872. [CrossRef]
28. Xu, J.; Xue, Y.-H.; Cui, F.-C.; Liu, H.; Lu, Z.-Y. Simultaneous polymer chain growth with the coexistence of bulk and surface initiators: Insight from computer simulations. *Phys. Chem. Chem. Phys.* **2018**, *20*, 22576–22584. [CrossRef]

29. Polanowski, P.; Hałagan, K.; Pietrasik, J.; Jeszka, J.K.; Matyjaszewski, K. Growth of polymer brushes by "grafting from" via ATRP–Monte Carlo simulations. *Polymer* **2017**, *130*, 267–279. [CrossRef]
30. Xue, Y.-H.; Quan, W.; Liu, X.-L.; Han, C.; Li, H.; Liu, H. Dependence of grafted polymer property on the initiator site distribution in surface-initiated polymerization: A computer simulation study. *Macromolecules* **2017**, *50*, 6482–6488. [CrossRef]
31. Arraez, F.J.; Van Steenberge, P.H.M.; Sobieski, J.; Matyjaszewski, K.; D'hooge, D.R. Conformational variations for surface-initiated reversible deactivation radical polymerization: From flat to curved nanoparticle surfaces. *Macromolecules* **2021**, *54*, 8270–8288. [CrossRef]
32. Arraez, F.J.; Van Steenberge, P.H.; D'hooge, D.R. The competition of termination and shielding to evaluate the success of surface-initiated reversible deactivation radical polymerization. *Polymers* **2020**, *12*, 1409. [CrossRef]
33. Li, W. Molecular dynamics simulations of ideal living polymerization: Terminal model and kinetic aspects. *J. Phy. Chem. B* **2023**, *127*, 7624–7635. [CrossRef]
34. Yang, B.; Liu, S.; Ma, J.; Yang, Y.; Li, J.; Jiang, B.-P.; Ji, S.; Shen, X.-C. A Monte Carlo simulation of surface-initiated polymerization: Heterogeneous reaction environment. *Macromolecules* **2022**, *55*, 1970–1980. [CrossRef]
35. Ma, J.-S.; Huang, Z.-N.; Li, J.-H.; Jiang, B.-P.; Liao, Y.-D.; Ji, S.-C.; Shen, X.-C. Simultaneous bulk- and surface-initiated living polymerization studied with a heterogeneous stochastic reaction model. *Chin. J. Polym. Sci.* **2023**, *42*, 364–372. [CrossRef]
36. Ma, J.; Li, J.; Yang, B.; Liu, S.; Jiang, B.-P.; Ji, S.; Shen, X.-C. A simple stochastic reaction model for heterogeneous polymerizations. *Polymers* **2022**, *14*, 3269. [CrossRef]
37. Larson, R. Monte Carlo lattice simulation of amphiphilic systems in two and three dimensions. *J. Chem. Phys.* **1988**, *89*, 1642–1650. [CrossRef]
38. Larson, R.; Scriven, L.; Davis, H. Monte Carlo simulation of model amphiphile-oil–water systems. *J. Chem. Phys.* **1985**, *83*, 2411–2420. [CrossRef]
39. Ji, S.; Ding, J. Rheology of polymer brush under oscillatory shear flow studied by nonequilibrium Monte Carlo simulation. *J. Chem. Phys.* **2005**, *123*, 144904. [CrossRef] [PubMed]
40. Farah, K.; Müller-Plathe, F.; Böhm, M.C. Classical reactive molecular dynamics implementations: State of the art. *ChemPhysChem* **2012**, *13*, 1127–1151. [CrossRef] [PubMed]
41. de Gennes, P.G. Conformations of polymers attached to an interface. *Macromolecules* **1980**, *13*, 1069–1075. [CrossRef]
42. Milner, S.T.; Witten, T.A.; Cates, M.E. Theory of the grafted polymer brush. *Macromolecules* **1988**, *21*, 2610–2619. [CrossRef]
43. Murat, M.; Grest, G.S. Structure of a grafted polymer brush: A molecular dynamics simulation. *Macromolecules* **1989**, *22*, 4054–4059. [CrossRef]
44. Jehser, M.; Zifferer, G.; Likos, C.N. Scaling and interactions of linear and ring polymer brushes via DPD simulations. *Polymers* **2019**, *11*, 541. [CrossRef]
45. Binder, K.; Milchev, A. Polymer brushes on flat and curved surfaces: How computer simulations can help to test theories and to interpret experiments. *J. Polym. Sci. Part B Polym. Phys.* **2012**, *50*, 1515–1555. [CrossRef]
46. Polanowski, P.; Sikorski, A. The structure of polymer brushes: The transition from dilute to dense systems: A computer simulation study. *Soft Matter* **2021**, *17*, 10516–10526. [CrossRef]
47. Whitmore, M.D.; Grest, G.S.; Douglas, J.F.; Kent, M.S.; Suo, T. End-anchored polymers in good solvents from the single chain limit to high anchoring densities. *J. Chem. Phys.* **2016**, *145*, 174904. [CrossRef]
48. Milner, S.T.; Witten, T.A.; Cates, M.E. Effects of polydispersity in the end-grafted polymer brush. *Macromolecules* **1989**, *22*, 853–861. [CrossRef]
49. Qi, S.; Klushin, L.I.; Skvortsov, A.M.; Schmid, F. Polydisperse polymer brushes: Internal structure, critical behavior, and interaction with flow. *Macromolecules* **2016**, *49*, 9665–9683. [CrossRef]

Disclaimer/Publisher's Note: The statements, opinions and data contained in all publications are solely those of the individual author(s) and contributor(s) and not of MDPI and/or the editor(s). MDPI and/or the editor(s) disclaim responsibility for any injury to people or property resulting from any ideas, methods, instructions or products referred to in the content.

Article

Polyurea–Graphene Nanocomposites—The Influence of Hard-Segment Content and Nanoparticle Loading on Mechanical Properties

Demetrios A. Tzelepis [1,2], Arman Khoshnevis [3], Mohsen Zayernouri [3,4] and Valeriy V. Ginzburg [1,*]

1. Department of Chemical Engineering and Materials Science, Michigan State University, East Lansing, MI 48824, USA; tzelepi1@msu.edu
2. Materials Division, US-Army, Ground Vehicle System Center, Warren, MI 48397, USA
3. Department of Mechanical Engineering, Michigan State University, East Lansing, MI 48824, USA; khoshne1@msu.edu (A.K.); zayern@msu.edu (M.Z.)
4. Department of Statistics and Probability, Michigan State University, East Lansing, MI 48824, USA
* Correspondence: ginzbur7@msu.edu

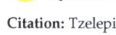

Citation: Tzelepis, D.A.; Khoshnevis, A.; Zayernouri, M.; Ginzburg, V.V. Polyurea–Graphene Nanocomposites—The Influence of Hard-Segment Content and Nanoparticle Loading on Mechanical Properties. *Polymers* **2023**, *15*, 4434. https://doi.org/10.3390/polym15224434

Academic Editor: Alexander Malkin

Received: 6 October 2023
Revised: 10 November 2023
Accepted: 13 November 2023
Published: 16 November 2023

Copyright: © 2023 by the authors. Licensee MDPI, Basel, Switzerland. This article is an open access article distributed under the terms and conditions of the Creative Commons Attribution (CC BY) license (https://creativecommons.org/licenses/by/4.0/).

Abstract: Polyurethane and polyurea-based adhesives are widely used in various applications, from automotive to electronics and medical applications. The adhesive performance depends strongly on its composition, and developing the formulation–structure–property relationship is crucial to making better products. Here, we investigate the dependence of the linear viscoelastic properties of polyurea nanocomposites, with an IPDI-based polyurea (PUa) matrix and exfoliated graphene nanoplatelet (xGnP) fillers, on the hard-segment weight fraction (HSWF) and the xGnP loading. We characterize the material using scanning electron microscopy (SEM) and dynamic mechanical analysis (DMA). It is found that changing the HSWF leads to a significant variation in the stiffness of the material, from about 10 MPa for 20% HSWF to about 100 MPa for 30% HSWF and about 250 MPa for the 40% HSWF polymer (as measured by the tensile storage modulus at room temperature). The effect of the xGNP loading was significantly more limited and was generally within experimental error, except for the 20% HSWF material, where the xGNP addition led to about an 80% increase in stiffness. To correctly interpret the DMA results, we developed a new physics-based rheological model for the description of the storage and loss moduli. The model is based on the fractional calculus approach and successfully describes the material rheology in a broad range of temperatures ($-70\ °C$–$+70\ °C$) and frequencies (0.1–$100\ s^{-1}$), using only six physically meaningful fitting parameters for each material. The results provide guidance for the development of nanocomposite PUa-based materials.

Keywords: polyurea; nanocomposite; graphene; elastomer; adhesive; DMA; SEM; fractional Maxwell model

1. Introduction

Polyurethanes (PUs), polyureas (PUas), and poly(urethaneureas) (PUUs), represent a class of polymers with a wide variety of applications [1–4]. Understanding the structure–property–performance relationship is critical in designing materials for specific applications. Critical parameters include the chemical structure of polymer constituents, extent of hydrogen bonding, and volume fraction of hard and soft segments [5–12]. In general, these classes of polymers are produced from a reaction between polyisocyanate (typically a diisocyanate) and a polyol in the case of pure PU and a polyamine in the case of pure PUa and both polyol and polyamine for PUU. In PUa, which will be the focus of this paper, the reaction of the diisocyanate and a polyamine forms the hard segments that have strong bidentate hydrogen bonds [1]. PUa can then be thought of as a multiblock copolymer in which the soft segment blocks alternate with the hard segment blocks. The strong hydrogen bonding within the hard segments drives microphase separation from the soft segments, resulting in a two-phase system—a percolated hard phase, consisting entirely of the hard segments, and a soft phase, consisting of the soft segments along with small amounts of non-percolated hard segments [10]. This microphase separation is similar to that of classical block copolymers, where various soft-crystalline phases (spherical, cylindrical, lamellar, etc.) are seen for different values of composition, f, and segregation strength, χN [13–15]. The morphology of the polymer, especially the total volume and the connectivity of the hard phase, has a decisive impact on the overall material properties (mechanical, transport, and thermophysical) [10,16–18].

Much of the PU and PUa literature has concentrated on the linear elasticity and especially the "temperature sweep" dynamic mechanical analysis (DMA), where the storage and loss moduli are measured as functions of temperature at a constant frequency (usually 1 Hz). However, many applications (automotive, electronics, etc.) require a good understanding of material performance in a wide range of temperatures and frequencies/strain rates. Thus, recently, Tzelepis et al. [19] used both temperature-sweep and frequency-sweep DMA to study the properties of PUa elastomers with different hard-segment weight fractions (HSWF). It was shown that the studied PUa materials obeyed the time–temperature superposition (TTS) principle. (We note that the application of TTS to PU and PUa was discussed earlier, e.g., by Velankar and Cooper [20] and Ionita et al. [21], but whether it is universally applicable to all PUs and PUas is still uncertain). The TTS shift factor, a_T, was successfully described using the TS2 function [22] that combines Arrhenius temperature dependence at high temperatures with a strong, but non-divergent, increase near the glass transition temperature. The storage and loss modulus master curves showed broad transition regions, indicating a wide distribution of relaxation times. Tzelepis et al. found that such a distribution was well-described by the so-called fractional Maxwell model (FMM) [23–29]—or, to be more precise, a sum of two fractional Maxwell gels (FMG), with one FMG element describing the continuous soft phase (with dispersed hard domains and dissolved hard segments) and the second FMG element representing the percolated hard phase. The plateau modulus of the first element was found to be nearly independent of the HSWF, while the plateau modulus of the second element was a strong function of HSWF, consistent with earlier experiments and theories [10].

In this study, we extend our previous work to investigate a set of PUa nanocomposites with exfoliated graphene nanoplatelets (xGnP), varying both HSWF and the xGnP weight fraction. The use of nanofillers, such as clay, talc, graphene, and graphene oxide, to modify the properties of polymers has been widespread since at least the 1980s [30–41]. The fillers are expected to significantly increase the modulus and strength of the material relative to the "neat" matrix polymer. For high-aspect-ratio nanoplatelets in rubbery polymers, the "reinforcement factor" (RF)—defined as the ratio of the nanocomposite modulus to the matrix modulus—can be as high as 2–4 at particle loadings of 1–4 weight percent [42,43]. Multiple models have been developed to predict reinforcement in simple two-component nanocomposites [42–44]. In general, the stiffness of the material increases strongly at the beginning, but often stays constant or even decreases as the filler loading is increased further—this is typically ascribed to the onset of nanofiller aggregation. Obviously, the problem becomes even more challenging when the matrix itself is multicomponent, like segmented polyurea. Are the nanofillers simply interacting with the pre-set domain nanostructure? Or are they modifying the arrangement of the hard domain itself—perhaps by nucleating their formation or by linking multiple domains? Here, we will attempt to address this problem by preparing multiple PUa-xGnP nanocomposites and investigating their structure and linear viscoelasticity. Starting with three neat PUa materials (20, 30, and 40 percent hard segment), we added xGnP nanofillers, with the xGnP weight percentage (wt%) varying from 0 to 1.5 wt% with increments of 0.5 wt%. We expected that this experimental design would capture the main reinforcement effect due to the nanofillers. On one hand, reinforcement effects are unlikely to be significant at loadings below 0.5 wt%, based on many earlier polymer nanocomposite studies (see, e.g., Pinnavaia and Beall [45]). On the other hand, as it will be seen later, at loadings of 1.5 wt% and above, the reinforcement effects diminish, possibly due to the nanoparticle interactions and transition from a fully to partially exfoliated morphology. For all twelve materials (neat PUa and nanocomposites), we measured the linear viscoelasticity and successfully fit it with the two-FMG model. The model parameters were then used to elucidate the structural details of the material and provide guidance for the impact of the design parameters (HSWF and wt% xGnP) on the nanocomposite properties.

2. Materials and Methods

2.1. Polymer Synthesis

The synthesis of the PUa-Neat materials is described in detail in our earlier paper [19]. We used isophorone diisocyanate (IPDI)-Vestanat from Evonik Corporation, Pisca-taway, NJ, USA; Jeffamine T5000 and D2000 polyetheramines from the Huntsman Corporation, The Woodlands, TX, USA; and the diethyltoluene diamine (DETDA) (Lonzacure) chain extender from Lonza, Morristown, NJ, USA. Toluene was purchased from Fisher Scientific, Hampton, NH, USA. All the materials were employed in our research "as received", with no further processing. The formulations for the three neat PUa-s having hard segment weight fractions of 20%, 30%, and 40%, are provided in Table 1. We produced polyurea prepolymer (A-side) by placing IPDI in the reactor, then adding toluene to prevent any possible gelling. Next, the amine blend for the prepolymer (comprised of Jeffamine D2000 and T5000) was mixed for 5 min in a separate 250 mL beaker at room temperature, subsequently degassed for approximately 10 min, and added to the IPDI–toluene mixture. Similarly, the amine blend for the B-side, comprised of Jeffamine D2000 and Lonzacure DETDA, was mixed for 5 min in a separate 250 mL beaker followed by vacuum degassing for approximately 10 min and then poured into a separate additional funnel. The reactor was assembled and then a vacuum was drawn for five minutes, followed by the addition of N_2 gas at a 0.3–0.4 L/min flow rate. The reactor temperature was increased to 80 °C and then A-side amine blend was added dropwise under mechanical stirring at 120 RPM. The mixture was subsequently stirred for another hour at 80 °C. Afterward, the reactor was cooled to 0 °C and the B-side amine blend was added dropwise, with mechanical stirring maintained at 120 RPM. Once all the B-side was added, the contents of the reactor were transferred into a 600 mL beaker, degassed for 5 min, and poured into molds. The molds were maintained at room temperature for 24 h to allow for gelation and solvent evaporation. After 24 h, the samples were placed in an oven at 40 °C for 12–24 h to accelerate the solvent evaporation. The curing of the PUa was then completed at 60 °C for 72 h.

Table 1. Summary of the constituents used in the synthesis of the model PUa-Neat. (Adapted with permission from Ref. [19]).

	Component	IPDI-2k-20HS	IPDI-2k-30HS	IPDI-2k-40HS
Isocyanate Prepolymer (A-Side)	IPDI	30.8 g	41.6 g	52.1 g
	T5000	14.5 g	12.1 g	10.1 g
	D2000	57.1 g	48.5 g	40.4 g
	Toluene	82.7 g (95 mL)	165.3 g (190 mL)	208.8 g (240 mL)
	%NCO	8.7%	12.9%	15.3%
Amine Blend (B-Side)	DETDA	10 g	19.4 g	29.3 g
	D2000	90 g	80.6 g	70.7 g

For each of the three PUa formulations described in Table 1, four nanocomposites were then prepared, with the xGnP weight percentage (wt%) varying from 0 to 1.5 wt% with increments of 0.5 wt%. For all the PUa-GnP nanocomposites, the process was identical to that for the neat systems with the following additional steps. Exfoliated nano-graphene (grade R-10, obtained from XG Sciences) was heat-treated at 400 °C for 1 h and allowed to furnace-cool. The required amount of xGnPs was placed in a 500 mL beaker, 190 mL of toluene was added to the beaker, and the slurry was simultaneously mechanically stirred and sonicated. The mechanical stirring was accomplished by magnetic stirring at 200 rpm. The sonication was accomplished using a Qsonica sonicator, manufac-tured by Qsonica L.L.C, Newtown, CT, USA. The amplitude was set to 20 and the process time was set to 30 min with a pulse time of 10 s on and 10 s off. The temperature of the slurry never exceeded 32 °C, and the total run time was ~1 h. The total amount of energy input was 38,610 J. The weight of xGnP added to the formulation is summarized as follows: for the 0.5 wt% xGnP formulations, 1.02 g of xGnP; for the 1.0 wt% xGnP formulations, 2.04 g of xGnP; and for the 1.5 wt% xGnP formulation, 3.06 g of xGnP.

2.2. PUa–xGnP Characterization

The surface chemistry of the top 50–80 Å was determined with X-ray photoelectron spectroscopy (XPS). The measurements were performed using a PHI 5400 ESCA system. The base pressure of the instrument was less than 10^{-8} Torr. A 1 cm^2 sample was mounted onto the sample holder with double-sided copper tape. The X-ray was a monochromatic Al source with a take-off angle of 45 degrees. Two types of scans were performed for each sample: a survey scan from 0–1100 eV taken with a pass energy of 187.85 eV and regional scans of each element at a pass energy of 23.70 eV. The data were fitted using the CASA XPS software package, version 2.3.15.

The xGnP particle size and dispersion were characterized using a Hitachi (Schaumburg, IL, USA) 3700 SEM. The acceleration was set to 5 keV to minimize charging effects. A 2-to-3-nanometer-thick gold coating was sputtered using a Quorum Q150R sputter coater. Geometric measurements of the xGnP were performed utilizing PCI software, version 9.0.

Dynamic mechanical analysis was conducted using a TA Instruments (New Castle, DE, USA) RSA-G2 rheometer. The curing of the polymer was determined by measuring the change in storage modulus with respect to time. All film samples were loaded in tension. Temperature sweeps, at a rate of 3 °C/min, were conducted from −95 °C to a maximum temperature depending on the polyurea formulation hard segment content. Six repeats per formulation were run for the temperature sweeps. The reference temperature for each material was set to equal its glass transition temperature, defined as the maximum of the loss modulus (see Table S1). All TTS shifts were completed with TA Instruments' TRIOS software package, version 5.1.1.46572.

2.3. Modeling
2.3.1. Fractional-Order Maxwell Gel Model

The Fractional Maxwell Model (FMM) can be employed in developing constitutive models for both soft solids and complex fluids. The FMM consists of two spring-pot elements in series, which describe the complex modulus, E^*, as presented in Equation (1) [25].

$$\frac{E^*(\omega)}{E_0} = \frac{(i\omega\tau_c)^\alpha}{1+(i\omega\tau_c)^{\alpha-\beta}} \quad (1)$$

where E_0 represents the characteristic modulus, τ_c denotes the characteristic relaxation time, and both α and β are fractional-order power-law exponents. The storage, E', and loss, E'', moduli were obtained by splitting the complex modulus into its real and imaginary components, respectively defined as

$$\frac{E'(\omega)}{E_0} = \frac{(\omega\tau_c)^\alpha \cos\left(\frac{\pi\alpha}{2}\right) + (\omega\tau_c)^{2\alpha-\beta}\cos\left(\frac{\pi\beta}{2}\right)}{1+(\omega\tau_c)^{\alpha-\beta}\cos\left(\frac{\pi(\alpha-\beta)}{2}\right)+(\omega\tau_c)^{2(\alpha-\beta)}} \quad (2a)$$

$$\frac{E''(\omega)}{E_0} = \frac{(\omega\tau_c)^\alpha \sin\left(\frac{\pi\alpha}{2}\right) + (\omega\tau_c)^{2\alpha-\beta}\sin\left(\frac{\pi\beta}{2}\right)}{1+(\omega\tau_c)^{\alpha-\beta}\cos\left(\frac{\pi(\alpha-\beta)}{2}\right)+(\omega\tau_c)^{2(\alpha-\beta)}} \quad (2b)$$

Within the FMM framework, one possible special case that can occur is the fractional Maxwell gel (FMG), denoted by β being set to 0, which models the material's elastic behavior past the gel point.

In our model of interest, two FMG elements were arranged in parallel, representing the *soft-phase* matrix and percolated *hard phase* of polyurea, as illustrated in Figure 1. Given the relatively low mass fraction of the added nanoparticles (only up to 1.5%), we assumed that they were ultimately dispersed within the two phases and not forming a new phase by themselves. Thus, no additional parallel branch was introduced, consistent with our prior work [19]. Consequently, each polymer was characterized by six model parameters, encompassing two characteristic moduli (often called plateau modulus) ($E_{0,1}$ and $E_{0,2}$), two relaxation characteristic times ($\tau_{c,1}$ and $\tau_{c,2}$), and two power-law exponents (α_1 and α_2). We expect these parameters—especially those related to the percolated hard phase (FMG2)—to depend on the material formulation, including HSWF and xGnP loading.

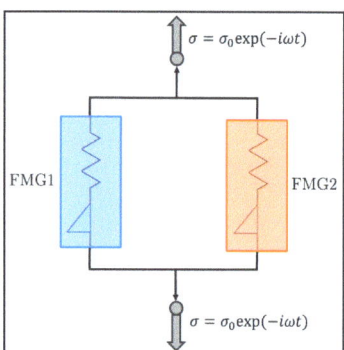

Figure 1. Schematic illustration of the two FMGs employed to model the polyurea. FMG1 represents the filled soft phase, whereas FMG2 corresponds to the percolated hard phase. No additional FMG element was considered for modeling nano-particles. Each FMG comprised an elastic spring and a spring-pot in a series arrangement.

To integrate the DMA data across various temperatures, the time–temperature superposition (TTS) principle was employed and the shift factor, denoted as a_T, was assumed to hold the same for both the soft and hard phases. As a result, the master curves could be described through the following equations:

$$E'(x) = \sum_{k=1}^{2} E_{0,k} \frac{(x\tau_k)^\alpha \cos\left(\frac{\pi\alpha}{2}\right) + (x\tau_k)^{2\alpha}}{1 + (x\tau_k)^\alpha \cos\left(\frac{\pi\alpha}{2}\right) + (x\tau_k)^{2\alpha}} \tag{3a}$$

$$E''(x) = \sum_{k=1}^{2} E_{0,k} \frac{(x\tau_k)^\alpha \sin\left(\frac{\pi\alpha}{2}\right)}{1 + (x\tau_k)^\alpha \cos\left(\frac{\pi\alpha}{2}\right) + (x\tau_k)^{2\alpha}} \tag{3b}$$

where $x = a_T \omega$. The equation for the shift factor as a function of temperature is discussed next.

2.3.2. The Shift Factor and the Two State, Two (Time) Scale (TS2) Model

In our previous paper [19], we applied three different functional forms to describe the shift factor—the Arrhenius, the Williams–Landel–Ferry (WLF) [46], and the TS2 [22] functions. The TS2 model describes the glass transition as the transition between the high-temperature and low-temperature Arrhenius regions,

$$ln(a_T) \equiv ln\left(\frac{\tau[T]}{\tau[T_0]}\right) = \frac{E_1}{RT} + \frac{E_2 - E_1}{RT}\left(\frac{1}{1 + exp\left\{\frac{\Delta S}{R}\left(1 - \frac{T^*}{T}\right)\right\}}\right) - \frac{E_1}{RT_0} + \frac{E_2 - E_1}{RT_0}\left(\frac{1}{1 + exp\left\{\frac{\Delta S}{R}\left(1 - \frac{T^*}{T_0}\right)\right\}}\right) \tag{4}$$

where E_1 and E_2 are activation energies (in J/mol), $\Delta S/R$ is the dimensionless transition entropy between the solid and liquid states of matter, T^* is the transition temperature (K) (typically, $T^* \approx T_g$), and T_o is the reference temperature of the TTS shifts. Equation (4) was shown to successfully describe the TTS of neat PUa polymers in the temperature range between $-70\,°C$ and $+70\,°C$ [19] and thus will be utilized here as well.

2.3.3. Optimization of the FMG Parameters

A global particle-swarm optimization (PSO) algorithm [47] was utilized to infer the fitting parameters in the two FMG branches, depicted in Figure 1. This is the same method that was employed in our previous paper [19]. Each optimization run maintained a constant population size of $N_{pop} = 200$ and performed $N_{it} = 6000$ iterations. Given the stochastic nature of the PSO algorithm, 50 optimization runs were conducted, and the expected values and standard deviations for each parameter of materials are reported.

The following parameter ranges are considered for all samples (20%, 30%, and 40% HS, with 0, 0.5%, 1%, and 1.5% GnP). The characteristic moduli are confined to the ranges of $0 \leq E_{0,1} \leq 10^4$ MPa and $0 \leq E_{0,2} \leq 10^3$ Mpa, the characteristic times are confined to the range of $10^{-3}\,s \leq \tau_{c,1(2)} \leq 10^2\,s$, and the fractional power law exponents α_1 and α_2 span from 0 to 1.

Equation (5) establishes a scalar multi-objective cost function via a weighted summation for the simultaneous fitting of storage and loss moduli.

$$\operatorname*{Min}_{\theta} \omega_1 f_1(\theta) + \omega_2 f_2(\theta) \tag{5}$$

where θ denotes the vector of fitting parameters, with $\omega_1 = 1/2$ and $\omega_2 = 1/2$, and the cost functions corresponding to both moduli are provided as follows:

$$\begin{aligned} f_1(\theta) &= \sum_{i=1}^{N_d} \left(\text{Log}\left(\frac{E'_{exp}}{E'_{model}}\right) \right)^2, \\ f_2(\theta) &= \sum_{i=1}^{N_d} \left(\text{Log}\left(\frac{E''_{exp}}{E''_{model}}\right) \right)^2 \end{aligned} \tag{6}$$

where N_d is the number of data points where our model is evaluated. The decision to employ a logarithmic difference between experimental data and model predictions arises from the significant variations in orders of magnitude for the storage and loss moduli across decades of frequency ranges. Moreover, the quality of the two-branch FMG model fits, if assessed by the relative error, is defined in Equation (7).

$$error = \frac{\omega_1 f_1(\theta) + \omega_2 f_2(\theta)}{\omega_1 \sum_{i=1}^{N_d} \left(\text{Log}\left(E'_{exp}\right) \right)^2 + \omega_2 \sum_{i=1}^{N_d} \left(\text{Log}\left(E''_{exp}\right) \right)^2} \tag{7}$$

Both the two-branch FMG model and PSO codes—similar to our previous paper [19]—were developed in MATLAB R2021b and executed in ICER MSU HPCC system with 1 node, 24 CPU, and 48 GB RAM.

Once again, we selected the GRG non-linear engine and imposed the following constraints: (1) $E_1 < 130$ kJ/mol, (2) $E_2 < 350$ kJ/mol, (3) $\Delta S/R < 25$, and (4) $T^* < 350$ K. The reference temperature (T_o) was set to match the glass transition temperature, defined as the maximum of the loss modulus ($-60 \pm 5\,°C$; for more details, see Table S1). The minimization function utilized is the average absolute value of the difference in the natural logarithm of the experimental and model shift factors, as defined in Equation (8).

$$\overline{error} = \left| ln(aT_{exp}) - ln(aT_{model}) \right| \tag{8}$$

This concludes the discussion of materials and methods; we now turn to the results.

3. Results

3.1. Experimental Results

Scanning electron microscopy (SEM) was used to determine the effect of sonication on the xGnP. Figure S1 shows SEM micrographs at various magnifications. Estimates of the particle diameter were produced by measuring the longest axis of the platelets, as shown in Figure S2. The average particle diameter before sonication was 15.4 +/− 6.3 μm (1σ). The average particle diameter after sonication was 15.0 +/− 4.5 μm (1σ). No change was seen in the morphology of the xGnP. The xGnP remained exfoliated throughout the sonication process and retained their shape and aspect ratio. Given that the technical data sheet for the R10 grade specified an average particle diameter size of approximately 10 μm, and accounting for the fact that the platelets in the images were at various angles, we conclude that there was no difference between the as-received and sonicated xGnP.

X-ray photoelectron spectroscopy (XPS) was used to evaluate the surface chemistry of the xGnP after heat treatment and after sonication. Figure 2 shows a survey of both a heat-treated sample and heat-treated and sonicated sample. Both spectra showed two peaks. The first at 281.6 eV and 282.4 eV for the heat-treated and sonicated samples, respectively, are associated with the C 1s position. The second at 530.4 eV for both heat-treated and sonicated samples is associated with the O 1s position. The atomic concentration was estimated and is presented in Figure 2.

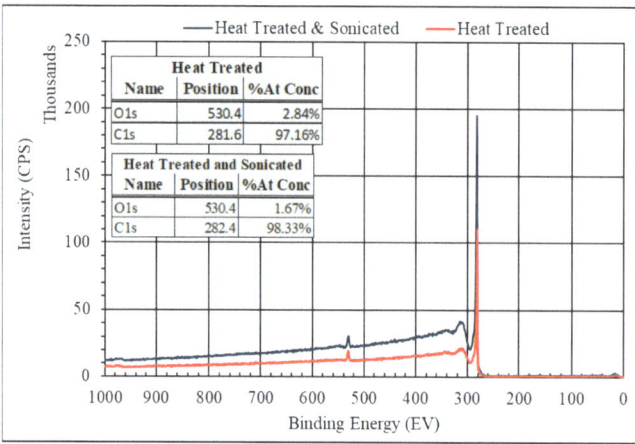

Figure 2. XPS survey of the xGnP after heat treatment (red spectrum), and after heat treatment and sonication. The sonication did not cause any change to the surface chemistry of the xGnP.

The atomic percentage of C was significantly higher than that of O for both heat-treated and heat-treated and sonicated samples. This is consistent with the expectation that the majority of the xGnP was carbon with very little oxygen-based functionalization on the edges of the basal plane. The approximate 1% difference seen between the two treatments was not considered significant. In order to explore the source of the oxygen peaks, a deconvolution of the XPS spectra for the heat-treated sample and the heat-treated and sonicated sample in the binding energy region for C and O is shown in Figure S3 and Figure S4, respectively. From Figure S3a, the peak at 283.2 eV, the largest peak, was associated with the C=C double bonds of graphene. The remaining C 1s peaks were associated with hydroxyl 284.7 eV. The C 1s peak at 288.0 eV was associated with the C=O, and the C 1s peak at 289.7 was associated with a COOH/COOR [48]. From Figure S3b, the O 1s peak at 531.2 eV was associated with COOH and the O 1s peak at 532.7 eV was associated with the –OH functional group [49]. The deconvolution of the heat-treated and sonicated samples is shown in Figure S4. Similar to the analysis for the heat-treated samples, the peak at 283.2 eV (the largest peak in the spectrum) was associated with the C=C double bonds of graphene.

The remaining C1s peak at 284.7 eV was associated with the –OH functional group. Likewise, the C 1s peak at 289.2 eV was associated with COOH/COOR functional groups (see Figure S4a,b). In Figure S4b, the O 1s peak at 531.0 eV was associated with COOH, and the O 1s peak at 532.4 eV was associated with the –OH functional group. From the analysis above, one concludes that there was very little –OH or –COOH functionalization on the xGnP; furthermore, the sonication process had very little effect on the chemistry, nor did it reduce the particle size.

In order to investigate particle dispersion at various concentrations of xGnP, tensile samples were placed in liquid nitrogen for about 5 min and then snapped in half. SEM micrographs of the fracture surface were then used to study the nanoparticle dispersion in the polymer matrix (see Figure 3 and also Figures S5–S7). In the lower-magnification micrographs, the xGnP was brighter, due to electron interaction with the jagged edges of the xGnP, than the polyurea matrix; examples of xGnP are highlighted by the arrows. No agglomeration or continuous networks of xGnP were found in any of the formulations. Figures S5f and S7f are higher-magnification micrographs (13 kX, and 10 kX respectively) of the xGnP. The edges of the individual nano-plates can be seen, suggesting the GnP remained exfoliated throughout the sample preparation process.

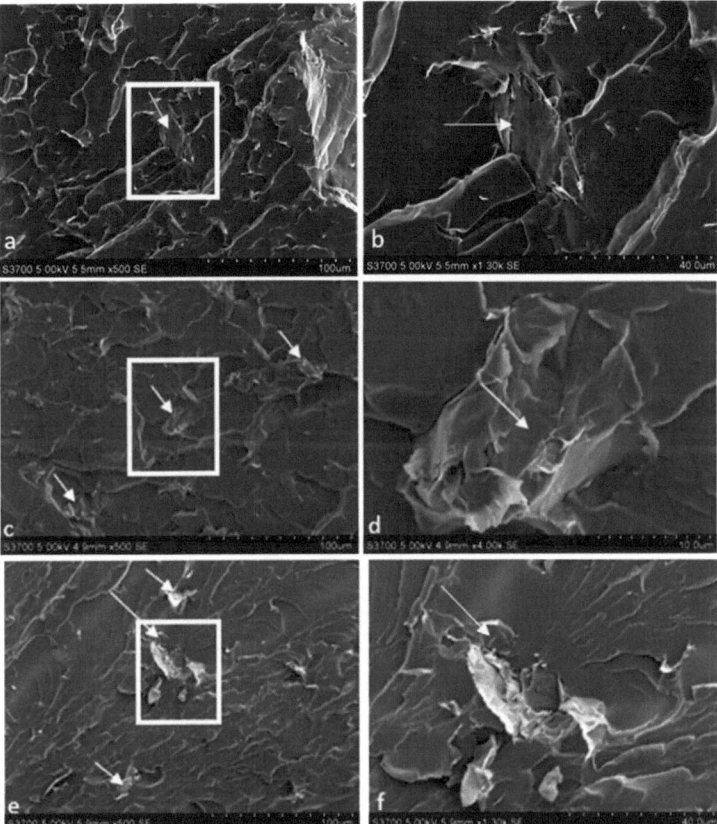

Figure 3. SEM photomicrographs of the fracture surface for PUa-xGnP nanocomposites with 0.5 wt% xGnP loading: (**a**) 20% HSWF at 500×, (**b**) photomicrograph of the white box in (**a**). (**c**) 30% HSWF at 500×, (**d**) photomicrograph of the white box in (**c**). (**e**) 40% HSWF at 500×, (**f**) photomicrograph of the white box in (**e**). In all photomicrographs, the arrows point to the xGnP.

Figure 4 shows the storage modulus curves for the DMA temperature sweeps (frequency 1 Hz) for the (a) IPDI-2k-20HS, (b) IPDI-2k-30HS, and (c) IPDI-2k-40HS formulations at various xGnP loadings. Note that the complete E', E'', and tan(δ) curves for all formulations are shown in the supplemental section, Figures S8–S10. For all formulations, the addition of xGnP did not have an appreciable effect on the T_g (as measured by E', E'' or tan(δ) curves) of the PUa formulations, nor did it have a significant effect on the glassy modulus. For these formulations, the T_g and glassy modulus were determined primarily by the soft phase [19]. This would tend to indicate the xGnP had little effect on the soft-phase microstructure, i.e., no crystallization or increase in the hydrogen bonding in the soft phase. For the IPDI-2k-20HS and IPDI-2k-30HS PUas, the addition of xGnP increased the plateau modulus and the temperature range of the plateau modulus. For the IPDI-2k-40HS formulations, the addition of xGnP had no effect on the temperature sweeps.

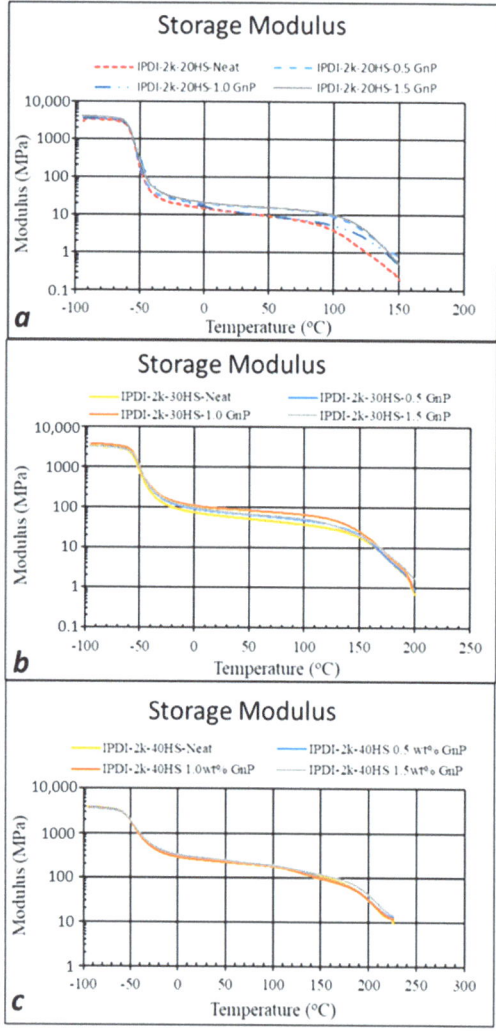

Figure 4. DMA temperature sweeps (tensile storage modulus, E') showing the effect of both an increase in % HS and an increase in xGnP loading: (**a**) IPDI-2k-20HS with 0, 0.5, 1.0, and 1.5 wt % xGNP; (**b**) the same for the IPDI-2k-30HS polyurea; (**c**) the same for the IPDI-2k-40HS polyurea.

3.2. Modeling Results

As previously discussed, the linear viscoelastic behavior of polyurea was described by a model comprising two parallel fractional Maxwell gel (FMG) branches representing the soft and hard phases. Even though, in the nanocomposites, there was a new phase (xGnPs), we continued to use the two-FMG model and expected that the impact of the nanofillers would be only in modifying the parameters of one or both of the FMGs, at least at sufficiently low loadings (<1.5 wt% in our case). This modeling approach enabled us to effectively capture the broad spectrum of relaxation times seen in these materials. The parameterization process was previously detailed, and we now present the results.

To begin with, in Figure 5, we plot the shift factor as a function of temperature for the IPDI-2k-20HS nanocomposites with (a) 0%, (b) 0.5%, (c) 1%, and (d) 1.5% xGnP. The symbols are the results of the TTS shift of the data (as outlined above), and the lines are the TS2 (Equation (4)) fits. Obviously, the addition of xGnPs did not have a qualitative impact on the TTS or the temperature dependence of the shift factor, although the model parameters (such as activation energies) changed slightly. Similar data and model fits for the IPDI-2k-30HS and IPDI-2k-40HS nanocomposites are presented in Figures S10 and S11, and the TS2 model parameters are summarized in Table S1.

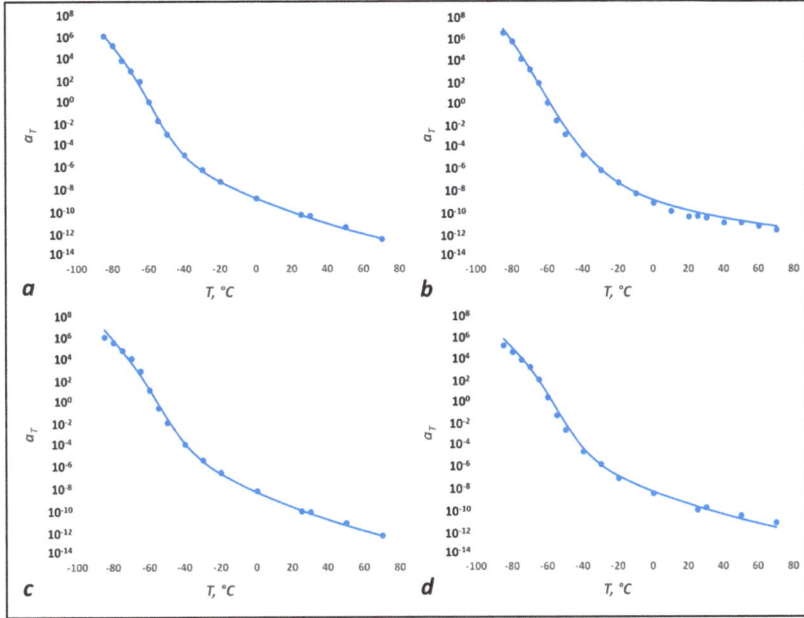

Figure 5. Experimental (symbols) and TS2 fit (lines) shift factors for 20HS polyureas with: (**a**) No added nanofillers; (**b**) 0.5 wt% xGnP; (**c**) 1.0 wt% xGnP; (**d**) 1.5 wt% xGnP.

In Figure 6, the storage and loss master curves are plotted for all nanocomposite systems: (a) IPDI-2k-20HS matrix, (b) IPDI-2k-30HS matrix; and (c) IPDI-2k-40HS matrix. Within each "family", all curves were very close to each other, with a possible exception of the IPDI-2k-20HS, 1% xGnP (blue symbols in Figure 6a). We will return to this system later to discuss the origins of its uniqueness.

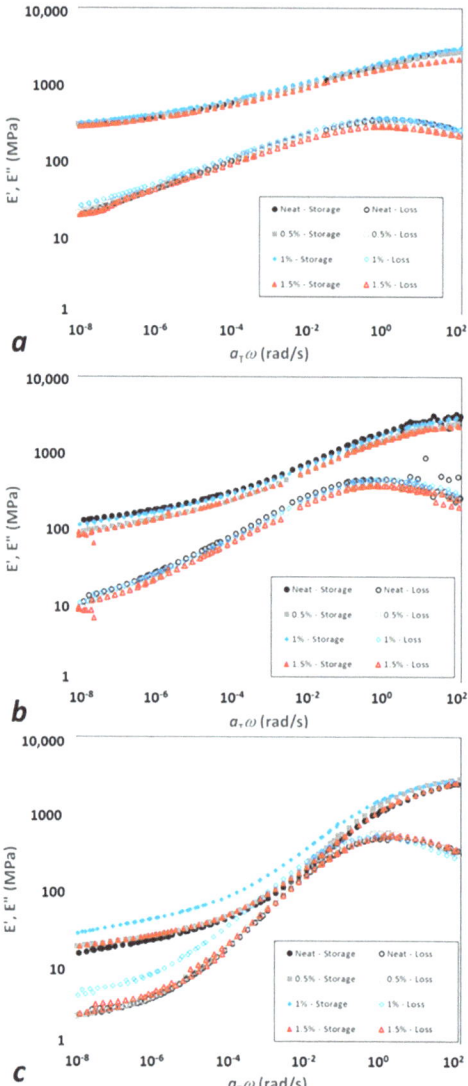

Figure 6. Master curves for the tensile storage (filled symbols) and loss (open symbols) moduli. (**a**) 20HS matrix with 0, 0.5, 1.0, and 1.5 wt % xGnP. (**b**) Same as (**a**) for the 30HS matrix. (**c**) Same as (**a**) for the 40HS matrix.

Next, let us consider the results of the two-FMG fitting to the master curves.

Table 2 provides the mean values and corresponding standard deviations for all six parameters in our two-FMG model. The optimization runs showed excellent convergence and reproducibility, as manifested in the low standard deviation values for all the systems considered.

Figure 7 presents the two-FMG model fits to the experimental shifted data for IPDI-2k-40HS nanocomposites; the results for IPDI-2k-20HS and IPDI-2k-30HS are depicted in Figures S12 and S13, respectively. All fitted curves were generated using the expected values for the model parameters, since the standard deviation of each model parameter was negligible. For all the formulations, the relative error between the model and data was less than 3.1%, with data spanning a broad range of frequencies (between 10^{-4} and 10^2 rad/s). However, for the 20 wt% hard segment sample at all nano-particle percentages, a minor deviation between the model and experimental data was observed above the glass transition point in the loss modulus, a phenomenon which was also noted in our prior work for the neat 20% HWSF case. It should be noted that the experimental data points exhibiting a high level of dispersion were excluded from the optimization and fitting process.

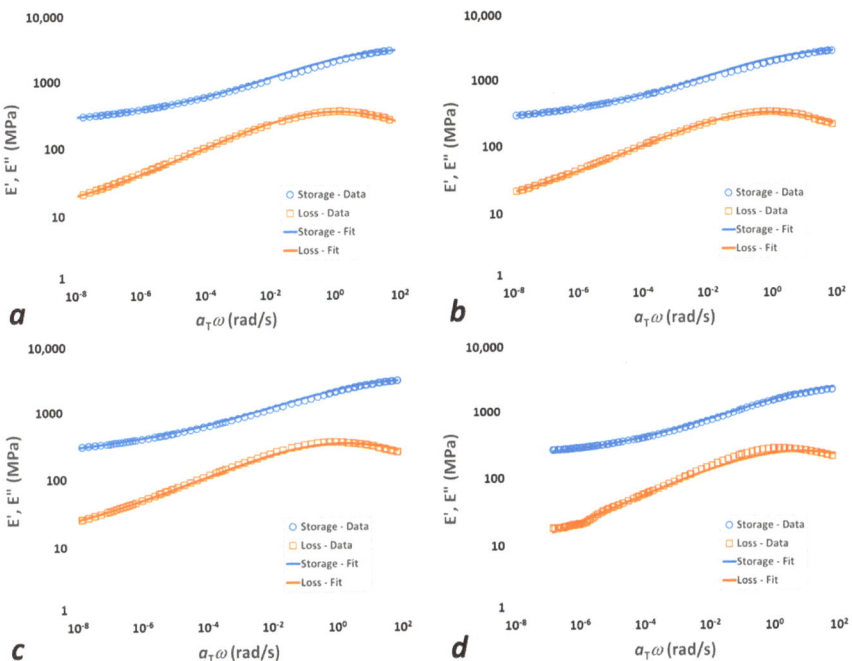

Figure 7. Experimental (symbols) and FMG-FMG fit (lines) master curves for 40HS polyureas with: (**a**) no added nanofillers; (**b**) 0.5 wt% xGnP; (**c**) 1.0 wt% xGnP; (**d**) 1.5 wt% xGnP. Blue open circles represent storage modulus data NS blue lines are the storage modulus model fits; orange open squares correspond to the loss modulus data and orange lines are the loss modulus model fits.

Figure 8a,b depicts the influence of the nanofiller content on the mean characteristic modulus of both branches. In general, the effect was very small, except for the significant increase in $E_{0,2}$ for the 1% xGnP in the IPDI-2k-20HS nanocomposite relative to the neat polymer. In that system, two factors contributed to the effect. First, the stiffness ratio between the filler and the matrix was the largest for the lower-HS polymers and became smaller as HSWF increased. Second, the impact of the fillers usually had a maximum as a function of filler loading. At low loadings, the effect was, obviously, very weak; at high loadings, on the other hand, the platelets aggregated, the aspect ratio decreased, and the overall effect decreased as well. Thus, 1% xGnP in the IPDI-2k-20HS represented the system corresponding to the maximum reinforcement in terms of both HSWF and %xGnP.

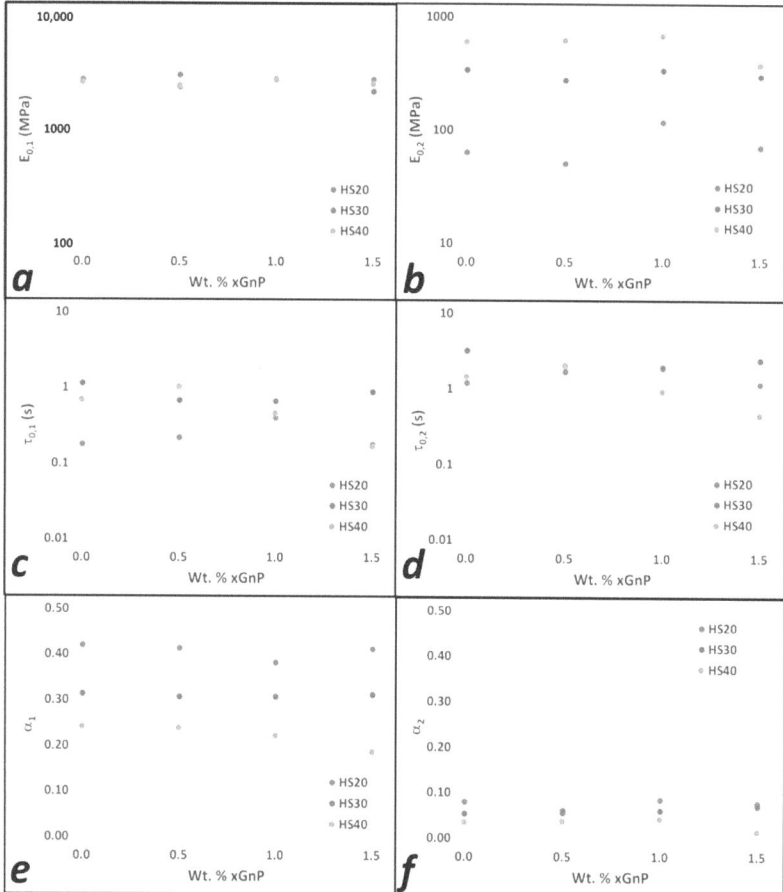

Figure 8. Effect of the nanofiller loading on the (**a**) characteristic modulus of the first branch ($E_{0,1}$) (**b**) characteristic modulus of the second branch ($E_{0,2}$), (**c**) characteristic time of the first branch ($\tau_{c,1}$), (**d**) characteristic time of the second branch ($\tau_{c,2}$), (**e**) power law exponent of the first branch (α_1), and (**f**) power law exponent of the first branch (α_2).

In Figure 8c,d, the variations in relaxation times for both branches with respect to the filler weight fraction are shown. These variations are also fairly small and do not show a clear dependence on the nanoparticle loading. Finally, Figure 8e,f shows the power law exponents, α, for both branches. Again, the dependence of α_1 on the xGnP loading was fairly weak. The soft-phase exponent, α_1, showed a strong dependence on HSWF, decreasing as HSWF increased. This is consistent with the material becoming "more elastic" and the average loss tangent decreasing. The hard-phase exponent, α_2, was quite small for all twelve neat and nanocomposite systems, indicating that they were almost always nearly perfectly elastic.

Table 2. FMG 1 and FMG 2 parameters. Rows represent hard segment weight fractions of 20%, 30%, and 40%, while columns correspond to xGnP weight fractions of 0.0%, 0.5%, 1%, and 1.5%.

	Model Parameters	0.0% GnP		0.5% GnP		1% GnP		1.5% GnP	
		FMG 1	FMG 2	FMG 1	FMG 2	FMG 1	FMG 2	FMG 1	FMG 2
IPDI-2k 20HS	$E_{0,i}$ (MPa)	2815 $\pm 3.3 \times 10^{-6}$	64 $\pm 3.8 \times 10^{-7}$	3097 $\pm 3.4 \times 10^{-6}$	51 $\pm 3.9 \times 10^{-7}$	2871 $\pm 2.6 \times 10^{-6}$	118 $\pm 6.2 \times 10^{-7}$	2837 $\pm 2.6 \times 10^{-6}$	71 $\pm 4.1 \times 10^{-7}$
	$\tau_{c,i}$ (s)	0.18 $\pm 8.2 \times 10^{-10}$	1.19 $\pm 5.9 \times 10^{-10}$	0.22 $\pm 6.8 \times 10^{-10}$	1.69 $\pm 7.1 \times 10^{-9}$	0.40 $\pm 1.4 \times 10^{-9}$	1.96 $\pm 8.1 \times 10^{-9}$	0.18 $\pm 6.7 \times 10^{-10}$	1.16 $\pm 4.5 \times 10^{-9}$
	α_i	0.42 $\pm 4.3 \times 10^{-10}$	0.080 $\pm 3.8 \times 10^{-10}$	0.413 $\pm 4.4 \times 10^{-10}$	0.056 $\pm 5.8 \times 10^{-10}$	0.382 $\pm 3.5 \times 10^{-10}$	0.085 $\pm 3.4 \times 10^{-10}$	0.412 $\pm 4.0 \times 10^{-10}$	0.077 $\pm 3.9 \times 10^{-10}$
IPDI-2k 30HS	$E_{0,i}$ (MPa)	2757 $\pm 1.8 \times 10^{-5}$	342 $\pm 1.7 \times 10^{-6}$	2405 $\pm 3.4 \times 10^{-6}$	278 $\pm 1.5 \times 10^{-6}$	2793 $\pm 2.5 \times 10^{-6}$	341 $\pm 1.2 \times 10^{-6}$	2203 $\pm 1.9 \times 10^{-5}$	301 $\pm 8.9 \times 10^{-6}$
	$\tau_{c,i}$ (s)	1.14 $\pm 4.1 \times 10^{-8}$	3.23 $\pm 1.0 \times 10^{-7}$	0.68 $\pm 5.8 \times 10^{-9}$	2.00 $\pm 1.2 \times 10^{-8}$	0.66 $\pm 2.7 \times 10^{-9}$	1.90 $\pm 6.3 \times 10^{-9}$	0.89 $\pm 2.7 \times 10^{-8}$	2.41 $\pm 3.0 \times 10^{-8}$
	α_i	0.314 $\pm 1.1 \times 10^{-9}$	0.054 $\pm 3.6 \times 10^{-10}$	0.307 $\pm 5.7 \times 10^{-10}$	0.061 $\pm 3.4 \times 10^{-10}$	0.307 $\pm 3.4 \times 10^{-10}$	0.061 $\pm 2.2 \times 10^{-10}$	0.312 $\pm 2.5 \times 10^{-9}$	0.071 $\pm 2.0 \times 10^{-9}$
IPDI-2k 40HS	$E_{0,i}$ (MPa)	2635 $\pm 2.6 \times 10^{-6}$	604 $\pm 2.0 \times 10^{-6}$	2476 $\pm 2.0 \times 10^{-6}$	622 $\pm 1.7 \times 10^{-6}$	2843 $\pm 2.8 \times 10^{-6}$	687 $\pm 2.8 \times 10^{-6}$	2567 $\pm 1.3 \times 10^{-5}$	378 $\pm 1.1 \times 10^{-5}$
	$\tau_{c,i}$ (s)	0.69 $\pm 2.8 \times 10^{-9}$	1.44 $\pm 5.2 \times 10^{-9}$	1.03 $\pm 3.4 \times 10^{-9}$	2.06 $\pm 8.8 \times 10^{-9}$	0.46 $\pm 2.1 \times 10^{-9}$	0.93 $\pm 4.9 \times 10^{-9}$	0.17 $\pm 2.7 \times 10^{-9}$	0.45 $\pm 3.7 \times 10^{-9}$
	α_i	0.242 $\pm 4.1 \times 10^{-10}$	0.035 $\pm 1.8 \times 10^{-10}$	0.239 $\pm 2.8 \times 10^{-10}$	0.038 $\pm 1.6 \times 10^{-10}$	0.222 $\pm 3.4 \times 10^{-10}$	0.043 $\pm 1.8 \times 10^{-10}$	0.187 $\pm 1.2 \times 10^{-9}$	0.015 $\pm 1.7 \times 10^{-9}$

4. Discussion

In this study, we investigated the structure and linear viscoelasticity of polyurea elastomers to be used in adhesive applications. The two main variables of interest were the polyurea hard segment (HS) weight fraction and the exfoliated graphene nanoplatelet (xGnP) loading. The hypothesis tested was that the polyurea hard segment and the nanofillers would interact strongly with each other and provide additional reinforcement by forming a "combined hard phase".

Using scanning electron microscopy (SEM), we verified that the heat treatment and the sonication in toluene resulted in no morphological changes in the xGnP. The average particle diameter did not change and the xGnPs remained exfoliated and well dispersed in the PUa matrix. Recall that the nanoparticles were placed on the isocyanate side (A-side) of the PUa reaction sequence; therefore, albeit small, there was a potential for the isocyanate to react with any hydroxyl or carboxylic acid functional groups located on the edges of the nano particles. However, XPS showed very few, if any, available reaction sites, whether they be hydroxyl or carboxylic acid that could potentially react with the isocyanate. Thus, we stipulate that the dispersed xGNPs had only weak physical interactions with the PUa matrix.

Given the complex structure of any polyurea nanocomposite (soft-phase matrix, hard-phase islands, percolated hard-phase domains, exfoliated nanofillers, aggregated nanofillers, etc.), the data from direct characterization, such as electron microscopy, are often inconclusive. Thus, here, we also concentrated on understanding the materials using linear viscoelasticity and inferring the information about the matrix–filler interaction from the DMA results.

Similar to the previous study [19], we observed that the DMA frequency sweeps in these systems are amenable to time–temperature superposition (TTS), with the TTS shift factors well-described by the TS2 [22] function. This was, in itself, a non-trivial result, since polyurea materials are multi-phase; understanding the reason why TTS works still requires additional analysis. We also found that the storage and loss master curves exhibited broad transition regions and thus could not be described with a single Maxwell model. Therefore, we used the fractional Maxwell model (FMM) approach [23–27] to quantify the viscoelastic response and fit the master curve. In particular, the material was well-described by the use of two fractional Maxwell gel (FMG) elements, one representing the soft phase and another one representing the percolated hard phase. We demonstrated that the plateau modulus of the percolated hard phase (FMG2) increased strongly with the hard segment weight fraction (HSWF), consistent with earlier studies [10]. Here, we used the same approach to determine the combined impact of HSWF and xGnP loading.

Based on the FMM analysis, we observed that the effect of the xGnP was significantly less pronounced than the effect of the HSWF change. The reinforcement factor (RF) was physically meaningful (significantly greater than 1) only for one nanocomposite system—HS20 with 1% xGnP. This result is consistent with expectations, as discussed above. Further increases in the xGnP loading likely resulted in at least some aggregation, thus blunting the effectiveness of the new fillers [50]. For polyureas with higher HSWF, the percolated hard-phase modulus is already quite high, and the contribution of the nanofillers becomes even less significant, regardless of their concentration. Thus, the addition of the nanofillers did not seem to offer a significant increase in the linear elastic properties of the polyureas studied here.

Of course, linear elasticity is not the only important property for adhesives—other properties of interest include tensile strength, ultimate elongation, fracture toughness, etc. The influence of nanofillers on those properties will be the subject of future work.

5. Conclusions

We investigated the structure and linear viscoelasticity of polyurea (PUa) elastomers and their nanocomposites with expanded graphene nanoplatelets (xGNPs) as a function of the hard segment weight fraction (HSWF) of the polyurea and the xGNP weight fraction in the overall nanocomposites. Experimentally, we found that the room-temperature modulus of the PUa-xGNP nanocomposites depended strongly on HSWF (about 10 MPa for the 20% HSWF to about 100 MPa for the 30% HSWF and about 250 MPa for the 40% HSWF polymer), but weakly on the xGNP weight fraction (for the weight fraction variations between 0 and 1.5 wt%, the modulus variations were generally within the experimental error, except for the 20% HSWF, 1% xGNP nanocomposite exhibiting nearly two-fold stiffening compared with the neat material).

Significantly, we have demonstrated that, despite their structural complexity, PUa-xGNP nanocomposites exhibit time–temperature superposition (TTS). For the first time, we demonstrated that the TTS master curves can be described by fractional calculus (FC)-based models with a small number of physically meaningful parameters (as opposed to the standard Prony-series modeling usually requiring twenty or more). The new model can be adapted to describe other polymers and nanocomposites for both linear and nonlinear mechanical tests.

Supplementary Materials: The following supporting information can be downloaded at: https://www.mdpi.com/article/10.3390/polym15224434/s1, Figure S1: SEM images at various magnifications of the xGnP before (left column) and after sonication (right column). The images indicate that there was no change in the morphology of the GnP with the sonication parameters used; Figure S2: SEM micrographs showing the measurement of the estimated diameter of the xGnP before sonication (a) and after sonication (b); Figure S3: Deconvolution of the XPS spectrum, for the heat-treated xGnP, in the binding energy region for C (a) and O (b); Figure S4: Deconvolution of the XPS spectrum for the heat-treated and sonicated xGnP and sonicated xGnP in the binding energy regions for C (a) and O (b); Figure S5: SEM photomicrographs of the fracture surface for all IPDI-2k-20HS xGnP loadings: (a) 0.5 wt% xGnP loading at 500×, (b) photomicrograph of the white box in (a). (c) 1.0 wt% xGnP loading at 500×, (b) photomicrograph of the white box in (c). (d) 1.5 wt% xGnP loading; Figure S6: SEM photomicrographs of the fracture surface for all IPDI-2k-30HS xGnP loadings: (a) 0.5 wt% xGnP loading at 500×, (b) photomicrograph of the white box in (a). (c) 1.0 wt% xGnP loading at 500×, (b) photomicrograph of the white box in (c). (d) 1.5 wt% xGnP loading at 650×. Photomicrograph of the white box in (e). In all photomicrographs, the arrows point to the xGnP; Figure S7: SEM photomicrographs of the fracture surface for all IPDI-2k-40HS xGnP loadings: (a) 0.5 wt% xGnP loading at 500×, (b) photomicrograph of the white box in (a). (c) 1.0 wt% xGnP loading at 500×, (b) photomicrograph of the white box in (c). (d) 1.5 wt% xGnP loading at 650×. Photomicrograph of the white box in (e). In all photomicrographs, the arrows point to the xGnP; Figure S8: DMA temperature sweep showing the storage, loss modulus, and the tan (δ) for the IPDI-2k-20HS; Figure S9: DMA temperature sweep showing the storage, loss modulus, and the tan (δ) for the IPDI-2k-30HS; Figure S10: DMA temperature sweep showing the storage, loss modulus, and the tan (δ) for the IPDI-2k-40HS; Figure S11: Experimental (symbols) and TS2 fit (lines) shift factors for IPDI-2k-30HS polyureas with: (a) no added nanofillers; (b) 0.5 wt% xGnP; (c) 1.0 wt% xGnP; (d) 1.5 wt% xGnP. Figure S12: Experimental (symbols) and TS2 fit (lines) shift factors for 40HS polyureas with: (a) no added nanofillers; (b) 0.5 wt% xGnP; (c) 1.0 wt% xGnP; (d) 1.5 wt% xGnP; Figure S13: Experimental (symbols) and FMG–FMG fit (lines) master curves for IPDI-2k-20HS polyureas with: (a) no added nanofillers; (b) 0.5 wt% xGnP; (c) 1.0 wt% xGnP; (d) 1.5 wt% xGnP. Blue open circles represent storage modulus data and blue lines are the storage modulus model fits; orange open squares correspond to the loss modulus data and orange lines are the loss modulus model fits; Figure S14: Experimental (symbols) and FMG–FMG fit (lines) master curves for 30HS polyureas with: (a) no added nanofillers; (b) 0.5 wt% xGnP; (c) 1.0 wt% xGnP; (d) 1.5 wt% xGnP. Blue open circles represent storage modulus data and blue lines are the storage modulus model fits; orange open squares correspond to the loss modulus data and orange lines are the loss modulus model fits; Table S1: TTS reference temperatures and TS2 fit parameters for all systems.

Author Contributions: Conceptualization, D.A.T., M.Z. and V.V.G.; methodology, D.A.T., M.Z. and V.V.G.; validation, D.A.T. and A.K.; formal analysis, D.A.T. and A.K.; investigation, D.A.T. and A.K.; resources, D.A.T. and M.Z.; data curation, D.A.T. and A.K.; writing—original draft preparation, D.A.T. and A.K.; writing—review and editing, D.A.T., M.Z. and V.V.G.; supervision, M.Z. and V.V.G.; project administration, D.A.T.; funding acquisition, D.A.T. and M.Z. All authors have read and agreed to the published version of the manuscript.

Funding: This research was supported by the US Army Ground Vehicle System Center. A.K. and M.Z. were funded by the ARO Young Investigator Program (YIP) award (W911NF-19-1-0444) and the NSF award (DMS-1923201).

Institutional Review Board Statement: Not applicable.

Data Availability Statement: The data and analysis procedures are available from the authors upon request.

Acknowledgments: We thank Peter Askeland (MSU) for support with XPS measurements.

Conflicts of Interest: The authors declare no conflict of interest. The funders had no role in the design of the study; in the collection, analyses, or interpretation of data; in the writing of the manuscript; or in the decision to publish the results.

References

1. Sonnenschein, M.F. *Polyurethanes: Science, Technology, Markets, and Trends*; Wiley Series on Polymer Engineering and Technology; Wiley: Hoboken, NJ, USA, 2020; ISBN 9781119669463.
2. Szycher, M. *Szycher's Handbook of Polyurethanes*, 2nd ed.; CRC Press: Boca Raton, FL, USA, 2013; ISBN 9781439863138/143986313X/9781523108022/1523108029.
3. Akindoyo, J.O.; Beg, M.; Ghazali, S.; Islam, M.R.; Jeyaratnam, N.; Yuvaraj, A.R. Polyurethane Types, Synthesis and Applications–a Review. *RSC Adv.* **2016**, *6*, 114453–114482. [CrossRef]
4. Petrović, Z.S.; Ferguson, J. Polyurethane Elastomers. *Prog. Polym. Sci.* **1991**, *16*, 695–836. [CrossRef]
5. Koberstein, J.T.; Stein, R.S. Small-angle X-ray Scattering Studies of Microdomain Structure in Segmented Polyurethane Elastomers. *J. Polym. Sci. Polym. Phys. Ed.* **1983**, *21*, 1439–1472. [CrossRef]
6. Koberstein, J.T.; Galambos, A.F. Multiple Melting in Segmented Polyurethane Block Copolymers. *Macromolecules* **1992**, *25*, 5618–5624. [CrossRef]
7. Leung, L.M.; Koberstein, J.T. Small-angle Scattering Analysis of Hard-microdomain Structure and Microphase Mixing in Polyurethane Elastomers. *J. Polym. Sci. Polym. Phys. Ed.* **1985**, *23*, 1883–1913. [CrossRef]
8. Koberstein, J.T.; Leung, L.M. Compression-Molded Polyurethane Block Copolymers. 2. Evaluation of Microphase Compositions. *Macromolecules* **1992**, *25*, 6205–6213. [CrossRef]
9. Christenson, C.P.; Harthcock, M.A.; Meadows, M.D.; Spell, H.L.; Howard, W.L.; Creswick, M.W.; Guerra, R.E.; Turner, R.B. Model MDI/Butanediol Polyurethanes: Molecular Structure, Morphology, Physical and Mechanical Properties. *J. Polym. Sci. B Polym. Phys.* **1986**, *24*, 1401–1439. [CrossRef]

10. Ginzburg, V.; Bicerano, J.; Christenson, C.P.; Schrock, A.K.; Patashinski, A.Z. Theoretical Modeling of the Relationship between Young's Modulus and Formulation Variables for Segmented Polyurethanes. *J. Polym. Sci. B Polym. Phys.* **2007**, *45*, 2123–2135. [CrossRef]
11. Garrett, J.T.; Siedlecki, C.A.; Runt, J. Microdomain Morphology of Poly (Urethane Urea) Multiblock Copolymers. *Macromolecules* **2001**, *34*, 7066–7070. [CrossRef]
12. Garrett, J.T.; Runt, J.; Lin, J.S. Microphase Separation of Segmented Poly (Urethane Urea) Block Copolymers. *Macromolecules* **2000**, *33*, 6353–6359. [CrossRef]
13. Matsen, M.W.; Bates, F.S. Unifying Weak-and Strong-Segregation Block Copolymer Theories. *Macromolecules* **1996**, *29*, 1091–1098. [CrossRef]
14. Drolet, F.; Fredrickson, G.H. Combinatorial Screening of Complex Block Copolymer Assembly with Self-Consistent Field Theory. *Phys. Rev. Lett.* **1999**, *83*, 4317. [CrossRef]
15. Benoit, H.; Hadziioannou, G. Scattering Theory and Properties of Block Copolymers with Various Architectures in the Homogeneous Bulk State. *Macromolecules* **1988**, *21*, 1449–1464. [CrossRef]
16. Qi, H.J.; Boyce, M.C. Stress–Strain Behavior of Thermoplastic Polyurethanes. *Mech. Mater.* **2005**, *37*, 817–839. [CrossRef]
17. Kolařk, J. Simultaneous Prediction of the Modulus, Tensile Strength and Gas Permeability of Binary Polymer Blends. *Eur. Polym. J.* **1998**, *34*, 585–590. [CrossRef]
18. Bicerano, J. *Prediction of Polymer Properties*, 3rd ed.; CRC Press: Boca Raton, FL, USA, 2002; ISBN 0203910117.
19. Tzelepis, D.A.; Suzuki, J.; Su, Y.F.; Wang, Y.; Lim, Y.C.; Zayernouri, M.; Ginzburg, V. V Experimental and Modeling Studies of IPDI-Based Polyurea Elastomers—The Role of Hard Segment Fraction. *J. Appl. Polym. Sci.* **2023**, *140*, e53592. [CrossRef]
20. Velankar, S.; Cooper, S.L. Microphase Separation and Rheological Properties of Polyurethane Melts. 1. Effect of Block Length. *Macromolecules* **1998**, *31*, 9181–9192. [CrossRef]
21. Ionita, D.; Cristea, M.; Gaina, C. Prediction of Polyurethane Behaviour via Time-Temperature Superposition: Meanings and Limitations. *Polym. Test* **2020**, *83*, 106340. [CrossRef]
22. Ginzburg, V. A Simple Mean-Field Model of Glassy Dynamics and Glass Transition. *Soft Matter* **2020**, *16*, 810–825. [CrossRef]
23. Jaishankar, A.; McKinley, G.H. A Fractional K-BKZ Constitutive Formulation for Describing the Nonlinear Rheology of Multiscale Complex Fluids. *J. Rheol.* **2014**, *58*, 1751–1788. [CrossRef]
24. Jaishankar, A.; McKinley, G.H. Power-Law Rheology in the Bulk and at the Interface: Quasi-Properties and Fractional Constitutive Equations. *Proc. R. Soc. A Math. Phys. Eng. Sci.* **2013**, *469*, 20120284. [CrossRef]
25. Rathinaraj, J.D.J.; McKinley, G.H.; Keshavarz, B. Incorporating Rheological Nonlinearity into Fractional Calculus Descriptions of Fractal Matter and Multi-Scale Complex Fluids. *Fractal Fract.* **2021**, *5*, 174. [CrossRef]
26. Suzuki, J.; Gulian, M.; Zayernouri, M.; D'Elia, M. Fractional Modeling in Action: A Survey of Nonlocal Models for Subsurface Transport, Turbulent Flows, and Anomalous Materials. *J. Peridynamics Nonlocal Model.* **2021**, *5*, 392–459. [CrossRef]
27. Suzuki, J.L.; Zayernouri, M.; Bittencourt, M.L.; Karniadakis, G.E. Fractional-Order Uniaxial Visco-Elasto-Plastic Models for Structural Analysis. *Comput. Methods Appl. Mech. Eng.* **2016**, *308*, 443–467. [CrossRef]
28. Suzuki, J.L.; Naghibolhosseini, M.; Zayernouri, M. A General Return-Mapping Framework for Fractional Visco-Elasto-Plasticity. *Fractal Fract.* **2022**, *6*, 715. [CrossRef]
29. Suzuki, J.L.; Tuttle, T.G.; Roccabianca, S.; Zayernouri, M. A Data-Driven Memory-Dependent Modeling Framework for Anomalous Rheology: Application to Urinary Bladder Tissue. *Fractal Fract.* **2021**, *5*, 223. [CrossRef]
30. Winey, K.I.; Vaia, R.A. Polymer Nanocomposites. *MRS Bull.* **2007**, *32*, 314–322. [CrossRef]
31. Kim, H.; Abdala, A.A.; Macosko, C.W. Graphene/Polymer Nanocomposites. *Macromolecules* **2010**, *43*, 6515–6530. [CrossRef]
32. Ray, S.S.; Okamoto, M. Polymer/Layered Silicate Nanocomposites: A Review from Preparation to Processing. *Prog. Polym. Sci.* **2003**, *28*, 1539–1641.
33. Lin, C.-L.; Li, J.-W.; Chen, Y.-F.; Chen, J.-X.; Cheng, C.-C.; Chiu, C.-W. Graphene Nanoplatelet/Multiwalled Carbon Nanotube/Polypyrrole Hybrid Fillers in Polyurethane Nanohybrids with 3D Conductive Networks for EMI Shielding. *ACS Omega* **2022**, *7*, 45697–45707. [CrossRef]
34. Kausar, A. Polyurethane Nanocomposite Coatings: State of the Art and Perspectives. *Polym. Int.* **2018**, *67*, 1470–1477. [CrossRef]
35. Chen, K.; Tian, Q.; Tian, C.; Yan, G.; Cao, F.; Liang, S.; Wang, X. Mechanical Reinforcement in Thermoplastic Polyurethane Nanocomposite Incorporated with Polydopamine Functionalized Graphene Nanoplatelet. *Ind. Eng. Chem. Res.* **2017**, *56*, 11827–11838. [CrossRef]
36. Shah, R.; Kausar, A.; Muhammad, B.; Shah, S. Progression from Graphene and Graphene Oxide to High Performance Polymer-Based Nanocomposite: A Review. *Polym. Plast. Technol. Eng.* **2015**, *54*, 173–183. [CrossRef]
37. Albozahid, M.; Naji, H.Z.; Alobad, Z.K.; Wychowaniec, J.K.; Saiani, A. Thermal, Mechanical, and Morphological Characterisations of Graphene Nanoplatelet/Graphene Oxide/High-Hard-Segment Polyurethane Nanocomposite: A Comparative Study. *Polymers* **2022**, *14*, 4224. [CrossRef] [PubMed]
38. Kausar, A. Shape Memory Polyurethane/Graphene Nanocomposites: Structures, Properties, and Applications. *J. Plast. Film Sheeting* **2020**, *36*, 151–166. [CrossRef]
39. Ginzburg, V.V.; Hall, L.M. *Theory and Modeling of Polymer Nanocomposites*; Springer: Berlin/Heidelberg, Germany, 2021; ISBN 303060442X.

40. Meng, Q.; Song, X.; Han, S.; Abbassi, F.; Zhou, Z.; Wu, B.; Wang, X.; Araby, S. Mechanical and Functional Properties of Polyamide/Graphene Nanocomposite Prepared by Chemicals Free-Approach and Selective Laser Sintering. *Compos. Commun.* **2022**, *36*, 101396. [CrossRef]
41. Su, X.; Wang, R.; Li, X.; Araby, S.; Kuan, H.-C.; Naeem, M.; Ma, J. A Comparative Study of Polymer Nanocomposites Containing Multi-Walled Carbon Nanotubes and Graphene Nanoplatelets. *Nano Mater. Sci.* **2022**, *4*, 185–204. [CrossRef]
42. Balazs, A.C.; Bicerano, J.; Ginzburg, V.V. Polyolefin/Clay Nanocomposites: Theory and Simulation. In *Polyolefin Composites*; Wiley: Hoboken, NJ, USA, 2007; pp. 415–448. ISBN 9780471790570.
43. Fornes, T.D.; Paul, D.R. Modeling Properties of Nylon 6/Clay Nanocomposites Using Composite Theories. *Polymer* **2003**, *44*, 4993–5013. [CrossRef]
44. Bicerano, J.; Douglas, J.F.; Brune, D.A. Model for the Viscosity of Particle Dispersions. *J. Macromol. Sci. Rev. Macromol. Chem. Phys.* **1999**, *39C*, 561–642. [CrossRef]
45. Pinnavaia, T.J.; Beall, G.W. *Polymer-Clay Nanocomposites*; John Wiley & Sons, Ltd.: Chichester, UK, 2000.
46. Williams, M.L.; Landel, R.F.; Ferry, J.D. The Temperature Dependence of Relaxation Mechanisms in Amorphous Polymers and Other Glass-Forming Liquids. *J. Am. Chem. Soc.* **1955**, *77*, 3701–3707. [CrossRef]
47. Kennedy, J.; Eberhart, R. Particle Swarm Optimization. In Proceedings of the ICNN'95-International Conference on Neural Networks, Perth, WA, Australia, 27 November–1 December 1995; Volume 4, pp. 1942–1948.
48. Chen, X.; Wang, X.; Fang, D. A Review on C1s XPS-Spectra for Some Kinds of Carbon Materials. *Fuller. Nanotub. Carbon Nanostructures* **2020**, *28*, 1048–1058. [CrossRef]
49. Kwan, Y.C.G.; Ng, G.M.; Huan, C.H.A. Identification of Functional Groups and Determination of Carboxyl Formation Temperature in Graphene Oxide Using the XPS O 1s Spectrum. *Thin. Solid. Film.* **2015**, *590*, 40–48. [CrossRef]
50. Brune, D.A.; Bicerano, J. Micromechanics of Nanocomposites: Comparison of Tensile and Compressive Elastic Moduli, and Prediction of Effects of Incomplete Exfoliation and Imperfect Alignment on Modulus. *Polymer* **2002**, *43*, 369–387. [CrossRef]

Disclaimer/Publisher's Note: The statements, opinions and data contained in all publications are solely those of the individual author(s) and contributor(s) and not of MDPI and/or the editor(s). MDPI and/or the editor(s) disclaim responsibility for any injury to people or property resulting from any ideas, methods, instructions or products referred to in the content.

Article

Numerical Simulation of Fatigue Life of Rubber Concrete on the Mesoscale

Xianfeng Pei [1], Xiaoyu Huang [1], Houmin Li [1,*], Zhou Cao [2], Zijiang Yang [2], Dingyi Hao [1], Kai Min [1], Wenchao Li [1], Cai Liu [1], Shuai Wang [3] and Keyang Wu [3]

[1] School of Engineering, Architecture and The Environment, Hubei University of Technology, Wuhan 430068, China; 102100805@hbut.edu.cn (X.P.); huangxiaoyu202203@163.com (X.H.); a424185749@163.com (D.H.); c1357962470@163.com (K.M.); cscecj319025@163.com (W.L.); zjsjygszh2023@163.com (C.L.)

[2] China Construction Third Bureau First Engineering Co., Ltd., Wuhan 430040, China; haodingyi1015@163.com (Z.C.); mk294588366@163.com (Z.Y.)

[3] Wuhan Construction Engineering Co., Ltd., Wuhan 430056, China; wangshuai@wceg.com.cn (S.W.); wukeyang@wceg.com.cn (K.W.)

* Correspondence: lihoumin2000@163.com

Citation: Pei, X.; Huang, X.; Li, H.; Cao, Z.; Yang, Z.; Hao, D.; Min, K.; Li, W.; Liu, C.; Wang, S.; et al. Numerical Simulation of Fatigue Life of Rubber Concrete on the Mesoscale. *Polymers* **2023**, *15*, 2048. https://doi.org/10.3390/polym15092048

Academic Editors: Alexey V. Lyulin and Valeriy V. Ginzburg

Received: 14 March 2023
Revised: 23 April 2023
Accepted: 23 April 2023
Published: 25 April 2023

Copyright: © 2023 by the authors. Licensee MDPI, Basel, Switzerland. This article is an open access article distributed under the terms and conditions of the Creative Commons Attribution (CC BY) license (https://creativecommons.org/licenses/by/4.0/).

Abstract: Rubber concrete (RC) exhibits high durability due to the rubber admixture. It is widely used in a large number of fatigue-resistant structures. Mesoscale studies are used to study the composition of polymers, but there is no method for fatigue simulation of RC. Therefore, this paper presents a finite element modeling approach to study the fatigue problem of RC on the mesoscale, which includes the random generation of the main components of the RC mesoscale structure. We also model the interfacial transition zone (ITZ) of aggregate mortar and the ITZ of rubber mortar. This paper combines the theory of concrete damage to plastic with the method of zero-thickness cohesive elements in the ITZ, and it is a new numerical approach. The results show that the model can simulate reasonably well the random damage pattern after RC beam load damage. The damage occurred in the middle of the beam span and tended to follow the ITZ. The model can predict the fatigue life of RC under various loads.

Keywords: numerical simulation; rubber concrete; fatigue life; three-point bending; polymer; mesoscale model

1. Introduction

With rapid economic development, the production of cars has increased, leading to the pollution of many waste tires, which are the primary source of waste rubber [1]. The combination of rubber, an excellent elastic material, and concrete, a brittle material, produces rubber concrete (RC), which has the advantages of low modulus of elasticity, high resistance to deformation, good crack resistance, good flexibility, and good wear resistance [2–4]. Liu et al. [2] found that RC improved concrete toughness and fatigue properties. Wang et al. [5] used the sounding technique to study the development of the fatigue damage process in RC at three stress levels—0.6, 0.7, and 0.8. It is a continuous process of the cumulative increase in damage, and it is divided into three processes: crack initiation, stable extension, and destabilization damage. With the rapid development of finite element theory and computer technology, concrete research is no longer limited to experimental studies. The method of finite element numerical simulation has become the primary research tool. Liu et al. [2] studied a mesoscale model of RC and analyzed its compressive properties. However, the model considered factors so simple that the results were unconvincing. Many scholars [6,7] have analyzed RC on microscopic, mesoscopic, and macroscopic scales, but no one has used a finite element model (FEM) to study RC fatigue. For this reason, this paper propose a finite element fatigue model of RC on the mesoscale.

Concrete mesoscale modeling studies have established aggregate, mortar, admixture, and interface transition zones (ITZ) over the last two decades [8]. Each component interacts with the others through mechanical relationships, thus influencing the strength of the overall structure. At the stage of concrete modeling, there are two ways of dealing with how to characterize concrete components. One technique is digital image technology recognition [9]. Zheng et al. [10] built a 2D concrete mesoscale model based on image recognition and investigated concrete's compressive strength and dimensional effects. He provided a reliable method for predicting compressive strength. Second, by analyzing the concrete composition and using computer programming to create a random aggregate model (RAM) [11] that meets the requirements, Sharif et al. [12] simulated samples of biphasic cubic concrete containing spherical aggregates embedded in homogeneous mortar and successfully demonstrated the failure modes of the pieces. After the characterization of the mesoscale aggregate composition method is completed, there are two methods of computational modeling: one is an FEM based on a RAM [11], and the other is a mechanical model based on a discrete element model (DEM) [13]. P.S.M. et al. [14] successfully modeled finite element RAM of ultrahigh-strength concrete fracture under uniaxial compression. It was found that damage initiation may occur in any of the three phases on the mesoscale, a degree that is difficult to achieve by experimental means and DEM. In addition, Zhou et al. [15] built a three-point bending-notched concrete beam as a model structure to discuss the mechanism of crack sprouting. However, realistic concrete beams have no prefabricated cracks. This paper uses an FEM with RAM to model three-point bent concrete without prefabricated cracks for fatigue simulation.

FEM calculations are primarily based on elasticity mechanics [16], plasticity mechanics [17], damage mechanics [18], and fracture mechanics [19] theories. The elastic model treats concrete as an elastomer and studies the mechanical properties of concrete in its elastic range. The disadvantage of the elastic model is that it is challenging to study the properties of concrete after large deformation or cracking. The concrete-smeared cracking model [20] uses a linear elastic model, which makes it difficult to calculate non-linear forms of damage. Kim et al. [21] presented a plasticity model that considers the form of concrete damage and the area of damage. The mechanical model of concrete damage first evolved through the study of metal fatigue [22], which considered a concrete failure as a process of quantitative damage triggered by microcracks in mesoscale structures. Ray et al. [23] found that the influencing factor for concrete fatigue is size through fracture mechanics models analyzed on a macroscopic structure. Concrete damage form is not determined by one mechanical behavior but by various mechanical methods. This paper used the concrete damaged plasticity (CDP) model, which combined concrete elasticity, plasticity, and damage. The CDP model was first proposed by J. et al. [24], and then B. Xu et al. [25] presented a damage model for the cyclic loading of concrete structures. B. Xu et al. [25] found that this model can simulate the inelastic behavior of RC beam–column members very well. In this paper, plastic damage theory is used, and the model conforms to the requirements by improvement.

In a 2D mesoscale study, the RC components are mortar, aggregate, rubber, aggregate-mortar ITZ, and rubber-mortar ITZ. The ITZ is complicated, and its thickness is usually 10–50 μm [26], which exceeds the minimum size for numerical simulations. With the development of research in recent years, a method called a cohesive element (CE) [27] for dealing with damage to very small-thickness elements has been proposed. Wang et al. [28] investigated the effect of cohesion models on the tensile behavior of concrete. Zhao et al. [29] developed a crystal plasticity model combining an extended finite element approach with a CE model. They analyzed fatigue cracking and found that the simulations were consistent with previous experimental observations.

The purpose of this paper is to present a fatigue damage model applicable to RC. This model uses a new numerical simulation method. Different RC peak loads of static pressure and fatigue life were simulated using the CDP and CE models. Based on the model's feasibility, the fatigue life of RC was predicted for different admixtures and loads, which can provide a basis for experimental reference in advance. Subsequent work can vary the load application methods, such as random and variable frequency loading, and can also consider a 3D mesoscale model study, which is of great significance.

2. Modeling Methods

The mesoscale model's modeling approach begins by considering the geometry of the model generation, which includes aggregate content, size, and location. Subsequently, the constitutive model of concrete and the ITZ model is considered to make the model feasible.

2.1. Mesoscale Model Geometry Generation

In mesoscale studies, concrete is usually considered a three-phase material consisting of aggregate, mortar, and ITZ. The rubber particles in RC replace part of the fine aggregates. Concrete is regarded as a homogeneous material in conventional macroscopic concrete FEM. This visual modeling approach makes it difficult to investigate how the concrete's inhomogeneity affects the macroscopic properties. This assumption of the homogeneity of concrete ignored several vital influences, such as aggregate size, particle size distribution, aggregate shape, and the effect of the ITZ. Zhong et al. [30] investigated the effect of aggregate shape (circular, elliptical, and polygonal) on the results of numerical analysis of the mesoscale model. They compared the stress–strain curves under different conditions with the experimental results. The results showed that the circular aggregate model is optimal for the numerical simulations. In this paper, circular aggregates are used so that meshing is easier and computer solutions are faster. In contrast, irregularly shaped aggregates are very complex to mesh and increase the computational burden.

In this paper, the coarse aggregates in the RC mesoscale model are aggregates of 5 mm or more in diameter and the fine aggregates are included in the mortar. The geometry of the mesoscale model needs to comply with three requirements: firstly, all the particles generated must be within the specified boundaries; secondly, none of the particles can overlap; and thirdly, there must be a gap between each particle, as the aggregates are wrapped in a layer of mortar and have no contact. Fuller's particle size [31] distribution curves are used in this paper. The Fuller curve is widely regarded as the grading curve, which provides an optimum particle size distribution for the working condition of the concrete. The Fuller curve equation is as follows:

$$P = 100\sqrt{\frac{D_0}{D_{max}}} \tag{1}$$

where P represents the percentage of aggregate passing through sieve hole diameter D_0, D_0 represents the diameter of the sieve hole, and D_{max} represents the diameter of the largest aggregate.

As this paper focuses on the 2D level, it is impossible to deal with the 2D problem directly with the help of the 3D Fuller set matching formula. It was used to obtain the best particle size distribution curve in 2D by applying the Walraven formula [32]. The formula is as follows:

$$P_c(D < D_0) = P_k\left(1.065 D_0^{0.5} D_{max}^{-0.5} - 0.053 D_0^4 D_{max}^{-4} - 0.012 D_0^6 D_{max}^{-6} - 0.0045 D_0^8 D_{max}^{-8} - 0.0025 D_0^{10} D_{max}^{-10}\right) \tag{2}$$

where P_c represents the percentage of the aggregate area, where size D is smaller than D_0. P_k represents the percentage of the aggregate area of the total area. In this paper, P_k is taken to be 0.7. D_{max} represents the diameter of the largest aggregate size, and the maximum diameter is taken as 20 mm.

The area of aggregate size distribution in the 550 mm × 150 mm area is listed by Formula (2) in Table 1.

Table 1. Size in 550 mm × 150 mm area occupied by different particle sizes.

Rubber Replacement Rate (%)	5–10 mm (mm^2)	10–15 mm (mm^2)	15–20 mm (mm^2)	Rubber (mm^2)
0	5612	8851	12,547	0
2.5	5612	8851	12,547	768
5	5612	8851	12,547	1537
7.5	5612	8851	12,547	2305
10	5612	8851	12,547	3074

The random generation is implemented in Python according to the aggregate area in Table 1, and the generated flowchart is shown in Figure 1.

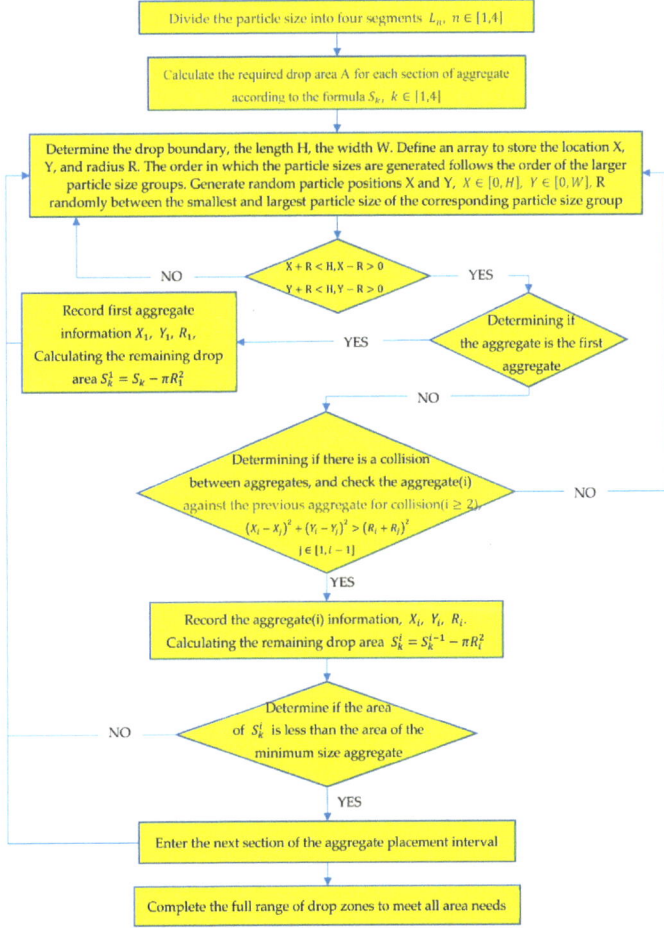

Figure 1. Flowchart of aggregate generation.

2.2. Constitutive Model of Concrete

The finite element software ABAQUS (2021 Version, Dassault systemes, France) and the programming language Python 3.8 are interconnected, and the code generated in Section 2.1 can be imported directly into ABAQUS. Aggregates are developed according to the program, and different property values are assigned to the different components to achieve the actual state of the mesoscale RC aggregates.

In this paper, coarse aggregate and rubber are considered homogeneous elastomers, and mortar is modeled numerically as a homogeneous continuum with elasticity. The mortar can be regarded as a lower-strength type of concrete, and its constitutive law uses the concrete–damage–plasticity (CDP) model. The CDP is a continuous, plasticity-based damage model that defines the concrete state by defining two mechanical behaviors: tensile cracking and compression damage. This model assumes that the concrete's uniaxial tensile and compressive response is characterized by plastic damage. The evolution of the yield surface is controlled by two hardening variables, the tensile equivalent plastic strain $\widetilde{\varepsilon}_t^{pl}$ and the compressive equivalent plastic strain $\widetilde{\varepsilon}_c^{pl}$, which are related to the damage mechanisms under tensile and compressive loading. The uniaxial tensile and compressive stress–strain response is shown in Figure 2.

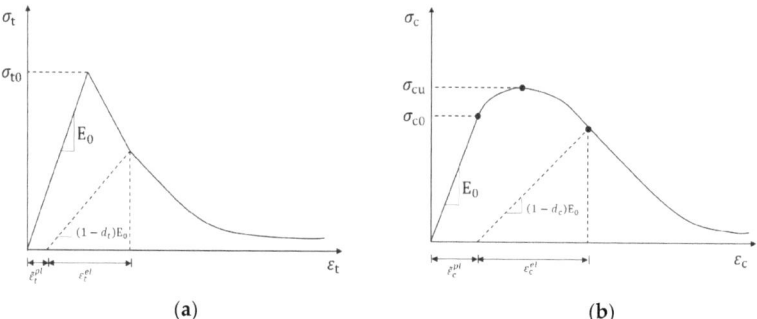

Figure 2. The uniaxial stress–strain response: (**a**) tensile; (**b**) compressive.

Where $\widetilde{\varepsilon}_t^{pl}$ and $\widetilde{\varepsilon}_c^{pl}$ represent the equivalent plastic strains in tension and compression, ε_t^{el} and ε_c^{el} represent the elastic strains corresponding to tension and compression, d_t and d_t represent the two damage variables for elastic stiffness degradation, with the damage variables taking values from 0 to 1, and $d = 0$ means the material is undamaged, $d = 1$ means the material is completely damaged, and E_0 represents the initial Young's modulus of the material.

In the case of uniaxial tension, the concrete stress–strain response obeys linear elastic variation up to the time of failure stress σ_{t0}, which is used to distinguish between the elastic and plastic phases of concrete. After σ_{t0}, the concrete enters the damage phase, and microcracking occurs in the macrostructure. In the case of uniaxial compression, the concrete stress–strain response obeys a linear elastic change to the compressive elastic ultimate stress, and σ_{c0}, and σ_{c0} used to distinguish the elastic phase from the plastic phase under uniaxial compression. Unlike uniaxial tension, there is a hardening phase to the ultimate compressive stress σ_{cu} after σ_{cu} where the concrete is softened and microcracked.

When a concrete specimen is unloaded from any point in the strain-softening branch of the stress–strain curve, the elastic stiffness of the material appears to be damaged. The stress–strain relationships for uniaxial tensile and compressive loading are (Equation (3)):

$$\sigma_t = (1 - d_t) E_0 \left(\varepsilon_t - \widetilde{\varepsilon}_t^{pl} \right) \tag{3}$$

$$\sigma_c = (1 - d_c) E_0 \left(\varepsilon_c - \widetilde{\varepsilon}_c^{pl} \right) \tag{4}$$

This paper deals with the numerical simulation of the three-point bending of concrete, where the general form of damage is tensile damage. After being subjected to cyclic loading, the tensile stiffness after damage needs to be redefined, as shown in Figure 3.

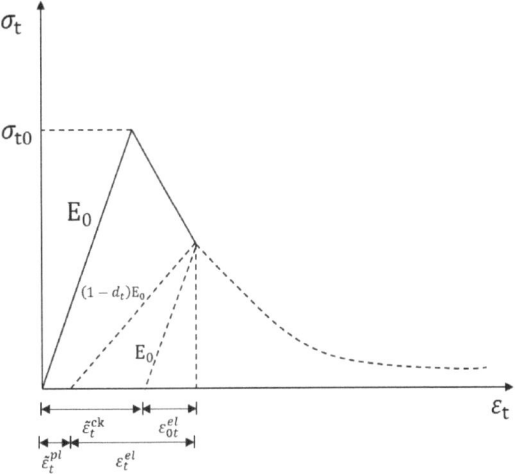

Figure 3. Redefinition of tensile stiffness after damage.

Where ε_{0t}^{el} represents elastic strain and $\tilde{\varepsilon}_t^{ck}$ represents cracking strain.

In the event of damage to the concrete, the cracking strain $\tilde{\varepsilon}_t^{ck}$ is defined by the following equation:

$$\tilde{\varepsilon}_t^{ck} = \varepsilon_t - \varepsilon_{0t}^{el} \tag{5}$$

$$\varepsilon_{0t}^{el} = \frac{\sigma_t}{E_0} \tag{6}$$

ABAQUS automatically converts cracking strain to plastic strain for use:

$$\tilde{\varepsilon}_t^{pl} = \tilde{\varepsilon}_t^{ck} - \frac{d_t}{(1-d_t)} \frac{\sigma_t}{E_0} \tag{7}$$

According to the Structural Design Code for Concrete [33], considering the tensile and compressive damage variables of the material, the specific concrete intrinsic model is determined by Young's modulus E and Poisson's ratio λ in the elastic phase and by the non-linear stress–strain equation in the inelastic phase. When the concrete structure is under pressure:

$$\sigma_c = (1 - d_c) E \varepsilon_c \tag{8}$$

$$d_c = \begin{cases} 1 - \frac{\rho_c n}{n - 1 + x^n} & x \leq 1 \\ 1 - \frac{\rho_c}{\alpha_c (x-1)^2 + x} & x > 1 \end{cases} \tag{9}$$

$$\rho_c = \frac{f_{c,r}}{E_c \varepsilon_{c,r}} \tag{10}$$

$$x = \frac{\varepsilon}{\varepsilon_{c,r}} \tag{11}$$

$$n = \frac{E_c \varepsilon_{c,r}}{E_c \varepsilon_{c,r} - f_{c,r}} \tag{12}$$

where α_c is the parameter value of the falling section of the uniaxial compressive stress–strain curve for concrete, $f_{c,r}$ is the representative value of the uniaxial compressive strength of concrete, $\varepsilon_{c,r}$ is the peak compressive strain corresponding to $f_{c,r}$, and d_c is the evolutionary parameter for uniaxial compressive damage to concrete.

When the concrete structure is in tension:

$$\sigma_t = (1 - d_t)E\varepsilon_t \tag{13}$$

$$d_t = \begin{cases} 1 - \rho_t \left[1.2 - 0.2x^5\right] x \leq 1 \\ 1 - \dfrac{\rho_t}{\alpha_t(x-1)^{1.7} + x} x > 1 \end{cases} \tag{14}$$

$$x = \frac{\varepsilon}{\varepsilon_{t,r}} \tag{15}$$

$$\rho_t = \frac{f_{t,r}}{E_c \varepsilon_{t,r}} \tag{16}$$

where α_t is the parameter value of the falling section of the uniaxial tension stress−strain curve for concrete, $f_{t,r}$ is the representative value of the uniaxial tension strength of concrete, $\varepsilon_{t,r}$ is the peak tension strain corresponding to $f_{t,r}$, and d_t is the evolutionary parameter for uniaxial tension damage to concrete.

In accordance with the Structural Design Code for Concrete [33], specific values are shown in Table 2.

Table 2. Parameters of CDP.

$f_{c,r}$ (MPa)	$\varepsilon_{c,r}$ ($\times 10^{-6}$)	α_c	$f_{t,r}$ (MPa)	$\varepsilon_{t,r}$ ($\times 10^{-6}$)	α_t
40.26	1790	1.947	3.01	84	2.831

In addition to this, the plasticity parameters for CDP are self-contained in ABAQUS, as shown in Table 3.

Table 3. Parameters of ABAQUS itself.

Dilation Angle	Eccentricity	f_{b0}/f_{c0}	K
30	0.1	1.16	0.666

The values in Table 3 have been verified by many academics to be generally consistent. This is a fixed value [10].

2.3. CE Model of the ITZ

After the aggregate model has been built, there are two approaches to the ITZ. One is establishing a solid FEM of the ITZ [34]. The advantage of this is that it can reflect the thickness relationship of the interface composed of concrete. However, the ITZ's actual thickness is 10–50 μm, which is difficult to achieve with FEM. Even if the thickness is expanded by a factor of 10 to a range that FEM can calculate, this will result in a dense and small mesh division and a significant increase in computational effort. Secondly, the ITZ is considered a zero-thickness element (ZTE) [35], which retains the relevant mechanical properties of the actual ITZ to achieve the accuracy of the simulation, and all ITZs in concrete can be represented by ZTE. In summary, we selected the ZTE.

The ZTE has three ways of simulating the behavior of the ITZ. Firstly, a layer of the ZTE can be inserted using a shared node, which can be used if the CE is on the same mesh as the surrounding element. Secondly, if the elements of the CE are divided differently from the surrounding mesh or if the CE uses a finer discretization than the adjacent parts, the tie constraint can be used to connect the CE to other parts. Thirdly, in some special cases where the requirements are met, a connected interaction can be added directly to the CE in contact without adding additional elements. Figure 4 shows the three methods of CE processing.

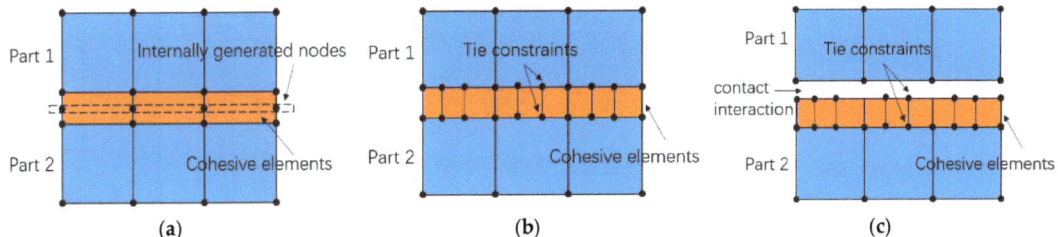

Figure 4. Three types of CE: (**a**) shared nodes; (**b**) tie constraint; (**c**) contact interaction.

Based on the random aggregates generated by the simulations in this paper, a cohesive zone will be added to the contact surface of the aggregate and mortar. The first ZTE (Figure 4a) is chosen to insert a layer of CE using a shared node. The size and location of each aggregate are uncertain, so choosing inserted CE is difficult. A Python program finds the node number of the aggregate place and copies the new node at the node number to create a zero-thickness CE. This fits perfectly with the shared node insertion approach. The ITZ generates CE, as shown in Figure 5.

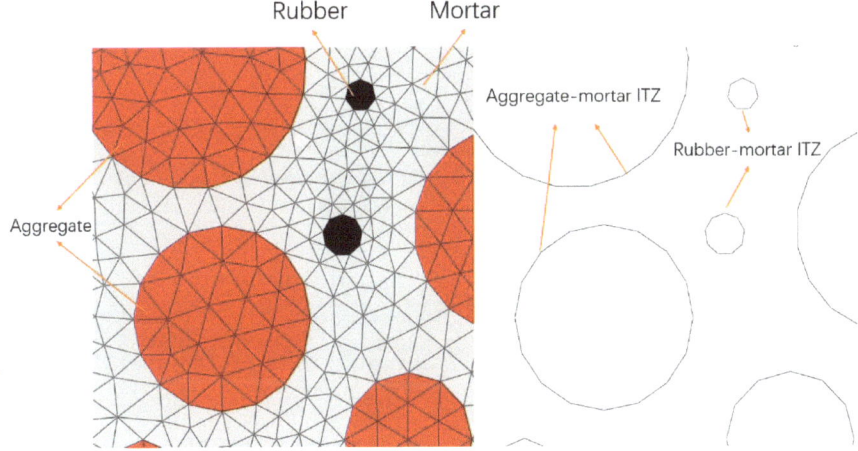

Figure 5. Aggregates and ITZ generated in ABAQUS.

Figure 5. Aggregates and ITZ generated in ABAQUS.

The CE damage is divided into four parts: the linear elastic phase, the damage initiation phase, the damage evolution phase, and complete damage.

The online resilience phase of the damage response of the CE is as follows:

$$\mathbf{t} = \begin{Bmatrix} t_n \\ t_s \\ t_t \end{Bmatrix} = \begin{bmatrix} E_{nn} & E_{ns} & E_{nt} \\ E_{ns} & E_{ss} & E_{st} \\ E_{nt} & E_{st} & E_{tt} \end{bmatrix} \begin{Bmatrix} \varepsilon_n \\ \varepsilon_s \\ \varepsilon_t \end{Bmatrix} = E\varepsilon \qquad (17)$$

where t_n is the nominal stress in the normal direction, t_s is the nominal stress in shear in the first direction, t_t is the nominal stress in shear in the second direction, E_{ij} is Young's modulus in each direction, and ε_i is the strain in the corresponding direction.

The quadratic stress criterion formula is used for damage initiation. Damage initiation occurs when the contact-stress ratio involved reaches 1:

$$\left\{\frac{t_n}{t_n^0}\right\}^2 + \left\{\frac{t_s}{t_s^0}\right\}^2 + \left\{\frac{t_t}{t_t^0}\right\}^2 = 1 \quad (18)$$

Damage evolution by way of traction separation is shown in Figure 6.

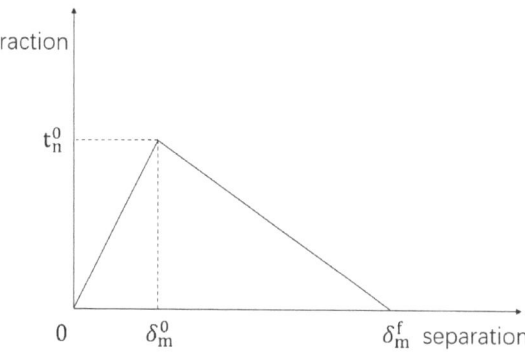

Figure 6. Traction separation relationship.

Where δ_m^0 is the separation displacement value at the onset of damage and δ_m^f is the separation displacement at the maximum damage.

The material enters the damage phase judged by the damage value D. When D is 1, the material is completely damaged by the following equation:

$$D = \frac{\delta_m^f \left(\delta_m^{max} - \delta_m^0\right)}{\delta_m^{max} \left(\delta_m^f - \delta_m^0\right)} \quad (19)$$

where D is the damage value and δ_m^{max} is the additional amount of maximum separation displacement during loading.

In this paper, the fracture energy is calculated using the Benzeggagh–Kenane (BK) criterion:

$$G^C = G_n^C + \left(G_s^C - G_n^C\right) \left\{\frac{G_s}{G_T}\right\}^\eta \quad (20)$$

where G^C is the hybrid fracture energy, G_n^C is the type I fracture energy of the cohesive element, G_s^C is the type II fracture energy of the cohesive element, G_s is the shear deformation energy, and G_T is the tensile deformation energy.

The performance of ITZ is difficult to test on the experimental scale, so the determination of simulation parameters for ITZ is difficult to determine. Usually, the performance of ITZ is approximated by the weak mortar composition, and researchers use the percentage of mortar to study and judge the performance of ITZ. Xiao et al. [36] considered the strength of ITZ to be 80% of the mortar. Kim et al. [37] considered the fracture energy of ITZ to be equivalent to 50% of the mortar. Li et al. [38] considered it to be 80%. It was obvious that different researchers have different opinions on determining the mechanical properties of ITZ. The ultimate purpose is to achieve unity between numerical simulations and experiments, so the performance parameters of the ITZ on numerical simulations are determined by trial and error to determine the optimum values of these relevant parameters. The parameters used in this paper are shown in Table 4.

Table 4. Parameters of ITZ.

	Normal Strength (MPa)	Tangential Strength (Mpa)	Normal Fracture Energy (N/mm)	Shear Fracture Energy (N/mm)
Aggregate-mortar ITZ [1]	3.1	9	0.03	0.09
Rubber-mortar ITZ [2]	2.8	8.4	0.028	0.084

[1] Data from [28], [2] trial values.

3. Verification of the Model

3.1. Experiment

The data for this summary test were obtained from Liu et al. [39]. He investigated the effect of rubber substitution rate and rubber particle size on the fatigue life of rubber concrete. The object of study was an RC beam with dimensions of 150 mm × 150 mm × 550 mm. A static load test was conducted under a three-point bending load, and a fatigue test was conducted under a cyclic load. The fatigue life and related fatigue life curves were obtained for different rubber substitution rates, particle sizes, and stress levels.

3.1.1. Experimental Materials

Material parameters are shown in Tables 5 and 6.

Table 5. Parameters of cement.

Type	Coagulation Time (min)		Compressive Strength (MPa)		Flexural Strength (MPa)	
	Initial Condensation	Final Condensation	3d	28d	3d	28d
P·O 42.5	180	270	26.9	50.1	5.62	8.3

Table 6. Parameters of aggregates.

Type	Gradation (mm)	Fineness Modulus	Apparent Density (kg/m^3)	Stacking Density (kg/m^3)	Water Absorption (%)	Mud Content (%)	Crushing Value (%)
Crushed stone	5–20	–	2775	1648	1.0	0.35	8.9
Sand	–	2.76	26.58	1736	1.3	1.9	–

Rubber: Crushed rubber granules from waste tyros 1–4 mm.

3.1.2. Experimental Test Methods

RC specimens with different rubber replacement rates are first tested by static loading to obtain the corresponding peak loads. The fatigue tests are carried out using models of the same material proportions. The maximum and minimum loads are applied to the RC beams using a load-controlled mode, which is an equal amplitude and uniform load mode.

3.2. Building Mesoscale Models

A 150 mm × 550 mm rubber concrete beam element is built according to Table 1, with different dosing of rubber concrete beams as shown in Figure 7, where gray means mortar, red means aggregate, and black means rubber.

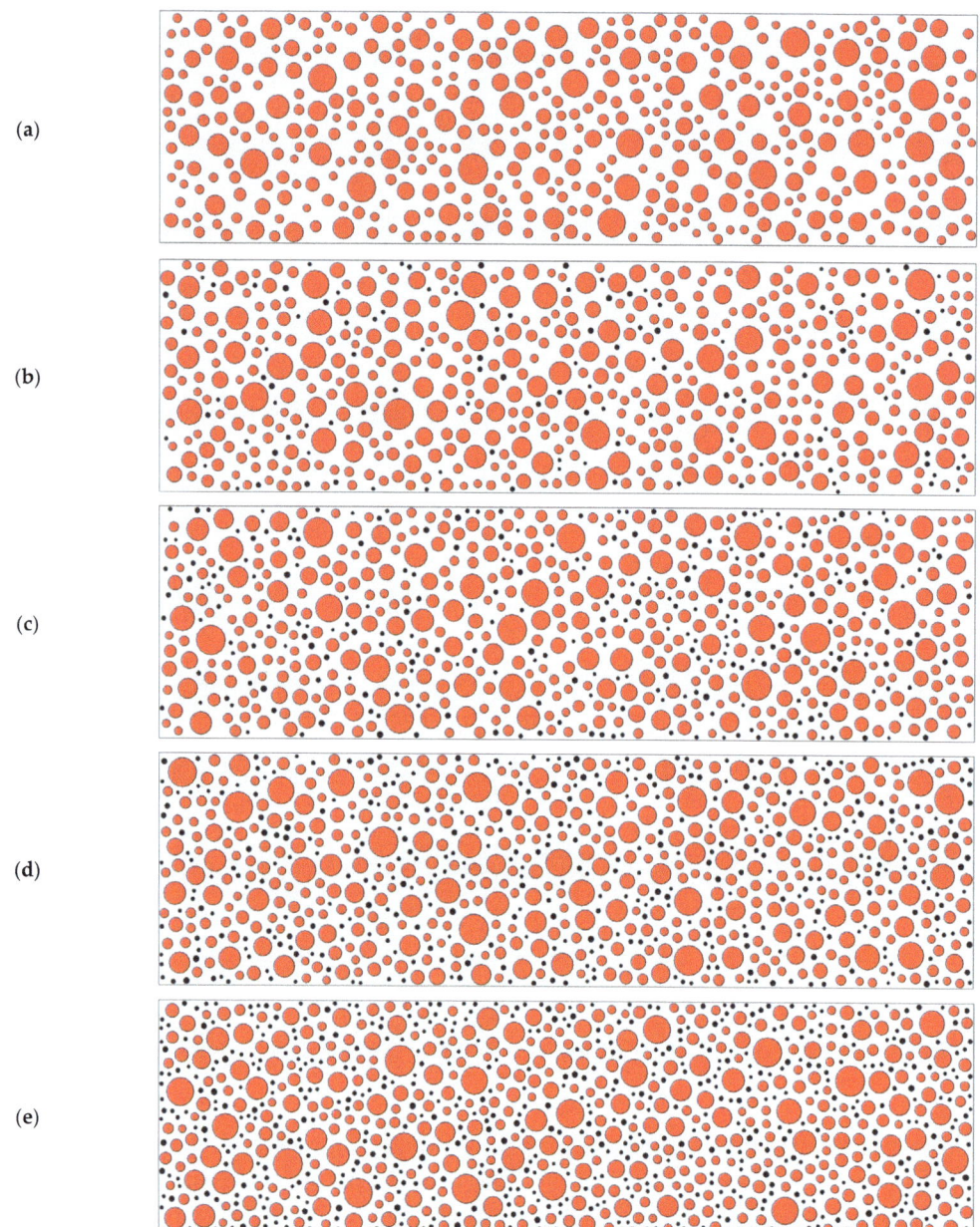

Figure 7. RC beams with different dosing: (**a**) 0; (**b**) 2.5%; (**c**) 5%; (**d**) 7.5%; (**e**) 10%.

The load loading point is in the middle of the upper part, with the bottom left constraint 100 mm from the left boundary and the right constraint 100 mm from the right, as in Figure 8.

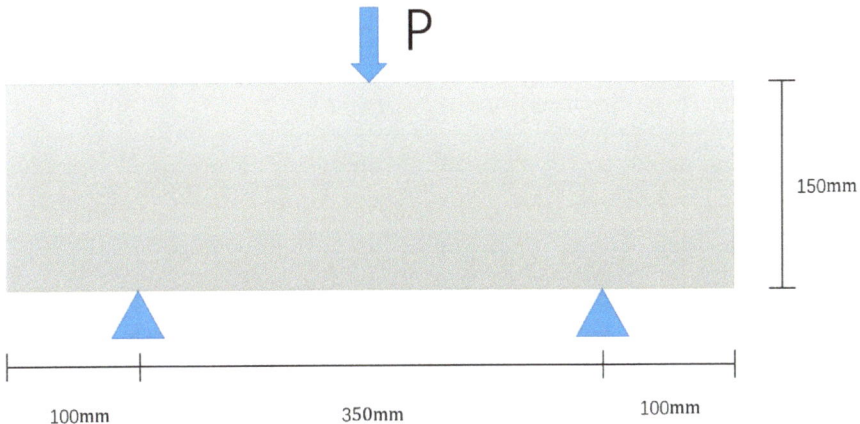

Figure 8. Load loading schematic.

The model is solved using the ABAQUS/Standard. First, apply a displacement constraint of 2 mm at the upper load loading point, stop the calculation when the model does not converge, and obtain the peak load F_{max} for the model. Subsequently, the fatigue life of the model is calculated at different stress levels, still using the same model with cyclic concentrated force constraints applied at the upper load loading points. Load application from minimum load P_{min} to maximum load P_{max}, where $P_{min}/P_{max} = 0.1$, fatigue load stress levels $S = P_{max}/F_{max}$. In this paper, S takes the values 0.9, 0.85, 0.8, and 0.75. When S is too high, the fatigue damage results are over in one go. When S is too small, the calculation is too large in the numerical simulation phase. The Fourier series method controls the equivalent mean amplitude load when fatigue loads are applied:

$$F(t) = \begin{cases} A_0 + \sum_n^N [A_n \cos n\omega(t-t_0) + B_n \sin n\omega(t-t_0)] & t \geq t_0 \\ A_0 & 0 \leq t \leq t_0 \end{cases} \quad (21)$$

where the period is T, circle frequency $\omega = 2\pi/T$, the loading initial time is A_0, and the number of steps parameters $A_1, B_1, A_2, B_2, \cdots, A_0, A_0$.

The parameters used for RC in this paper are shown in Table 7.

Table 7. Parameters of RC.

Type	Young's Modulus (GPa)	Poisson's Ratio
Mortar	36	0.2
Aggregate	72	0.16
Rubber	7	0.49

3.3. Experimental Versus Simulation

By comparing the results of this study with the three-point bending static load peak load results and fatigue load results from the literature [39], the feasibility of the model is verified.

3.3.1. Peak Load

The peak load tests and simulation results for this model under three-point bending loads at different stress levels are summarized in Table 8. It can be seen that the magnitude of the peak load decreases as the rubber content increases, which is in line with the researchers' judgment on the performance of RC. Comparing the test and simulation for peak loads at the same stress levels proved that the simulation and test results agree well. The maximum absolute error is 3.6%. The new numerical model proved reliable for peak loads under three-point bending loads.

Table 8. Experimental and simulated peak loads.

	RC-0	RC-2.5	RC-5	RC-7.5	RC-10
Experimental peak loads (KN)	28.15	26.32	25.05	24.1	23.06
Simulated peak loads (KN)	28.62	25.94	25.52	24.44	22.23
Error (%)	1.67	−1.44	1.88	1.41	3.60

3.3.2. Fatigue Life

The results of tests and simulations with different dosing levels of rubber concrete at stress levels S = 0.85 and S = 0.75 are summarized in Table 9. The increase in rubber admixture can improve the fatigue resistance of RC and extend the fatigue life. Due to the large dispersion of the fatigue life results, only the minimum and maximum lives are taken as a reference in the test results. Moreover, the overall life trend improves with increasing rubber doping. As shown in Figure 9, the results obtained from the numerical model of rubber concrete in this paper are all between the maximum and minimum values of the test results and meet the feasibility requirements of the model. This proves the reliability of the new numerical model in fatigue life calculation. There are some differences between the expected life and the experiment, but this is acceptable. Because the experiment phase is a one-off for each test beam, the RC is already destroyed after the experiments with peak load. Although each beam is made to the same size and aggregate content, the mechanical properties are not the same. The different mechanical behaviors of the concrete beam can be observed in [37].

Table 9. Experimental and simulated fatigue life.

		RC-0	RC-2.5	RC-5	RC-7.5	RC-10
S = 0.85	experiment min/max	1615/4236	2281/3459	1485/5883	2261/6781	3827/6832
	simulation	1742	2678	3824	5018	6779
S = 0.75	experiment min/max	8654/15,432	9876/19,536	14,876/23,654	15,245/31,132	20,268/34,538
	simulation	11,812	14,208	19,081	24,085	32,742

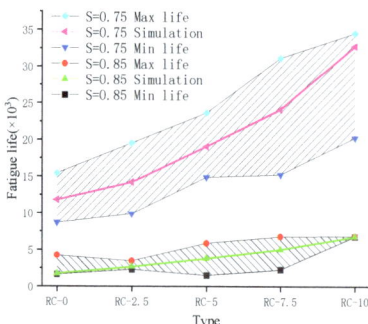

Figure 9. Experimental and simulated fatigue life.

4. Analysis of Variables

Finite element static pressure simulations of three-point bending were carried out for RC rubber admixtures of 0, 2.5%, 5%, 7.5%, and 10% to obtain the corresponding peak load and displacement relationships. Based on the stress levels S = 0.85 and S = 0.75 above, add stress levels S = 0.9 and S = 0.8 to the loading method for the RC fatigue simulation to analyze the effect of different rubber doping and stress levels on damage form and fatigue life.

4.1. Force–Deflection Curves

The force–deflection curves for RC at different admixtures were obtained from numerical simulations, as shown in Figure 10. The concrete deflection increases as the rubber admixture increases and the peak load tends to decrease significantly. The rising and falling phases of the curve for ordinary concrete are steeper than the gentle curve for RC. The trend becomes more subdued as the amount of rubber added increases. The comparison of the trends of the two curves RC-0 and RC-10 in Figure 10 is exceptionally different. It confirmed the effect of rubber particles on concrete in the mesoscale study. Rubber was able to reduce the extension of concrete damage and increase the toughness of concrete, reducing the brittleness of concrete. This reflects the actual validity of the new numerical model.

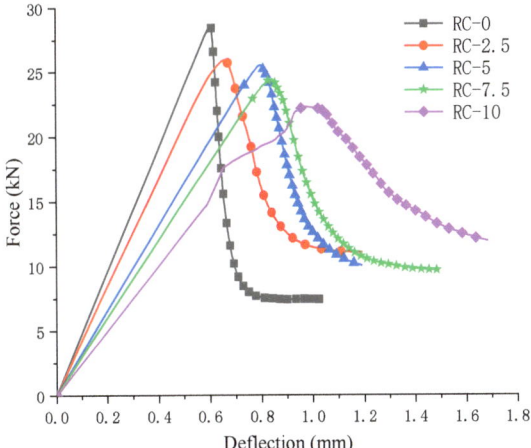

Figure 10. Load–deflection curve.

4.2. Types of Damage

In this paper, the damage to the RC is shown through stiffness in the form of two factors: one is static compression, and the other is fatigue. The visual form of the damage is represented by the SDEG cloud map output by ABAQUS (SDEG = 0 means no damage to the structure, and SDEG = 1 means complete damage to the structure). Figure 11 shows the damage to an ordinary concrete beam of 150 mm × 550 mm without rubber admixture after a three-point bending static load. As the beam damage occurs in the middle of the span, for ease of observation, the structure is taken in the middle of the beam, as shown in the black box in Figure 11, with a size of 150 mm × 150 mm. The following are screenshots of the damage obtained by this method.

The SDEG damage clouds for 0, 2.5%, 5%, 7.5%, and 10% rubber doping after damage are shown in Figure 12. A form of static pressure damage to rubber concrete was observed in the mesoscale study. The damage was mainly at the mid-span of the beam, with an irregular damage zone extending from the bottom to the top. The damage course follows the edges of the aggregate and rubber and is consistent with existing fracture and damage mechanics theories. The point of damage to the zero rubber-doped concrete is only at the opening of the damage zone, with no damage to the surrounding concrete aggregate, as shown in Figure 12a. Damage points occur not only at the opening of the damage zone but also minor damage to the rubber around the opening, as shown in Figure 12b–e. On the mesoscale, it is observed that the rubber particles take up a small part of the load-bearing capacity under load. Furthermore, with the increase of rubber admixture, the damage point at the bottom of the concrete increases, and the damage zone is influenced by the surrounding rubber particles in the middle of the extension. The 10% and 7.5% rubber-doped concrete leads particularly well, with multiple damage points at the bottom and tiny branches of the damage zone midway through, as shown in Figure 12d,e. Various forms of damage indicate that adding rubber particles to concrete helps to retard concrete damage, which also provides the basis for research into the fatigue resistance of RC.

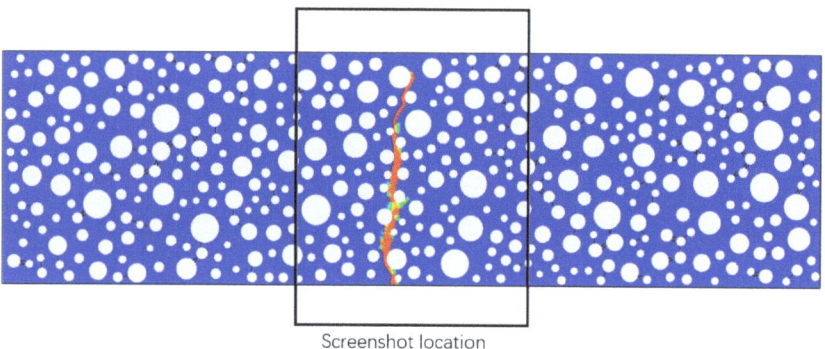

Figure 11. Damage to ordinary concrete and location of damage sampling.

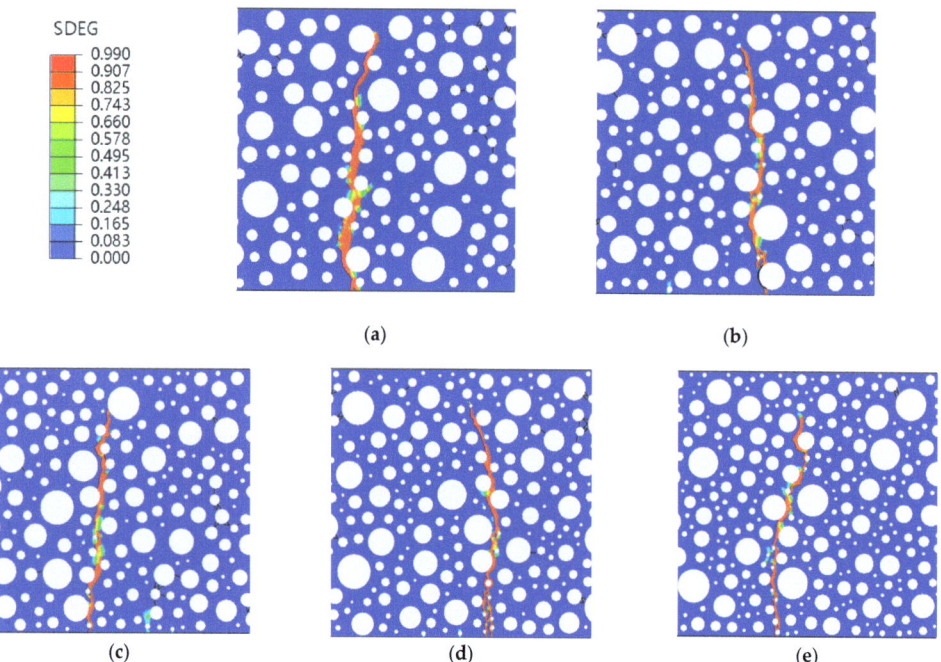

Figure 12. RC SDEG with different mixes for static pressure loading: (**a**) 0; (**b**) 2.5%; (**c**) 5%; (**d**) 7.5%; (**e**) 10%.

The fatigue simulation of the same RC beam at different stress levels is a unique advantage of the fine-view simulation. The stress levels are guaranteed to be the same peak load each time, something that cannot be achieved experimentally. This paper uses fatigue simulations for four stress levels of 0.9, 0.85, 0.8, and 0.75, with each stress level corresponding to five rubber doping levels. Shown in Figure 13 are four forms of stress level fatigue damage for ordinary concrete. It can be observed that fatigue damage to ordinary concrete at different stress levels takes the same form, with the damage zone starting at the same point of failure at the bottom of the concrete. In ordinary concrete, from the start of the damage to the end, only the weakest point within the concrete bears the load, regardless of the force acting. Its fatigue damage is also essentially the same as static pressure damage (Figures 12a and 13), proving that ordinary concrete is relatively homogeneous regarding internal forces when damaged, with the same place bearing the load.

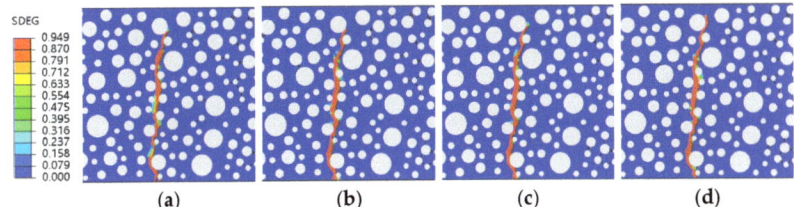

Figure 13. SDEG of ordinary concrete under different stress levels: (**a**) S = 0.9; (**b**) S = 0.85; (**c**) S = 0.8; (**d**) S = 0.75.

Figures 14–17 show fatigue damage at four stress levels for four doped RC. Unlike ordinary concrete, the fatigue loads do not take the same form of damage at different stress levels when rubber is added. As shown in Figure 14, the damage zone for 2.5% admixture stress levels of 0.85, 0.8, and 0.75 differ, and the damage point is also different at the bottom of the concrete. As shown in Figure 15, the location of the damage zone is different for 5% doping stress levels of 0.9, 0.85, and 0.8, but the location of the bottom damage point is the same for stress levels of 0.85, 0.8, and 0.75. As shown in Figure 16, the 7.5% doping stress level only differs in the damage zone and damage at a stress level of 0.9; the damage zone and damage point are essentially the same at other stress levels. As shown in Figure 17, the orientation of the damage zone and the location of the initial damage point at the bottom stabilize and remain the same when the doping level reaches 10%. The rubber dosing ranges from 0 to 10%, with the damage zone orientation and initial damage point location stabilizing from the beginning, through the disorder of the intermediate dosing, and to final stability. The fatigue damage of RC is different from hydrostatic damage, which is also different from ordinary concrete. The addition of the rubber creates a fragile ITZ between the rubber and the mortar, even weaker than the ITZ between the aggregate and the mortar. The model successfully simulated the effect of rubber doping.

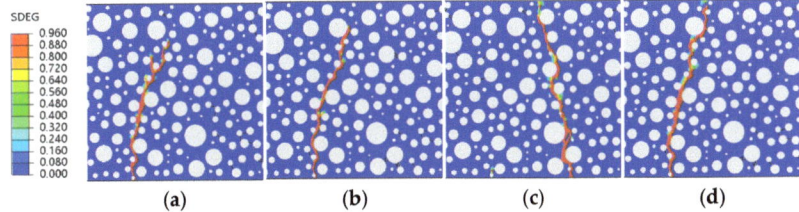

Figure 14. SDEG of RC-2.5 under different stress levels: (**a**) S = 0.9; (**b**) S = 0.85; (**c**) S = 0.8; (**d**) S = 0.75.

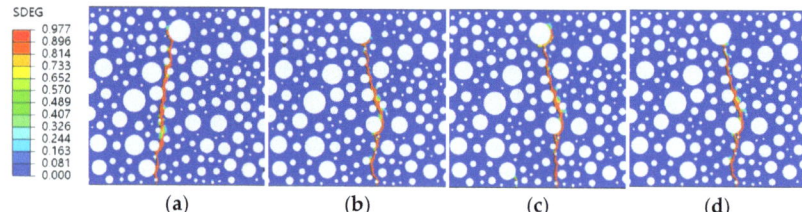

Figure 15. SDEG of RC-5 under different stress levels: (**a**) S = 0.9; (**b**) S = 0.85; (**c**) S = 0.8; (**d**) S = 0.75.

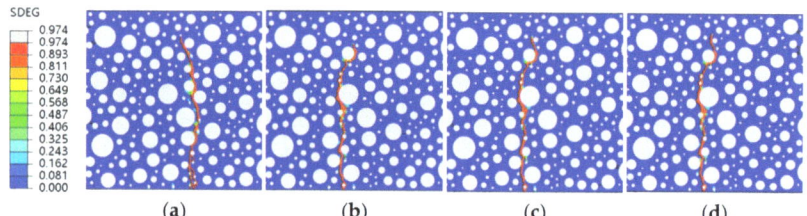

Figure 16. SDEG of RC-7.5 under different stress levels: (**a**) S = 0.9; (**b**) S = 0.85; (**c**) S = 0.8; (**d**) S = 0.75.

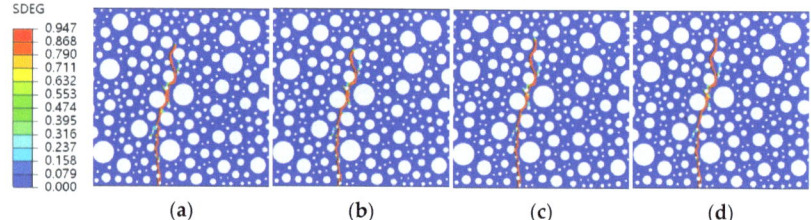

Figure 17. SDEG of RC-10 under different stress levels: (**a**) S = 0.9; (**b**) S = 0.85; (**c**) S = 0.8; (**d**) S = 0.75.

4.3. Fatigue Life

Table 10 shows the peak loads and the life of the fatigue loads at the corresponding four stress levels for the five doped rubber concretes. At the same stress level, the fatigue life increases as the rubber content increases, indicating that rubber concrete carries higher cyclic loads than ordinary concrete for a given cyclic load. In the case of rubber concrete, this result is because the rubber particles act as an energy absorber and load cushion in the concrete. Rubber particles have better elastic properties on the mesoscale level than concrete particles. In the case of concrete suffering from tension and compression, part of the energy is converted into the elastic energy of the rubber particles. The fatigue life of 10% rubber is 7.3, 3.89, 4.45, and 2.77 times greater than that of ordinary concrete at four stress levels.

Table 10. Fatigue life of different rubber doping at different stress levels.

	RC-0	RC-2.5	RC-5	RC-7.5	RC-10
S = 0.9	135	243	581	792	986
S = 0.85	1742	2678	3824	5018	6779
S = 0.8	4212	5385	9821	12036	18735
S = 0.75	11,812	14,208	19,081	24,085	32,742

The relationship between rubber doping, stress level, and fatigue life is shown in Figure 18. It is obvious that, within a specific range, an increase in rubber content and a decrease in stress level increase the fatigue life of RC. The increase in fatigue life is a non-linear relationship, as shown in Figure 19.

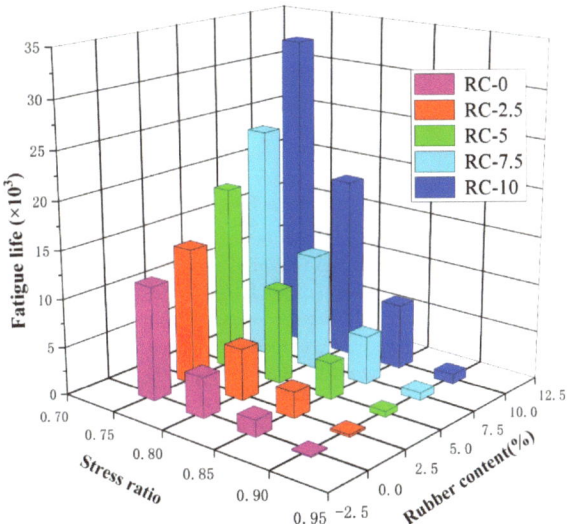

Figure 18. Fatigue life of different rubber doping at different stress levels.

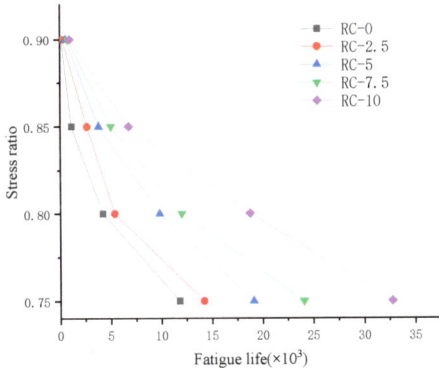

Figure 19. Relationship between stress level and fatigue life of different rubber doping.

5. Discussion

This model simulation study generates the mesoscale structure of RC through random aggregates, applies the improved properties of CDP to mortar, and combines modeling of the aggregate-mortar ITZ and rubber-mortar ITZ to achieve the mesoscale structure of actual RC. It is a new numerical approach. The model was subjected to a series of three-point bending fatigue loads to analyze the causes of damage forms and fatigue life from a mesoscale.

5.1. Causes of Damage Types

The structural form of the model after damage by static pressure and fatigue loading is consistent with reality, and the appearance of concrete damage on the mesoscale can be accurately observed from the mesoscale structure. Subsequent damage develops along the weakness of the ITZ around the aggregate and rubber particles. There are two main reasons for producing an irregular damage band consistent with reality. First, the model adds the damage theory in the CE model, setting a zero-thickness damage zone in the aggregate-mortar and rubber-mortar layers. The material's mechanical properties in the CE are less than those of the mortar. When subjected to forces, the ITZ is more easily damaged than the mortar aggregate. Secondly, the mesoscale model is randomly generated for aggregate size and location, which aligns with the actual aggregate distribution of concrete materials and better reflects the model's realism. The rubber particles are smaller than the coarse aggregate, and it is easier for damage to occur around the rubber than around the coarse aggregate.

5.2. Factors Influencing Fatigue Life

The addition of rubber benefits the fatigue properties of rubber concrete. When the model is damaged, there is some damage around the rubber particles not in the damage zone. These share some of the fatigue load, confirming the effect of the rubber particles on the mesoscale level. The mesoscale model can be applied to complex fatigue loads. The model shows fatigue life agreement at stress levels of 0.75 to 0.9 and can simulate the effects of fatigue life due to different doping levels of rubber. Rubber particles have better elastic properties on the mesoscale level than concrete particles. When the concrete is loaded, part of the energy is converted into the elastic energy of the rubber particles. RC life increases with increasing rubber and decreases with increasing stress ratio. As a rule of thumb, the magnitude of the stress ratio is related to the logarithm of the fatigue life [40]. After trying various fitting formulae, the following relationship is assumed:

$$S = A + B\ln(N) + C\ln^2(N) \quad \left(N > e^{-\frac{B}{2C}}\right) \tag{22}$$

where S is the concrete stress ratio, A, B, and C are constants whose magnitude is related to the concrete admixture, and N is the concrete fatigue life.

According to Formula (22). for curve fitting, as shown in Figure 20, the specific formula and correlation coefficient R results are shown in Table 11, and the fitted results meet the requirements. It is found that the stress level is related to the quadratic function of the logarithm of fatigue life. In addition, the results of this study allow for reasonable extrapolation of the three-point bending fatigue life of rubber concrete at dosing levels between 0.75 and 0.9. This provides a corresponding reference for the test, a model for calculating fatigue life correctly on the mesoscale.

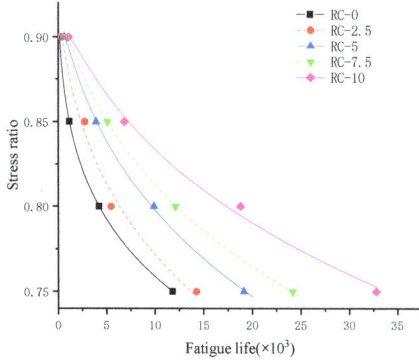

Figure 20. Fitting relationship between different rubber doping stress levels and fatigue life.

Table 11. Fitting formulae for different rubber doping.

Type	Fitting Formula	R^2
RC-0	$S = 0.865 + 0.028\ln(N) - 0.0043\ln^2(N)$ (N > 26)	0.999
RC-2.5	$S = 0.671 + 0.087\ln(N) - 0.0083\ln^2(N)$ (N > 189)	0.989
RC-5	$S = 0.519 + 0.126\ln(N) - 0.0104\ln^2(N)$ (N > 361)	0.997
RC-7.5	$S = 0.475 + 0.135\ln(N) - 0.0107\ln^2(N)$ (N > 550)	0.999
RC-10	$S = 0.384 + 0.152\ln(N) - 0.0112\ln^2(N)$ (N > 886)	0.996

5.3. Potential Applications and Developments

The mesoscale model proposed in this study can accurately represent the fatigue life of rubber concrete under three-point bending fatigue loading. In addition, analysis of the static pressure load's damage form and the RC's ultimate load reveals that the model is also accurately represented. The study of RC is equally informative for ordinary concrete and other polymer admixture concrete. The difference lies in the polymer's shape, size, location, and properties. Shape, size, and position are solved by Python code, and performance could be solved by setting properties on the polymer. However, it is often difficult to achieve the desired effect, and an interface layer is needed to change the mechanical relationship between the different substances. This study could also be applied to reinforced RC to simulate the location of damage and structural life of specific damaged structures in macrostructures to provide an initial structural performance judgment for actual structures.

The limitations of this research method lie in the 2D structure. When considering the mesoscale design in the 3D structure, the lack of computer performance is challenging to resolve, and the vast number of calculations leads to increased calculation time. Future work could be improved to develop a 3D mesoscale concrete model to calculate fatigue life, achieving the desired accuracy and computational efficiency requirements.

6. Conclusions

This paper proposed an RAM on the scope of a mesoscale study. The model used plastic damage theory and the insertion of cohesive elements in the ITZ, and is a new numerical model. This paper verifies the model's correctness in peak load and fatigue life. Peak loads were verified for five doping levels of 0, 2.5%, 5%, 7.5%, and 10%, and fatigue life was verified for stress levels of 0.75 and 0.85. After this, the results for stress levels of 0.8 and 0.9 were simulated and analyzed. The peak static pressure load in three-point bending was successfully modeled on a mesoscale as decreasing with increasing rubber doping, and the resulting deflection increased with increasing rubber doping. Static pressure and fatigue forms of damage could be observed in the mesoscale, where the point of damage produced by RC damage is not unique and increases with the amount of rubber admixture. The damage element produced by RC damage shows an order–disorder–order process as the rubber dosing increases. It was observed from the model that when the damage occurred to the RC, the internal rubber took the load. In addition, the model could simulate the three-point bending fatigue life at different stress levels for various rubber doping on a mesoscale. A quadratic function relating stress levels with different rubber doping to fatigue life was fitted, which can predict fatigue life for stress levels between 0.75 and 0.9, providing some reference value for the test. The mesoscale model in this paper satisfied the fatigue life simulation requirements perfectly. The method, with model improvements, could be applied to all RC fatigue structures in the future.

Author Contributions: Conceptualization, X.P. and X.H.; methodology, X.P. and K.W.; software, X.P. and D.H.; validation, X.P., X.H. and D.H.; formal analysis, X.P.; investigation, X.P. and D.H.; resources, X.P. and D.H.; data curation, X.P.; writing—original draft preparation, X.P.; writing—review and editing, H.L.; visualization, X.P., Z.C., Z.Y. and K.M.; supervision, H.L., C.L. and W.L.; project administration, H.L.; funding acquisition, H.L., S.W. and K.W. All authors have read and agreed to the published version of the manuscript.

Funding: This research was funded by the 2022 Hubei Construction Science and Technology Program Project (grant 90) and the 2021 Hubei Construction Science and Technology Program Project (grant number 43).

Institutional Review Board Statement: Not applicable.

Data Availability Statement: The data used in the article can be obtained from the author here.

Acknowledgments: The authors would like to acknowledge Hubei University of Technology.

Conflicts of Interest: The authors declare no conflict of interest.

References

1. Xiao, Z.; Pramanik, A.; Basak, A.K.; Prakash, C.; Shankar, S. Material Recovery and Recycling of Waste Tyres—A Review. *Clean. Mater.* **2022**, *5*, 100115. [CrossRef]
2. Liu, F.; Zheng, W.; Li, L.; Feng, W.; Ning, G. Mechanical and Fatigue Performance of Rubber Concrete. *Constr. Build. Mater.* **2013**, *47*, 711–719. [CrossRef]
3. Sukontasukkul, P.; Chaikaew, C. Properties of Concrete Pedestrian Block Mixed with Crumb Rubber. *Constr. Build. Mater.* **2006**, *20*, 450–457. [CrossRef]
4. Liu, F.; Wang, B.; Yuan, X.; Wang, Y.; Li, J. Experimental Study on the Toughness of Concrete Mixed with Waste Rubber Particles. *Concrete* **2019**, *353*, 78–81+85.
5. Wang, L.; Wang, C.; Zhang, Y.; Ma, A. Study on Fatigue Damage Process of Rubberized Cement Concrete by Acoustic Emission Technique. *J. Southeast Univ. Nat. Sci. Ed.* **2009**, *39*, 3.
6. Kachkouch, F.Z.; Noberto, C.C.; Babadopulos, L.F.D.A.L.; Melo, A.R.S.; Machado, A.M.L.; Sebaibi, N.; Boukhelf, F.; El Mendili, Y. Fatigue Behavior of Concrete: A Literature Review on the Main Relevant Parameters. *Constr. Build. Mater.* **2022**, *338*, 127510. [CrossRef]
7. Xu, J.; Yao, Z.; Yang, G.; Han, Q. Research on Crumb Rubber Concrete: From a Multi-Scale Review. *Constr. Build. Mater.* **2020**, *232*, 117282. [CrossRef]
8. He, J.; Liu, D.; Gao, Z.; Zhu, F.; Bai, P. Research Progress on Mechanical Properties Testing Methods of Interfacial Transition Zone in Concrete. *J. Exp. Mech.* **2022**, *37*, 805–820.
9. Huang, Y. Overview of Research Progress of Digital Image Processing Technology. *J. Phys. Conf. Ser.* **2022**, *2386*, 012034. [CrossRef]
10. Zheng, Z.; Tian, C.; Wei, X.; Zeng, C. Numerical Investigation and ANN-Based Prediction on Compressive Strength and Size Effect Using the Concrete Mesoscale Concretization Model. *Case Stud. Constr. Mater.* **2022**, *16*, e01056. [CrossRef]
11. Yu, L.; Liu, C.; Mei, H.; Xia, Y.; Liu, Z.; Xu, F.; Zhou, C. Effects of Aggregate and Interface Characteristics on Chloride Diffusion in Concrete Based on 3D Random Aggregate Model. *Constr. Build. Mater.* **2022**, *314*, 125690. [CrossRef]
12. Yu, Y.; Zheng, Y.; Guo, Y.; Hu, S.; Hua, K. Mesoscale Finite Element Modeling of Recycled Aggregate Concrete under Axial Tension. *Constr. Build. Mater.* **2021**, *266*, 121002. [CrossRef]
13. Abu-Haifa, M.; Lee, S.J. Image-Based Modeling-to-Simulation of Masonry Walls. *J. Arch. Eng.* **2022**, *28*, 06022001. [CrossRef]
14. Thilakarathna, P.; Baduge, K.K.; Mendis, P.; Chandrathilaka, E.; Vimonsatit, V.; Lee, H. Understanding Fracture Mechanism and Behaviour of Ultra-High Strength Concrete Using Mesoscale Modelling. *Eng. Fract. Mech.* **2020**, *234*, 107080. [CrossRef]
15. Zhou, R.; Lu, Y.; Wang, L.-G.; Chen, H.-M. Mesoscale Modelling of Size Effect on the Evolution of Fracture Process Zone in Concrete. *Eng. Fract. Mech.* **2021**, *245*, 107559. [CrossRef]
16. Lurie, A.I.; Belyaev, A. *Theory of Elasticity. Foundations of Engineering Mechanics*; Springer: Berlin/Heidelberg, Germany, 2005; ISBN 978-3-540-24556-8.
17. Abu-Lebdeh, T.M.; Voyiadjis, G.Z. Plasticity-Damage Model for Concrete under Cyclic Multiaxial Loading. *J. Eng. Mech.* **1993**, *119*, 7.
18. Nguyen, G.D. A Thermodynamic Approach to Non-Local Damage Modelling of Concrete. *Int. J. Solids Struct.* **2008**, *45*, 1918–1934.
19. González-Velázquez, J.L. Chapter 2—Linear Elastic Fracture Mechanics. In *A Practical Approach to Fracture Mechanics*; González-Velázquez, J.L., Ed.; Elsevier: Amsterdam, The Netherlands, 2021; pp. 35–74. ISBN 978-0-12-823020-6.
20. AHillerborg, A.; Modéer, M.; Petersson, P.E. Analysis of Crack Formation and Crack Growth in Concrete by Means of Fracture Mechanics and Finite Elements. *Cem. Concr. Res.* **1976**, *6*, 773–781. [CrossRef]
21. Jae, Y.; Hong, G.; Seong, T. Plasticity Model for Directional Nonlocality by Tension Cracks in Concrete Planar Members, *Eng. Struct.* **2011**, *33*, 1001–1012.
22. Kachanov, L.M. Rupture Time Under Creep Conditions. *Int. J. Fract.* **1999**, *97*, 11–18. [CrossRef]
23. Sonalisa, R.; Kishen, J.M.C. Fatigue Crack Propagation Model and Size Effect in Concrete Using Dimensional Analysis. *Mech. Mater.* **2011**, *43*, 75–86.
24. Lubliner, J.; Oliver, J.; Oller, S.; Oñate, E. A Plastic-Damage Model for Concrete. *Int. J. Solids Struct.* **1987**, *25*, 299–326. [CrossRef]
25. Xu, B.; Bompa, D.V.; Elghazouli, A.Y.; Ruiz-Teran, A.M.; Stafford, P.J. Numerical Assessment of Reinforced Concrete Members Incorporating Recycled Rubber Materials. *Eng. Struct.* **2020**, *204*, 110017. [CrossRef]
26. Gaedicke, C.; Roesler, J.; Evangelista, F. Three-Dimensional Cohesive Crack Model Prediction of the Flexural Capacity of Concrete Slabs on Soil. *Eng. Fract. Mech.* **2012**, *94*, 1–12. [CrossRef]

27. Zhang, B.; Nadimi, S.; Eissa, A.; Rouainia, M. Modelling Fracturing Process Using Cohesive Interface Elements: Theoretical Verification and Experimental Validation. *Constr. Build. Mater.* **2023**, *365*, 130132. [CrossRef]
28. Wang, J.; Jivkov, A.P.; Engelberg, D.L.; Li, Q. Parametric Study of Cohesive ITZ in Meso-Scale Concrete Model. *Procedia Struct. Integr.* **2019**, *23*, 167–172. [CrossRef]
29. Zhao, Q.; Wahab, M.A.; Ling, Y.; Liu, Z. Fatigue Crack Propagation within Al-Cu-Mg Single Crystals Based on Crystal Plasticity and XFEM Combined with Cohesive Zone Model. *Mater. Des.* **2021**, *210*, 110015. [CrossRef]
30. Zhong, G.Q.; Guo, Y.C.; Li, L.J.; Liu, F. Analysis of Mechanical Performance of Crumb Rubber Concrete by Different Aggregate Shape under Uniaxial Compression on Mesoscopic. *Key Eng. Mater.* **2011**, *462–463*, 219–222. [CrossRef]
31. Schlangen, E.; van Mier, J.G.M. Simple Lattice Model for Numerical Simulation of Fracture of Concrete Materials and Structures. *Mater. Struct.* **1992**, *25*, 534–542. [CrossRef]
32. Wang, Y.; Peng, Y.; Kamel, M.M.A.; Ying, L. Mesomechanical Properties of Concrete with Different Shapes and Replacement Ratios of Recycled Aggregate Based on Base Force Element Method. *Struct. Concr.* **2019**, *20*, 1425–1437. [CrossRef]
33. Mohurd. *Code for Design of Concrete Structures*; China Architecture and Building Press: Beijing, China, 2010.
34. Song, Z.; Lu, Y. Mesoscopic Analysis of Concrete under Excessively High Strain Rate Compression and Implications on Interpretation of Test Data. *Int. J. Impact Eng.* **2012**, *46*, 41–55. [CrossRef]
35. Víctor, R.; Marta, S.; Flora, F.; Juan, M. Self-Compacting Concrete Manufactured with Recycled Concrete Aggregate: An Overview. *J. Clean. Prod.* **2020**, *262*, 121362.
36. Xiao, J.; Li, W.; Sun, Z.; David, A.L.; Surendra, P.S. Properties of Interfacial Transition Zones in Recycled Aggregate Concrete Tested by Nanoindentation. *Cem. Concr. Compos.* **2013**, *37*, 276–292. [CrossRef]
37. Kim, S.-M.; Abu Al-Rub, R.K. Meso-Scale Computational Modeling of the Plastic-Damage Response of Cementitious Composites. *Cem. Concr. Res.* **2011**, *41*, 339–358. [CrossRef]
38. Li, W.; Xiao, J.; Corr, D.J.; Shah, S.P. Numerical Modeling on the Stress-Strain Response and Fracture of Modeled Recycled Aggregate Concrete. *Civil Environ. Eng.* **2013**, *1*, 749–759.
39. Liu, Y. Fatigue Test and Life Prediction of Pre-Treated. Master's Thesis, North China University of Water Resources and Electric Power, Zhengzhou, China, 2020.
40. Mohammadi, Y.; Kaushik, S.K. Flexural Fatigue-Life Distributions of Plain and Fibrous Concrete at Various Stress Levels. *J. Mater. Civil Eng.* **2005**, *17*, 650e8. [CrossRef]

Disclaimer/Publisher's Note: The statements, opinions and data contained in all publications are solely those of the individual author(s) and contributor(s) and not of MDPI and/or the editor(s). MDPI and/or the editor(s) disclaim responsibility for any injury to people or property resulting from any ideas, methods, instructions or products referred to in the content.

Article

The Light-Fueled Self-Rotation of a Liquid Crystal Elastomer Fiber-Propelled Slider on a Circular Track

Lu Wei, Yanan Chen, Junjie Hu, Xueao Hu, Yunlong Qiu and Kai Li *

School of Civil Engineering, Anhui Jianzhu University, Hefei 230601, China; weilu@ahjzu.edu.cn (L.W.); cyn@stu.ahjzu.edu.cn (Y.C.); jjhu@stu.ahjzu.edu.cn (J.H.); hxa@stu.ahjzu.edu.cn (X.H.); ylqiu@stu.ahjzu.edu.cn (Y.Q.)
* Correspondence: kli@ahjzu.edu.cn

Citation: Wei, L.; Chen, Y.; Hu, J.; Hu, X.; Qiu, Y.; Li, K. The Light-Fueled Self-Rotation of a Liquid Crystal Elastomer Fiber-Propelled Slider on a Circular Track. *Polymers* **2024**, *16*, 2263. https://doi.org/10.3390/polym16162263

Academic Editors: Valeriy V. Ginzburg and Alexey V. Lyulin

Received: 18 June 2024
Revised: 3 August 2024
Accepted: 7 August 2024
Published: 9 August 2024

Copyright: © 2024 by the authors. Licensee MDPI, Basel, Switzerland. This article is an open access article distributed under the terms and conditions of the Creative Commons Attribution (CC BY) license (https://creativecommons.org/licenses/by/4.0/).

Abstract: The self-excited oscillation system, owing to its capability of harvesting environmental energy, exhibits immense potential in diverse fields, such as micromachines, biomedicine, communications, and construction, with its adaptability, efficiency, and sustainability being highly regarded. Despite the current interest in track sliders in self-vibrating systems, LCE fiber-propelled track sliders face significant limitations in two-dimensional movement, especially self-rotation, necessitating the development of more flexible and mobile designs. In this paper, we design a spatial slider system which ensures the self-rotation of the slider propelled by a light-fueled LCE fiber on a rigid circular track. A nonlinear dynamic model is introduced to analyze the system's dynamic behaviors. The numerical simulations reveal a smooth transition from the static to self-rotating states, supported by ambient illumination. Quantitative analysis shows that increased light intensity, the contraction coefficient, and the elastic coefficient enhance the self-rotating frequency, while more damping decreases it. The track radius exhibits a non-monotonic effect. The initial tangential velocity has no impact. The reliable self-rotating performance under steady light suggests potential applications in periodic motion-demanding fields, especially in the construction industry where energy dissipation and utilization are of utmost urgency. Furthermore, this spatial slider system possesses the ability to rotate and self-vibrate, and it is capable of being adapted to other non-circular curved tracks, thereby highlighting its flexibility and multi-use capabilities.

Keywords: self-rotation; liquid crystal elastomer; light fueled; slider; curved track

1. Introduction

A self-oscillating system [1–3] refers to the phenomenon where a system relies on fixed environmental stimuli to trigger and induce continuous and stable periodic motion [4–6] without external drives. The system absorbs energy from external stimuli, inducing the periodic amplification of its energy. This results in periodic energy conversion within the system, compensating for the energy dissipation caused by damping in the process of motion. Through this positive feedback mechanism, the system is propelled to generate nonlinear responses [7] and amplify the control effect of the stimuli on its components or motion, enabling the system to reach a novel steady state, and consequently perform spontaneous periodic motion with invariant frequency [8–10]. The previous information implies that a self-rotating system does not require a sophisticated controller [11], which will result in more convenient and straightforward operation. Currently, the existing feedback mechanisms often involve multi-process coupling and internal adaptive feedback to achieve the purpose of energy compensation, such as the multi-process coupling of droplet evaporation with membrane deformation and movement [12], the coupling of air expansion with liquid column movement [13], the coupling mechanism in plate buckling and chemical reactions [14], the coupling of bridge vibrations with electrical energy [15], and photo-induced thermo-surface tension gradients [16,17]. Additionally, the period and amplitude in the state of self-rotating system equilibrium are dictated solely by the inherent properties of the system, regardless of the initial conditions. This unique characteristic guarantees the internal stability of the feedback system, enhancing its robustness [18–20] and facilitating the deeper comprehension of non-equilibrium thermodynamic theories. Given the unique characteristics and advantages of self-rotating systems, numerous self-rotating machines [21–23] have been created to explore their practical application potential. The commonly available self-rotating machines include miniaturized autonomous robots [24], nano-generators [25], energy harvesting and capture systems [26], active machines [27–29], autonomous separators [30], and mass transport equipment [31].

Moreover, the achievement of self-rotation necessitates the use of active materials capable of responding to external stimuli. Such materials include dielectric elastomers [32], liquid crystal elastomers, stimulation-responsive ionic gels [33], polyelectrolyte gels [34], shape memory polymers, smart polymer hydrogels [35], and thermotropic liquid crystalline polymer composites [36]. They possess immense application potential in diverse fields, including biomedicine [37], agriculture, transportation, corrosion prevention, and material science. LCEs, one of the emerging environmentally responsive materials [38–40], are polymeric networks formed by the crosslinking of liquid crystal monomers, possessing the characteristics of both liquid crystals and elastomers. Upon being exposed to external environmental stimuli, the liquid crystal units within LCEs transition from anisotropy to isotropy at the microscopic level, with the liquid crystal monomers undergoing rotation or phase changes [41]. This leads to the conversion of *trans-isomers* in the liquid crystal elastomer fibers to *cis-isomers*, causing the nonlinear unidirectional contraction of the LCEs macroscopically [42]. Once the stimulus is removed, the LCEs exhibit their inherent reversibility, enabling them to return to their original state, which contributes significantly to the realization of non-driven spontaneous periodic motion.

Different types of LCE fibers demonstrate distinct responses to a variety of external stimuli, such as light, heat, electricity, magnetism, and humidity. With continuous development and in-depth research on LCE fibers, other researchers have been able to fabricate LCEs that respond to a single or multiple coupled stimuli, including magnetically responsive LCEs [43,44], light-responsive LCEs [45,46], thermally responsive LCEs [47], humidity-responsive LCEs [48], electrically responsive LCEs [49,50], and multi-responsive LCEs [51,52]. The self-rotating modes based on the stimulus-responsive behavior of LCEs are constantly diversifying. They encompass various forms, such as vibration [53], bending [54,55], self-rotation [56], torsion [57,58], rolling [59], buckling [60], chaos [61,62], the sitting up of LCE thin film [44], eversion [63], and inversion, as well as the synchronized movement of multiple coupled self-oscillators [64].

Among the diverse stimulus-responsive LCEs, light emerges as a highly attractive clean energy source due to its ease of accessibility and controllability [65]. Meanwhile, light-fueled LCEs exhibit unique advantages [66–68], including significant strain resistance, reversible deformation, a rapid response, silence, etc. [69,70]. Therefore, the light–mechanical coupled self-rotating system stands out prominently in terms of both its application potential and value. The existing self-oscillating systems of LCEs have been applied in self-rotating engines [71], self-oscillating flexible circuits [72], self-fluttering aircraft [73], self-paddling boats [74], soft robots [75], self-moving automobiles [76], and other fields.

In the last few years, the intense focus on LCEs has been driven by extensive research into self-sliding systems [77–79], particularly with regard to track-mounted sliders [80,81]. However, the potential of these systems is still constrained by the limitations in in-plane motion, such as the difficulty in achieving self-rotation due to necessary deformation. To address this challenge, this paper proposes a spatial LCE fiber-propelled slider system that reduces the deformation of LCEs during energy conversion, enabling self-rotation. The main contents of this paper are outlined as follows: In Section 2, we establish a theoretical model of the system based on the proposed light–mechanical coupling dynamics and the deformation mechanism of light-responsive LCE fibers. In Section 3, we explore the system's behavior in both the static and self-rotating states under constant illumination.

Analysis reveals spontaneous periodic motion modes and the underlying mechanisms governing these modes. In Section 4, we quantitatively investigate the impact of critical dimensionless parameters on the cyclic frequency of the system through mathematical modeling and data calculations. Finally, we conclude the key findings of this paper.

2. Theoretical Model and Formulation

In this section, we first describe a newly designed light-fueled self-rotating dynamic system that comprises an LCE fiber, a slotted slider, and a rigid circular track. Subsequently, based on the dynamic mechanical model of the LCE optical fiber and the theorem of momentum moment, combined with spatial analytic geometry derivations, the dynamic control equation of the periodic self-rotating system is calculated. Finally, to address the dimensional impact among the parameters in the control equation, we normalize and standardize the parameters using the dimensionless method and introduce the process of numerical analytical calculation.

2.1. Dynamics of Self-Rotating System

Figure 1 and Video S1 illustrate the spatial structural model of a light-fueled self-rotating system of an LCE fiber propelling a slider. This system consists of an LCE fiber, a slotted slider, and a rigid circular track with a radius of r. The LCE fiber has an original length of L_0, with one end attached to the horizontal fixed end and the other end connected to a slotted slider of mass m. The slider is firmly nested into the circular track through the slot, while the circular track itself is fastened stably to horizontal fixed supports. Under the given initial tangential velocity v_0 and the stimulation of the designed light, the slider propelled by the LCE fiber can perform self-sustaining periodic motion on the circular track. Given that the mass of the LCE fiber is significantly smaller than that of the slider of m, the influence of its mass on motion can be disregarded. In addition, while the slider is moving on the circular track, the slider is subjected to damping force, acting in the opposite direction of the slider's motion. We took the center of the circular track as the origin o, with the vertical upward direction as the positive z axis, the horizontal rightward direction as the positive y axis, and the perpendicular outward direction in the paper as the positive x axis. We define point A as the intersection of the negative y axis and the circular track, which serves as the original position for movement. The connection point between the LCE fiber and the horizontal fixed end is labeled as point M, where the projection of M onto the xoy plane falls on the negative y axis, referred to as point N. Let P be the instantaneous position of the mass block, with the angles $\angle NPO = \alpha$, $\angle MPN = \beta$, and $\angle AOP = \theta$.

As shown in Figure 1a,b, the yellow area represents the illuminated zone with an angle θ_0, ranging from π to 2π, while the colorless area represents the non-illuminated zone in this paper. Driven by the initial tangential velocity, the slider moves in a tangential direction within the plane, continuing its rotation until the slider reaches zero velocity within the illuminated area. At this moment, the *trans-isomers* in the LCE fibers transforms into the *cis-isomers* upon UV light irradiation, leading to the unidirectional contraction of the LCE fiber. This contraction propels the slider to continue moving in a counterclockwise direction, subsequently exiting the illuminated zone. Due to inertia, the slider continues to move in the non-illuminated zone at a decelerated speed, until it re-enters the illuminated zone and repeats the process. The repetition of this cycle results in spontaneous circular motion, termed self-rotation.

As depicted in Figure 1c, with the vertical forces acting on the slider cancelling each other out at all times, there is no chance of vertical displacement, allowing the slider to move solely within the plane of the circular track. Therefore, we only consider the dynamic response of the slider within the xoy plane. In Figure 1, θ is the angle which is the projection of the LCE string rotating angular displacement on the horizontal plane. When

θ falls within the interval from 0 to π, based on the theorem of momentum moment, the mechanical control equation of the system is given as follows [82–84]:

$$mr^2\ddot{\theta} = -rF_L\cos\beta\sin\alpha - rF_D \quad (1)$$

where $\ddot{\theta}$ refers to the angular acceleration of the slider at its instantaneous position, F_L represents the tension of the LCE fiber, and F_D denotes the damping force.

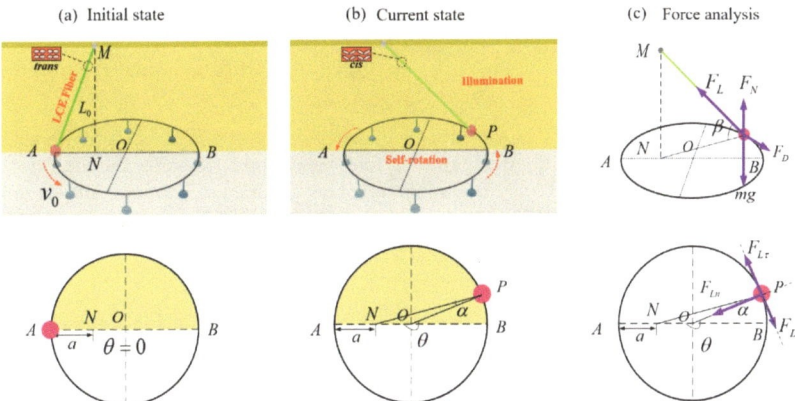

Figure 1. The side and top views of light-fueled self-rotating system with an LCE fiber, a slotted slider, and a rigid circular track: (**a**) initial state; (**b**) current state; and (**c**) force analysis. Under stable illumination, the slotted slider propelled by the LCE fiber can undergo spontaneous and continuous periodic motion on the circular track.

When θ lies between π and 2π, the equation is given as follows:

$$mr^2\ddot{\theta} = rF_L\cos\beta\sin\alpha - rF_D \quad (2)$$

The explanations of the variables $\ddot{\theta}$, F_L, and F_D in Equation (2) are the same as those above in Equation (1). According to spatial geometric relationship of the structure in Figure 1c, we can determine that $\cos\beta = \dfrac{\sqrt{(r-a)^2+r^2-2r(r-a)\cos\theta}}{\sqrt{h^2+(r-a)^2+r^2-2r(r-a)\cos\theta}}$ and $\sin\alpha = \dfrac{(r-a)\sin\theta}{\sqrt{(r-a)^2+r^2-2r(r-a)\cos\theta}}$, where r is the radius of circular track, a represents the horizontal distance of the LCE fiber on the plane at the initial location, and h denotes the height of the LCE fiber in the z direction.

The tensile force in the LCE fiber is assumed to be directly proportional to the elastic strain and can be formulated as follows [85,86]:

$$F_L = KL_0\varepsilon_e(t) \quad (3)$$

where K signifies the elastic coefficient in the LCE fiber, and $\varepsilon_e(t)$ designates the elastic strain present within the LCE fiber. To simplify analysis, the elastic strain $\varepsilon_e(t)$ in the case of small deformations can be approximated as a linear sum of the total strain $\varepsilon_{tot}(t)$ and the strain due to light-activated contraction $\varepsilon_L(t)$, namely, $\varepsilon_{tot}(t) = \varepsilon_e(t) + \varepsilon_L(t)$. Hence, the expression for the tension of the LCE fiber in Equation (3) can be rewritten as follows:

$$F_L = KL_0(\varepsilon_{tot}(t) - \varepsilon_L(t)) \quad (4)$$

For simplicity, we define the total strain $\varepsilon_{tot}(t)$ as the change in length from the original length L_0, expressed as $\varepsilon_{tot}(t) = \frac{L-L_0}{L_0}$. Consequently, the tension F_L in Equation (4) can be reformulated as follows:

$$F_L = K(L - L_0(1 + \varepsilon_L(t))) \tag{5}$$

where L is the instantaneous length of the LCE fiber, which can be mathematically expressed as $L = \sqrt{h^2 + (r-a)^2 + r^2 - 2r(r-a)\cos\theta}$ using the cosine theorem of a triangle.

To simplify analysis, under the condition of low velocity, the damping force is typically modeled as a quadratic function, always acting in the opposite direction of the motion.

$$F_D = \beta_1 \dot\theta r + \beta_2 \left(\dot\theta r\right)^2 \tag{6}$$

where $\dot\theta$ denotes the angular velocity of the slider within a horizontal plane, β_1 represents the first damping coefficient, and β_2 signifies the second damping coefficient.

After integrating both Equations (5) and (6) into Equations (1) and (2), we can deduce the corresponding Equation (7) for the range of θ spanning from 0 to π and Equation (8) for the range between θ π and 2π.

$$mr^2\ddot\theta = -Kr\left(1 - \frac{L_0(1+\varepsilon_L(t))}{\sqrt{h^2+(r-a)^2+r^2-2r(r-a)\cos\theta}}\right)(r-a)\sin\theta - \beta_1\dot\theta r^2 - \beta_2\left(\dot\theta\right)^2 r^3 \tag{7}$$

$$mr^2\ddot\theta = Kr\left(1 - \frac{L_0(1+\varepsilon_L(t))}{\sqrt{h^2+(r-a)^2+r^2-2r(r-a)\cos(2\pi-\theta)}}\right)(r-a)\sin(2\pi-\theta) - \beta_1\dot\theta r^2 - \beta_2\left(\dot\theta\right)^2 r^3 \tag{8}$$

2.2. Dynamic Model of LCE

This section primarily focuses on describing the dynamic characteristics of the contraction strain induced by light in the LCE fibers. To simplify analysis, the light-induced contraction strain in the LCE fibers under small-scale deformation is considered to be directly correlated with the numerical fraction $\varphi(t)$ of *cis-isomer* within the LCE fibers, i.e.,

$$\varepsilon_L(t) = -C\varphi(t) \tag{9}$$

where C is the coefficient that characterizes the contraction of the LCE fiber.

Yu et al. [87] discovered that LCE fibers integrated with azobenzene moieties absorb UV light around 360 nm, enabling repeatable deformation without fatigue. Upon light exposure, molecular rearrangement leads to *trans–cis* isomerization and contraction. Azobenzene moieties convert light energy to a mechanical force, allowing for optical-to-mechanical coupling. When the LCE fiber is not illuminated, $\varphi(t)$ remains zero, resulting in no contraction strain. However, upon light exposure, the *cis-isomer* $\varphi(t)$ increases, triggering unidirectional contraction. This highlights that the *cis-isomer* fraction in the LCE fiber determines the degree of contraction strain under light stimulus.

Given the negligible effect of strain on the LCE's *cis-trans* isomerization, we disregard it. The findings [88,89] shows that the fraction of *cis-isomers* is influenced by thermal excitation, thermally driven relaxation, and light-responsive isomerization. However, the thermal excitation's impact is minor, so we omit it. Consequently, the governing equation for the *cis-isomer* fraction is simplified as follows:

$$\frac{\partial\varphi(t)}{\partial t} = \eta_0 I(1-\varphi(t)) - \frac{\varphi(t)}{T_0} \tag{10}$$

where η_0 denotes the light absorption constant, T_0 refers to the thermally driven relaxation time from the *cis* to the *trans* state, and I signifies the light intensity. By solving Equation (10), we can obtain the number fraction of the *cis*-isomer:

$$\varphi(t) = \frac{\eta_0 T_0 I}{\eta_0 T_0 I + 1} + \left(\varphi_0 - \frac{\eta_0 T_0 I}{\eta_0 T_0 I + 1}\right) exp\left[-\frac{t}{T_0}(\eta_0 T_0 I + 1)\right] \quad (11)$$

where φ_0 represents the initial number fraction of *cis* photochromic molecules in the non-illuminated zone. For simplicity, we assume that φ_0 initially takes a value of zero upon entering the illuminated zone; thus Equation, (11) can be simplified as follows:

$$\varphi(t) = \frac{\eta_0 T_0 I}{\eta_0 T_0 I + 1}\left\{1 - exp\left[-\frac{t}{T_0}(\eta_0 T_0 I + 1)\right]\right\} \quad (12)$$

In the non-illuminated zone, by setting the value of I to zero, the *cis* number fraction of photosensitive molecules can be obtained as follows:

$$\varphi(t) = \varphi_0 exp\left(-\frac{t}{T_0}\right) \quad (13)$$

In Equation (12), at the initial time $t = 0$, the maximum possible value of φ_0 is denoted as $\varphi_{0max} = \frac{\eta_0 T_0 I}{\eta_0 T_0 I + 1}$. Substituting this value into Equation (12) yields the following:

$$\varphi(t) = \frac{\eta_0 T_0 I}{\eta_0 T_0 I + 1} exp\left(-\frac{t}{T_0}\right) \quad (14)$$

2.3. Nondimensionalization

It is evident that the numerical calculations in this study involve multiple parameters. To reveal the characteristic properties of the system and simplify the equations, the following dimensionless parameters are introduced: $\bar{\dot{\theta}} = \dot{\theta} T_0$, $\bar{\ddot{\theta}} = \ddot{\theta} T_0^2$, $\bar{t} = t/T_0$, $\bar{K} = K T_0^2/m$, $\bar{I} = \eta_0 T_0 I_0$, $\bar{\beta_1} = \beta_1 T_0/m$, $\bar{\beta_2} = \beta_2 r/m$, $\bar{\varphi}(t) = \varphi(t)\frac{\eta_0 T_0 I + 1}{\eta_0 T_0 I}$. Substituting the dimensionless parameters into Equations (7) and (8), respectively, we can obtain Equations (15) and (16) in a dimensionless form:

$$\bar{\ddot{\theta}} = -\bar{K}\left(1 - \frac{L_0}{r}\frac{1 + \varepsilon_L(t)}{\sqrt{1 + \frac{h^2}{r^2} + \left(1 - \frac{a}{r}\right)^2 - 2\left(1 - \frac{a}{r}\right)\cos\theta}}\right)\left(1 - \frac{a}{r}\right)\sin\theta - \bar{\beta_1}\bar{\dot{\theta}} - \bar{\beta_2}\left(\bar{\dot{\theta}}\right)^2 \quad (15)$$

$$\bar{\ddot{\theta}} = \bar{K}\left(1 - \frac{L_0}{r}\frac{1 + \varepsilon_L(t)}{\sqrt{1 + \frac{h^2}{r^2} + \left(1 - \frac{a}{r}\right)^2 - 2\left(1 - \frac{a}{r}\right)\cos(2\pi - \theta)}}\right)\left(1 - \frac{a}{r}\right)\sin(2\pi - \theta) - \bar{\beta_1}\bar{\dot{\theta}} - \bar{\beta_2}\left(\bar{\dot{\theta}}\right)^2 \quad (16)$$

When entering the illuminated zone, Equation (12) can be simplified as follows:

$$\bar{\varphi}(t) = 1 - exp\left[-(1 + \bar{I})\bar{t}\right] \quad (17)$$

When exiting the illuminated zone, Equation (14) can be simplified as follows:

$$\bar{\varphi}(t) = exp(-\bar{t}) \quad (18)$$

Simultaneously, the horizontal tangential component F_{LT} of the tension of the LCE fiber, as defined in Equation (5), and the damping force, as stated in Equation (6), can be expressed in a dimensionless form as follows:

When θ ranges from 0 to π, we can derive Equation (19):

$$\overline{F_{L\tau}} = -\overline{K}\left(1 - \frac{L_0}{r}\frac{1+\varepsilon_L(t)}{\sqrt{1+\frac{h^2}{r^2}+\left(1-\frac{a}{r}\right)^2-2\left(1-\frac{a}{r}\right)\cos\theta}}\right)\left(1-\frac{a}{r}\right)\sin\theta \quad (19)$$

When θ ranges from π to 2π, we can derive Equation (20):

$$\overline{F_{L\tau}} = \overline{K}\left(1 - \frac{L_0}{r}\frac{1+\varepsilon_L(t)}{\sqrt{1+\frac{h^2}{r^2}+\left(1-\frac{a}{r}\right)^2-2\left(1-\frac{a}{r}\right)\cos(2\pi-\theta)}}\right)\left(1-\frac{a}{r}\right)\sin(2\pi-\theta) \quad (20)$$

$$\overline{F_D} = \overline{\beta_1}\dot{\overline{\theta}} + \overline{\beta_2}\left(\dot{\overline{\theta}}\right)^2 \quad (21)$$

As observed from Equations (15) and (16), both the equations are second-order nonlinear differential equations, which makes it impossible to find precise solutions. Consequently, aiming for precision, we choose the fourth-order Runge–Kuttamethod to iteratively solve the nonlinear high-order ordinary differential equations, with MATLAB R2021a software facilitating numerical computations and analyses. By adjusting the relevant parameters in Equations (15) and (16), including the mean values of $\overline{I}$, $\overline{K}$, C, $\overline{\beta_1}$, and $\overline{\beta_2}$, we can attain the self-rotation of the system. At the same time, we can obtain the tensile force, the damping force, the contraction strain, the angular velocity, and the position of the LCE fiber light–mechanical coupling system under instantaneous conditions.

3. Two Dynamic States and Mechanism of Self-Rotation

In this section, utilizing the control equations outlined in Section 2, we analyze the dynamic response of the light-fueled self-rotating system when it is subjected to constant illumination. Initially, we present two characteristic dynamic modes of the static state and the self-rotating state. Following this, we describe the underlying mechanisms that enable self-rotation.

3.1. Two Dynamic States

Before investigating the self-rotating dynamic behavior and photoresponsive characteristics of the system, it is necessary to obtain the range of actual typical values for the dimensionless parameters. Based on the existing experimental verifications and research results [90,91], the specific property parameter values of the materials and structure are presented in Table 1. The corresponding dimensionless parameter values required in this study are shown in Table 2.

The time–history graph and phase trajectory plot for the system are attainable through the numerical solution of Equations (15) and (16), presented in Figure 2. The findings reveal the existence of two characteristic dynamic states of the system, namely, the static state and the self-rotating state, during constant exposure to light of $\overline{I} = 0.2$ and $\overline{I} = 0.8$. During the numerical simulation, we establish the following dimensionless variables for the system: $C = 0.3$, $\overline{K} = 1.0$, $\overline{v_0} = 1.3$, $\overline{\beta_1} = 0.015$, $\overline{\beta_2} = 0.005$, $\overline{r} = 1.5$, $\overline{L_0} = 5$, $\overline{a} = 0.5$, $\theta_0 = \pi \sim 2\pi$. When $\overline{I} = 0.2$, initially, the mass block rotates counterclockwise for two revolutions. Subsequently, it begins to rotate clockwise and counterclockwise in an alternating manner, with the rotating angle and angular velocity gradually decreasing, and ultimately settling at zero as a result of the damping force, indicating that it has reached a static state. Time–history curves of the vibrational response during this process are depicted in Figure 2a,b. The corresponding phase trajectory plot in Figure 2c shows that the motion trajectory eventually stabilizes at a single point. When $\overline{I} = 0.8$, the angular velocity of the slider gradually stabilizes, indicating that the system has entered a self-rotating state, as shown in Figure 2d,e. Eventually, the maintenance of a limit cycle, resembling the phase trajectory in Figure 2f, exemplifies a periodically stable operational mode.

Table 1. Material properties and geometric parameters.

Parameter	Definition	Value	Unit
I	light intensity	0–80	kW/m^2
C	contraction coefficient of LCE fiber	0–0.4	/
K	elastic coefficient of LCE fiber	20–40	N/m
T_0	*Cis* to *trans* thermal relaxation time	0.02–0.45	s
η_0	light absorption constant	0.002	m^2/(s·W)
m	mass of the slider	0–0.02	kg
β_1	the first damping coefficient	0–0.3	kg/s
β_2	the second damping coefficient	0–0.15	kg/m
v_0	initial tangential velocity	0–5	m/s
θ_0	range of illuminated zone	0–2π	rad
r	radius of circular track	0.01–5	m
a	horizontal projection distance of LCE string	0.01–2.5	m
L_0	original length of LCE fiber	0.1–5	m

Table 2. Dimensionless parameters.

Parameter	$\bar{I}$	C	$\bar{K}$	$\bar{v_0}$	θ_0	$\bar{\beta_1}$	$\bar{\beta_2}$
Value	0–1	0–0.4	0–10	0–3	π–2π	0–0.2	0–0.1

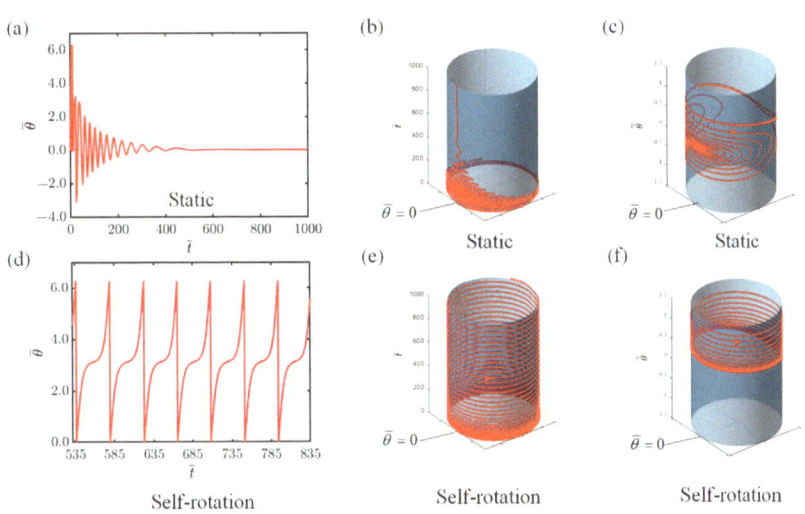

Figure 2. The two characteristic dynamic states of the system during constant exposure to light: the static state and the self-rotating state. (**a,b**) Time–history graph of angular displacement when $\bar{I} = 0.2$. (**c**) Phase trajectory plot when $\bar{I} = 0.2$. (**d,e**) Time–history graph of angular displacement when $\bar{I} = 0.8$. (**f**) Phase trajectory plot when $\bar{I} = 0.8$.

3.2. Mechanism of Self-Rotation

In the investigation of the self-rotating mechanism, we particularly focus on how the system counters energy loss stemming from the damping forces. To further clarify this intricate dynamic, we utilize the visual aid of relationship curves to highlight the intricate connections between the critical variables that contribute to the self-rotating process, as visualized in Figure 3. For the purpose of analysis, we choose the following dimensionless variables of $\bar{I} = 0.8$, $C = 0.3$, $\bar{K} = 1.0$, $\bar{v_0} = 1.3$, $\bar{\beta_1} = 0.015$, $\bar{\beta_2} = 0.005$, $\bar{r} = 1.5$, $\bar{L_0} = 5$,

$\bar{a} = 0.5$, with θ_0 ranging from π to 2π. In Figure 3a, we observe the changes in the rotating angle of the system as time progresses. The illuminated region, highlighted in yellow, indicates where the LCE fiber absorbs light. It is noticeable that the self-rotating system exhibits a consistent pattern, with the slider rotating repeatedly between the illuminated and non-illuminated sections. In Figure 3b, the fluctuation of the LCE fiber's number fraction over time is revealed in relation to light exposure. When the rotating angle of the mass exceeds π, the LCE fiber comes into the illuminated areas, triggering a gradual rise in its number fraction towards a defined maximum. However, as the slider shifts from the illuminated to the non-illuminated regions, the LCE fiber's number fraction drops sharply to zero. This recurring pattern of the system's traversal between the illuminated and non-illuminated zones results in the periodic variations observed in the LCE fiber's number fraction.

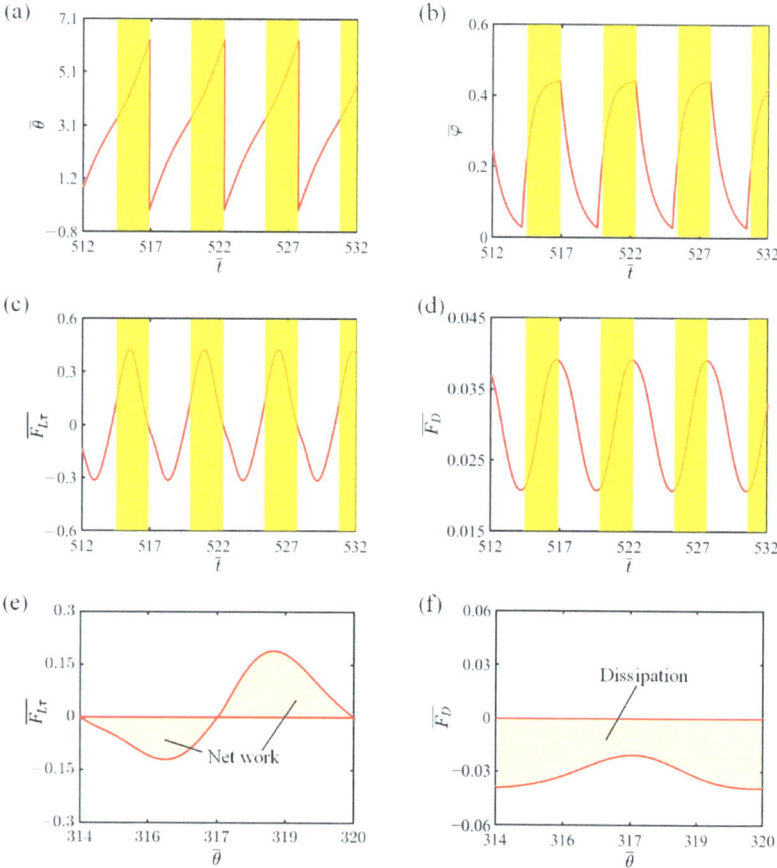

Figure 3. The self-rotating mechanism of the system. (**a**) The variation in rotating angle with time. (**b**) The variation in the number fraction of *cis-isomers* in the LCE fiber with time. (**c**) A time–history curve of horizontal tangential tension of LCE fiber. (**d**) A time–history curve of damping force. (**e**) Rotating angle-dependent horizontal tangential tension in the LCE fiber. (**f**) The rotating angle-dependent damping force.

Figure 3c demonstrates the temporal evolution of tension in the LCE fiber. The cyclical self-rotating motion of the system is responsible for the periodic changes in tension. As the LCE fiber moves into the illuminated areas, the increased number fraction of the LCE fiber leads to a corresponding rise in contraction strain, accompanied by an augmentation in elastic strain. This ultimately results in an increment in tension within the LCE fiber. Conversely, when the system exits the illuminated regions, the tension decreases due to the reversal of light-induced contraction. As illustrated in Figure 3c, the variation in horizontal tangential tension of the LCE fiber is consistent with the theoretical framework presented in this study, particularly Equation (5). Figure 3d shows a time–history curve of the damping force, which also follows a period cycle. In the non-illuminated region, the damping force decreases, while in the illuminated region, it increases. This is due to the fact that the damping force is directly proportional to velocity, and as depicted in Figure 2f, the velocity initially decreases before subsequently increasing over time.

To gain a deeper understanding of the system's energy absorption and compensation mechanism, we chart the dependency of horizontal tangential tension on a rotating angle, as depicted in Figure 3e, and we also represent the relationship of damping force with the rotating angle in Figure 3f. The hysteresis loop in Figure 3e shows LCE fiber's net work (0.913) in a rotating cycle, balancing the absorbed energy from light-responsive contraction and released energy during recovery. The loop in Figure 3f quantifies the damping force's energy consumption (also 0.913). These two balance, indicating the damping force losses are precisely compensated by the LCE fiber's energy differences. This demonstrates that the LCE fiber-propelled slider system maintains its periodic rotation effectively.

4. Parameter Study

In the previous section, we analyze the dynamic behavior of the slider propelled by the light-fueled LCE fiber based on Equations (15)–(21) and the following dimensionless physical parameters: $\bar{I}, C, \bar{K}, \bar{v_0}, \bar{\beta_1}, \bar{\beta_2}, \bar{r}, \bar{L_0}, \bar{a}$, and θ_0. In this section, under the condition that $\bar{L_0} = 5$, $\bar{a} = 0.5$, and $\bar{\beta_2} = 0.005$, and with the illumination region remaining stable within the range θ_0 of π to 2π, we proceed to conduct quantitative analysis on the dynamic impact of each of the six major dimensionless parameters, i.e., $\bar{I}, C, \bar{K}, \bar{v_0}, \bar{\beta_1}$, and $\bar{r}$, specifically focusing on how they affect the self-rotating frequency, denoted as f.

4.1. Effect of Light Intensity

Given the specified dimensionless variables, $C = 0.3$, $\bar{K} = 1.0$, $\bar{v_0} = 1.3$, $\bar{\beta_1} = 0.015$, and $\bar{r} = 1.5$, Figure 4 illustrates how the intensity of light affects the self-rotating mechanism of the slider propelled by the light-fueled LCE fiber. As shown in Figure 4a, there is direct proportionality between the light intensity and its impact on frequency, indicating that as the intensity of light rises, the frequency also increases. This is due to the fact that higher light intensities empower the LCE fiber to absorb a larger quantity of energy and convert it into kinetic energy, which enables the system to cycle through a full revolution more quickly. As evident from Figure 4a, the key intensity of light that divides the static state and the self-rotating state is $\bar{I} = 0.25$. Below this intensity of 0.25, the LCE fiber fails to absorb enough light energy to counter damping dissipation, leading to the transition into a static state due to its inability to maintain motion. Conversely, when the light intensity surpasses 0.25, the LCE fiber absorbs sufficient energy to overcome damping dissipation, enabling it to sustain a continuous and stable self-rotation, which defines the self-rotating state. In Figure 4b, the respective limit cycles for self-rotation are exhibited for various $\bar{I}$ values, including 0.3, 0.8, and 1.3. It is evident that as the light intensity rises at any given point on the circular ring, the velocity of the slider's rotation increases significantly. This observation strongly suggests that boosting the light intensity plays a pivotal role in improving the energy utilization efficiency of the LCE fiber-propelled slider system.

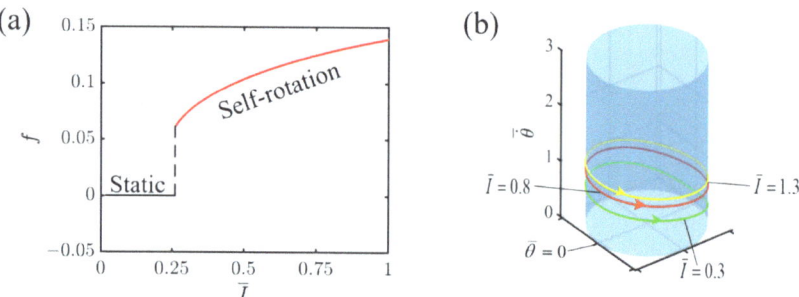

Figure 4. Effect of light intensity on self-rotating frequency. (**a**) Frequency variations with light intensities. (**b**) Depictions of limit cycles at $\bar{I} = 0.3$, 0.8, and 1.3.

4.2. Effect of Contraction Coefficient of LCE

Given the specified dimensionless variables, $\bar{I} = 0.8$, $\bar{K} = 1.0$, $\bar{v_0} = 1.3$, $\bar{\beta_1} = 0.015$, $\bar{r} = 1.5$, Figure 5 illustrates how the contraction coefficient of the LCE affects the self-rotating mechanism of the slider propelled by the light-fueled LCE fiber. As depicted in Figure 5a, there is a clear limiting value for the contraction coefficient, mathematically identified as 0.12, marking a critical point for initiating self-rotation. Below this value of 0.12, the slider remains stationary. Nonetheless, upon exceeding 0.12, the system transitions into a state of self-rotation. Moreover, there is a tendency for the frequency to rise as C increases, which stems from the decrease in the LCE fiber's capacity to absorb light, triggered by a reduction in the contraction coefficient, ultimately causing a decrease in the kinetic energy and frequency of the system. In Figure 5b, for various C values, including 0.2, 0.3, and 0.4, the corresponding limit cycles for self-rotation are displayed. In addition, at any fixed position on the circular ring, the increase in the contraction coefficient is accompanied by a marked augmentation in the slider's rotational velocity. The observation indicates that augmenting the contraction coefficient of an LCE fiber can enhance the efficient transformation of light energy into mechanical energy.

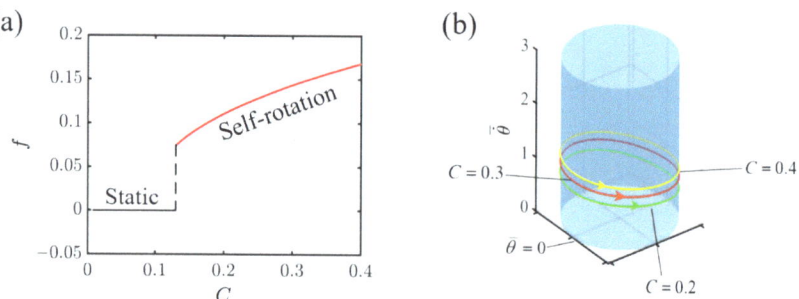

Figure 5. Effect of contraction coefficient on self-rotating frequency. (**a**) Frequency variations with contraction coefficient. (**b**) Depictions of limit cycles at $C = 0.2, 0.3,$ and 0.4.

4.3. Effect of Elastic Coefficient of LCE

Given the specified dimensionless variables, $\bar{I} = 0.8$, $C = 0.3$, $\bar{v_0} = 1.3$, $\bar{\beta_1} = 0.015$, $\bar{r} = 1.5$, Figure 6 illustrates how the elastic coefficient of the LCE affects the self-rotating mechanism of the slider propelled by the light-fueled LCE fiber. As illustrated in Figure 6a, the elastic coefficient serves as a crucial factor in determining the frequency of self-rotation. As the elastic coefficient rises, so does the frequency of self-rotation. This is attributed to the fact that a higher elastic coefficient yields a stronger elastic force from the LCE fiber. Consequently, the system accumulates more elastic potential energy, which is then converted into kinetic energy, ultimately resulting in a greater frequency of self-rotation. As seen in Figure 6a, an elastic coefficient of 0.42 acts as the vital value between the static and self-rotating modes for the system. Under continuous illumination, if the elastic coefficient falls below 0.42, the LCE fiber cannot harvest enough light energy to overcome the damping force, resulting in a static mode. Conversely, when the coefficient exceeds 0.42, the LCE fiber accumulates sufficient energy to counter the damping force and maintain continuous self-rotation. Figure 6b shows the respective limit cycles of self-rotation corresponding to the elastic coefficients of $\overline{K} = 0.5$, 1.0, and 1.5. Notably, when considering a specific point on the circular track, an increase in the elastic coefficient $\overline{K}$ leads to a corresponding acceleration in the slider's velocity, thereby enhancing the frequency. Therefore, when designing an LCE propelling system, selecting the appropriate elastic coefficient is crucial to achieving a superior performance.

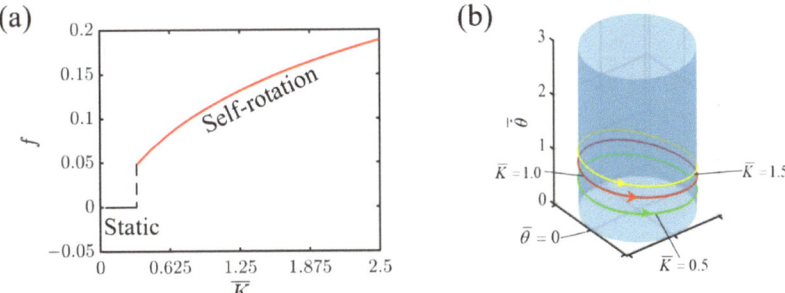

Figure 6. Effect of elastic coefficient on self-rotating frequency. (**a**) Frequency variations with elastic coefficient. (**b**) Depictions of limit cycles at $\overline{K} = 0.5$, 1.0, and 1.5.

4.4. Effect of Initial Tangential Velocity

Given the specified dimensionless variables, $\bar{I} = 0.8$, $C = 0.3$, $\overline{K} = 1.0$, $\overline{\beta_1} = 0.015$, $\bar{r} = 1.5$, Figure 7 illustrates how the initial tangential velocity affects the self-rotating mechanism of the slider propelled by the light-fueled LCE fiber.

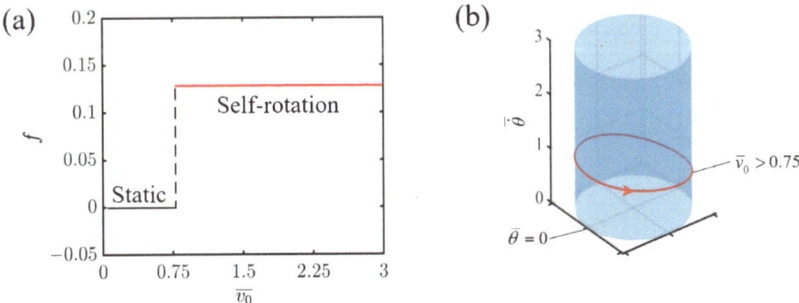

Figure 7. Effect of initial tangential velocity on self-rotating frequency. (**a**) Frequency variations with initial tangential velocity. (**b**) Depictions of limit cycles at $\overline{v_0} = 1.1$, 1.3, and 1.5.

Figure 7a depicts the relationship between the frequency of self-rotation and the initial tangential velocity. It is clearly shown that under constant illumination, the initial tangential velocity does not affect the system's frequency. This is because the frequency of self-rotation is primarily determined by the interaction between the energy dissipated by the damping force and the net work generated by the light-fueled LCEs. These internal dynamics, together with the material properties, constitute the inherent characteristics of the system. It can be seen that when the initial tangential velocity is less than 0.75, the system attains a static state. This is attributed to the fact that with such a low initial tangential velocity, the LCE fiber fails to enter the illumination zone, thus preventing it from capturing sufficient light energy to sustain its dynamic movement. Conversely, when the initial tangential velocity surpasses 0.75, specifically at $\overline{v_0} = 1.1, 1.3$, and 1.5, the system transitions into a self-rotating state. Furthermore, the corresponding limit cycle remains the same for $\overline{v_0} = 1.1, 1.3,$ and 1.5, as depicted in Figure 7b. The results show that when designing an LCE propelling self-rotating system, the initial velocity has little impact on the system performance as long as it can trigger self-rotation.

4.5. Effect of the First Damping Coefficient

Given the specified dimensionless variables, $\overline{I} = 0.8$, $C = 0.3$, $\overline{K} = 1.0$, $\overline{v_0} = 1.3$, $\overline{r} = 1.5$, Figure 8 illustrates how the first damping coefficient affects the self-rotating mechanism of the slider propelled by the light-fueled LCE fiber. As can be observed from Figure 8a, with the increase in the first damping coefficient, the system frequency gradually decreases. When the damping coefficient exceeds the critical value of 0.04, the system changes from a self-rotating state to a static state. The reason for this is that as the damping coefficient increases, the dissipative energy generated by the damping force also increases. When the slider propelled by the light-fueled LCE fiber enters the illuminated area, the energy collected becomes insufficient to overcome the increased dissipative energy, ultimately leading the system to enter a static state. When the system is in a self-rotating state, numerical calculations are performed with different values of $\overline{\beta_1}$, specifically 0.005, 0.015, and 0.025. The results indicate that as the first damping coefficient increases, the corresponding limit cycle shifts downwards in the depiction. Conversely, for smaller damping coefficients, the limit cycle is positioned higher, as illustrated in Figure 8b. Consequently, decreasing the damping coefficient of the medium facilitates the efficient transformation of light energy into mechanical energy.

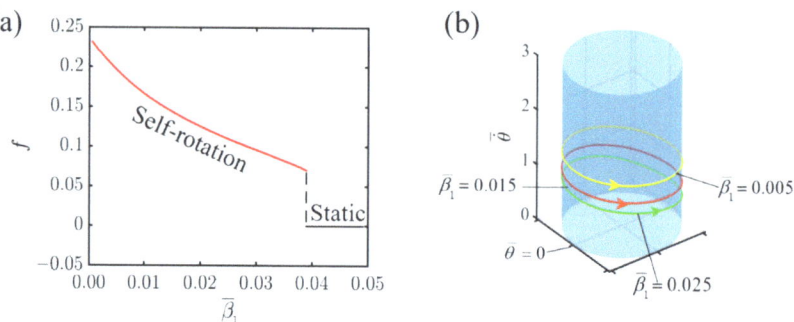

Figure 8. Effect of the first damping coefficient on self-rotating frequency. (**a**) Frequency variations with the first damping coefficient. (**b**) Depictions of limit cycles at $\overline{\beta_1} = 0.005, 0.015$, and 0.025.

4.6. Effect of Radius of Circular Track

Given the specified dimensionless variables, $\overline{I} = 0.8$, $C = 0.3$, $\overline{K} = 1.0$, $\overline{v_0} = 1.3$, $\overline{\beta_1} = 0.015$, Figure 9 illustrates how the radius of circular track affects the self-rotating mechanism of the slider propelled by the light-fueled LCE fiber. As shown in Figure 9a, as the radius increases, the system's rotational frequency also increases. Owing to the increase in radius, there is more conversion of light energy from the light-fueled LCE into mechanical energy, which leads to an increase in the system's internal kinetic energy. In the self-rotating state, a larger radius enables the system to complete more rotation cycles per unit time, thus increasing the rotational frequency. However, when the radius exceeds the threshold value of 2.0, the situation changes. At this point, as the radius continues to increase, the system's rotational frequency begins to decrease until it finally enters a static state. The reason for this is that as the radius increases, the damping force experienced by the system also increases, and these damping forces consume the energy of the system's rotation. When the dissipated energy reaches a certain level, the system may no longer be able to maintain its self-rotating state, and eventually enters a static state. Furthermore, as the radius continues to increase, it becomes increasingly challenging for the slider propelled by the light-fueled LCE fiber to enter the illuminated area. Consequently, the LCE fiber is unable to gather sufficient light energy to overcome the damping-induced energy losses, which ultimately leads to the gradual transition into a stationary state. It can also be observed from Figure 9a that the radius starts at 0.5 due to the assumption that the horizontal projected distance $\bar{a}$ of the LCE string is less than the radius. Figure 9b presents the corresponding limit cycles of self-rotation for $\bar{r} = 1.0, \bar{r} = 1.5$, and $\bar{r} = 2.0$. It is evident that in the self-rotating state, the limit cycle with a larger radius experiences faster velocity variation. This is due to the negative work completed by the tensile force of the LCE before entering the illuminated region. The larger the radius is, the longer the elastic LCE becomes, coupled with the dissipative capability caused by damping, resulting in a more significant impact on the velocity. The findings suggest that the selection of the radius of circular track is essential for enhancing the energy efficiency of the LCE fiber-propelled slider system.

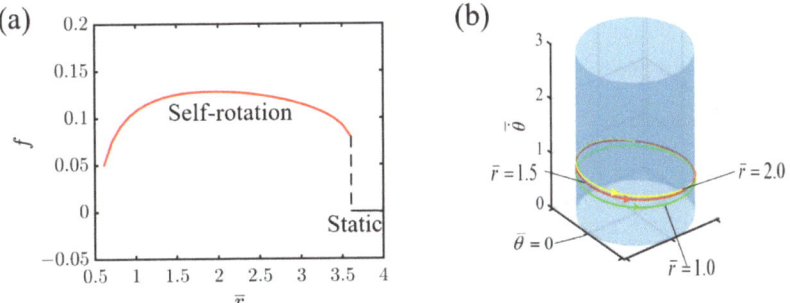

Figure 9. Effect of radius of circular track on self-rotating frequency. (**a**) Frequency variations with radius of circular track. (**b**) Depictions of limit cycles at $\bar{r} = 1.0, 1.5,$ and 2.0.

5. Conclusions

Despite the current interest in track sliders incorporating LCE fibers and their adaptability, efficiency, and sustainability in self-vibrating systems, their movement within a two-dimensional space is severely restricted. Specifically, self-rotation, a process requiring the intricate shaping deformation of the sliders and the LCE fiber-based system, poses a significant challenge. This limitation highlights the need for developing more flexible and mobile track slider designs with enhanced kinematic capabilities. To address these challenges, we present a novel light-fueled spatial system comprising an LCE fiber, a slotted slider, and a rigid track. This innovative design ensures the smooth self-rotation of the slider on a circular track under constant illumination, overcoming deformation issues during operation. Based on the dynamic mechanical model of the LCE optical fiber and the theory of momentum moment, combined with spatial analytic geometry derivations, we have derived the dimensionless dynamic control equation for the periodic self-rotating system. Utilizing the established fourth-order Runge–Kutta method and MATLAB R2021a software, we numerically solved the dynamic control equations. Our findings reveal two distinct motion states of the self-rotating slider system: the static state and the self-sliding state. Notably, we elaborate on the self-rotating process and its accompanying energy balancing mechanism. Here, the consistent external energy source compensates for the dissipation caused by system damping, thereby maintaining the dynamic equilibrium of the system.

In addition, quantitative analysis was carried out on the light intensity, the contraction coefficient, the elastic coefficient, the initial tangential velocity, the damping coefficient, and the radius of the circular track. The numerical calculation results show that the increase in the light intensity, the contraction coefficient, and the elastic coefficient lead to an increase in the self-rotating frequency. In contrast, the increase in the damping coefficient results in a significant decrease in the self-rotating frequency. Compared with the monotonic influence of the other parameters, the effect of the track radius on the frequency is non-monotonic. As the radius increases, the frequency first increases, and then decreases, and finally tends to be static. It is worth noting that the initial tangential velocity has no effect on the frequency of the system.

Although the simplicity, flexibility, and diverse motion capabilities of the proposed LCE fiber-propelled slider system promise widespread adoption, limitations persist. Notably, the small-scale deformation assumption oversimplifies the mechanical behavior, and the exclusion of viscoelastic effects in the LCE fibers, crucial under large actuation strains, introduces challenges. Viscoelasticity's time-dependent nature causes hysteresis and energy dissipation, reducing energy efficiency by hindering mechanical work extraction from photomechanical coupling. Delayed deformation recovery further threatens motion stability in dynamic environments with fluctuating illumination. To enhance the system's full potential, future research will integrate viscoelasticity into the model, enabling a deeper understanding of its impact on energy efficiency and stability. Additionally, exploring the system's dynamic behavior under variable illumination and non-circular trajectories will enhance its robustness and adaptability in complex environments.

Supplementary Materials: The following supporting information can be downloaded at: https://www.mdpi.com/article/10.3390/polym16162263/s1, Video S1: The process of self-rotation of the system.

Author Contributions: Conceptualization, K.L.; methodology, L.W.; software, Y.Q. and J.H.; validation, K.L.; investigation, Y.C. and J.H.; data curation, Y.C. and X.H.; drafting the original manuscript, Y.C. and J.H.; reviewing and revising the manuscript, L.W. and K.L.; supervision, K.L. All authors have read and agreed to the published version of the manuscript.

Funding: This research is supported by the University Natural Science Research Project of Anhui Province (Grant Nos. KJ2021A0609 and 2022AH020029), the National Natural Science Foundation of China (Grant No. 12172001), the Anhui Provincial Natural Science Foundation (No. 2208085Y01), and the Housing and Urban-Rural Development Science and Technology Project of Anhui Province (Grant No. 2023-YF036).

Institutional Review Board Statement: Not applicable.

Data Availability Statement: The original contributions presented in the study are included in the article/Supplementary Material, further inquiries can be directed to the corresponding author/s.

Conflicts of Interest: The authors declare no conflicts of interest.

References

1. Ding, W. *Self-Excited Vibration*; Tsing-Hua University Press: Beijing, China, 2009.
2. Thomson, W. *Theory of Vibration with Applications*; CRC Press: Boca Raton, FL, USA, 2018.
3. Jenkins, A. Self-oscillation. *Phys. Rep.* **2013**, *525*, 167–222. [CrossRef]
4. Zhang, Z.; Duan, N.; Lin, C.; Hua, H. Coupled dynamic analysis of a heavily-loaded propulsion shafting system with continuous bearing-shaft friction. *Int. J. Mech..Sci.* **2020**, *172*, 105431. [CrossRef]
5. Hara, Y.; Jahan, R.A. Influence of initial substrate concentration of the Belouzov-Zhabotinsky reaction on transmittance self-oscillation for a nonthermoresponsive polymer chain. *Polymers* **2011**, *3*, 330–339. [CrossRef]
6. Lu, X.; Zhang, H.; Fei, G.; Yu, B.; Tong, X.; Xia, H.; Zhao, Y. Liquid-crystalline dynamic networks doped with gold nanorods showing enhanced photocontrol of actuation. *Adv. Mater.* **2018**, *30*, 1706597. [CrossRef] [PubMed]
7. Parrany, A.M. Nonlinear light-induced vibration behavior of liquid crystal elastomer beam. *Int. J. Mech. Sci.* **2018**, *136*, 179–187. [CrossRef]
8. Li, M.H.; Keller, P.; Li, B.; Wang, X.; Brunet, M. Light-driven side-on nematic elastomer actuators. *Adv. Mater.* **2003**, *15*, 569–572. [CrossRef]
9. Yu, Y.; Zhou, L.; Du, C.; Zhu, F.; Dai, Y.; Ge, D.; Li, K. Self-galloping of a liquid crystal elastomer catenary cable under a steady temperature field. *Thin-Walled Struct.* **2024**, *202*, 112071. [CrossRef]

10. Zhou, L.; Chen, H.; Li, K. Optically-responsive liquid crystal elastomer thin film motors in linear/nonlinear optical fields. *Thin-Walled Struct.* **2024**, *202*, 112082. [CrossRef]
11. Rothemund, P.; Ainla, A.; Belding, L.; Preston, D.J.; Kurihara, S.; Suo, Z.; Whitesides, G.M. A soft, bistable valve for autonomous control of soft actuators. *Sci. Robot.* **2018**, *3*, eaar7986. [CrossRef]
12. Lv, X.; Yu, M.; Wang, W.; Yu, H. Photothermal pneumatic wheel with high loadbearing capacity. *Compos. Commun.* **2021**, *24*, 100651. [CrossRef]
13. Lendlein, A.; Jiang, H.; Jünger, O.; Langer, R. Light-induced shape-memory polymers. *Nature* **2005**, *434*, 879–882. [CrossRef] [PubMed]
14. Bubnov, A.; Domenici, V.; Hamplová, V.; Kašpar, M.; Zalar, B. First liquid single crystal elastomer containing lactic acid derivative as chiral co-monomer: Synthesis and properties. *Polymer* **2011**, *52*, 4490–4497. [CrossRef]
15. Zhou, Z.; Huang, H.; Cao, D.; Qin, W.; Zhu, P.; Du, W. Harvest more bridge vibration energy by nonlinear multi-stable piezomagnetoelastic harvester. *J. Phys. D Appl. Phys.* **2024**, *57*, 135501. [CrossRef]
16. Hauser, A.W.; Sundaram, S.; Hayward, R.C. Photothermocapillary oscillators. *Phys. Rev. Lett.* **2018**, *121*, 158001. [CrossRef] [PubMed]
17. Kim, H.; Sundaram, S.; Kang, J.-H.; Tanjeem, N.; Emrick, T.; Hayward, R.C. Coupled oscillation and spinning of photothermal particles in Marangoni optical traps. *Proc. Natl. Acad. Sci. USA* **2021**, *118*, e2024581118. [CrossRef]
18. Chatterjee, S. Self-excited oscillation under nonlinear feedback with time-delay. *J. Sound Vib.* **2011**, *330*, 1860–1876. [CrossRef]
19. Hines, L.; Petersen, K.; Lum, G.Z.; Sitti, M. Soft actuators for small-scale robotics. *Adv. Mater.* **2017**, *29*, 1603483. [CrossRef] [PubMed]
20. Sangwan, V.; Taneja, A.; Mukherjee, S. Design of a robust self-excited biped walking mechanism. *Mech. Mach. Theory* **2004**, *39*, 1385–1397. [CrossRef]
21. Qiu, Y.; Li, K. Self-rotation-eversion of an anisotropic-friction-surface torus. *Int. J. Mech. Sci.* **2024**, *281*, 109584. [CrossRef]
22. Hu, W.; Lum, G.Z.; Mastrangeli, M.; Sitti, M. Small-scale soft-bodied robot with multimodal locomotion. *Nature* **2018**, *554*, 81–85. [CrossRef]
23. Wu, H.; Ge, D.; Chen, J.; Xu, P.; Li, K. A light-fueled self-rolling unicycle with a liquid crystal elastomer rod engine. *Chaos Solitons Fractals* **2024**, *186*, 115327. [CrossRef]
24. Kageyama, Y.; Ikegami, T.; Satonaga, S.; Obara, K.; Sato, H.; Takeda, S. Light-Driven Flipping of Azobenzene Assemblies—Sparse Crystal Structures and Responsive Behaviour to Polarised Light. *Chem.–A Eur. J.* **2020**, *26*, 10759–10768. [CrossRef]
25. Chun, S.; Pang, C.; Cho, S.B. A micropillar-assisted versatile strategy for highly sensitive and efficient triboelectric energy generation under in-plane stimuli. *Adv. Mater.* **2020**, *32*, 1905539. [CrossRef]
26. Tang, R.; Liu, Z.; Xu, D.; Liu, J.; Yu, L.; Yu, H. Optical pendulum generator based on photomechanical liquid-crystalline actuators. *ACS Appl. Mater. Interfaces* **2015**, *7*, 8393–8397. [CrossRef] [PubMed]
27. Serak, S.; Tabiryan, N.; Vergara, R.; White, T.J.; Vaia, R.A.; Bunning, T.J. Liquid crystalline polymer cantilever oscillators fueled by light. *Soft Matter* **2010**, *6*, 779–783. [CrossRef]
28. Zeng, H.; Lahikainen, M.; Liu, L.; Ahmed, Z.; Wani, O.M.; Wang, M.; Yang, H.; Priimagi, A. Light-fuelled freestyle self-oscillators. *Nat. Commun.* **2019**, *10*, 5057. [CrossRef]
29. White, T.J.; Tabiryan, N.V.; Serak, S.V.; Hrozhyk, U.A.; Tondiglia, V.P.; Koerner, H.; Vaia, R.A.; Bunning, T.J. A high frequency photodriven polymer oscillator. *Soft Matter* **2008**, *4*, 1796–1798. [CrossRef]
30. Akbar, F.; Rivkin, B.; Aziz, A.; Becker, C.; Karnaushenko, D.D.; Medina-Sánchez, M.; Karnaushenko, D.; Schmidt, O.G. Self-sufficient self-oscillating microsystem driven by low power at low Reynolds numbers. *Sci. Adv.* **2021**, *7*, eabj0767. [CrossRef]
31. Yang, L.; Miao, J.; Li, G.; Ren, H.; Zhang, T.; Guo, D.; Tang, Y.; Shang, W.; Shen, Y. Soft tunable gelatin robot with insect-like claw for grasping, transportation, and delivery. *ACS Appl. Polym. Mater.* **2022**, *4*, 5431–5440. [CrossRef]
32. Wu, J.; Yao, S.; Zhang, H.; Man, W.; Bai, Z.; Zhang, F.; Wang, X.; Fang, D.; Zhang, Y. Liquid crystal elastomer metamaterials with giant biaxial thermal shrinkage for enhancing skin regeneration. *Adv. Mater.* **2021**, *33*, 2106175. [CrossRef]
33. Boissonade, J.; De Kepper, P. Multiple types of spatio-temporal oscillations induced by differential diffusion in the Landolt reaction. *Phys. Chem. Chem. Phys.* **2011**, *13*, 4132–4137. [CrossRef] [PubMed]
34. Cicconofri, G.; Damioli, V.; Noselli, G. Nonreciprocal oscillations of polyelectrolyte gel filaments subject to a steady and uniform electric field. *J. Mech. Phys. Solids* **2023**, *173*, 105225. [CrossRef]
35. Kim, Y.; van den Berg, J.; Crosby, A.J. Autonomous snapping and jumping polymer gels. *Nat. Mater.* **2021**, *20*, 1695–1701. [CrossRef]
36. Rešetič, A.; Milavec, J.; Domenici, V.; Zupančič, B.; Bubnov, A.; Zalar, B. Deuteron NMR investigation on orientational order parameter in polymer dispersed liquid crystal elastomers. *Phys. Chem. Chem. Phys.* **2020**, *22*, 23064–23072. [CrossRef] [PubMed]
37. Charroyer, L.; Chiello, O.; Sinou, J.-J. Self-excited vibrations of a non-smooth contact dynamical system with planar friction based on the shooting method. *Int. J. Mech. Sci.* **2018**, *144*, 90–101. [CrossRef]
38. Warner, M.; Terentjev, E.M. *Liquid Crystal Elastomers*; Oxford University Press: Oxford, UK, 2007; Volume 120.
39. Wang, Y.; Dang, A.; Zhang, Z.; Yin, R.; Gao, Y.; Feng, L.; Yang, S. Repeatable and reprogrammable shape morphing from photoresponsive gold nanorod/liquid crystal elastomers. *Adv. Mater.* **2020**, *32*, 2004270. [CrossRef]
40. Yang, H.; Zhang, C.; Chen, B.; Wang, Z.; Xu, Y.; Xiao, R. Bioinspired design of stimuli-responsive artificial muscles with multiple actuation modes. *Smart Mater. Struct.* **2023**, *32*, 085023. [CrossRef]

41. Bai, R.; Bhattacharya, K. Photomechanical coupling in photoactive nematic elastomers. *J. Mech. Phys. Solids* **2020**, *144*, 104115. [CrossRef]
42. Yang, L.; Chang, L.; Hu, Y.; Huang, M.; Ji, Q.; Lu, P.; Liu, J.; Chen, W.; Wu, Y. An autonomous soft actuator with light-driven self-sustained wavelike oscillation for phototactic self-locomotion and power generation. *Adv. Funct. Mater.* **2020**, *30*, 1908842. [CrossRef]
43. Zhang, J.; Guo, Y.; Hu, W.; Soon, R.H.; Davidson, Z.S.; Sitti, M. Liquid crystal elastomer-based magnetic composite films for reconfigurable shape-morphing soft miniature machines. *Adv. Mater.* **2021**, *33*, 2006191. [CrossRef]
44. Espíndola-Pérez, E.R.; Campo, J.; Sánchez-Somolinos, C. Multimodal and Multistimuli 4D-Printed Magnetic Composite Liquid Crystal Elastomer Actuators. *ACS Appl. Mater. Interfaces* **2023**, *16*, 2704–2715. [CrossRef]
45. Wang, Y.; Yin, R.; Jin, L.; Liu, M.; Gao, Y.; Raney, J.; Yang, S. 3D-Printed Photoresponsive Liquid Crystal Elastomer Composites for Free-Form Actuation. *Adv. Funct. Mater.* **2023**, *33*, 2210614. [CrossRef]
46. Ferrantini, C.; Pioner, J.M.; Martella, D.; Coppini, R.; Piroddi, N.; Paoli, P.; Calamai, M.; Pavone, F.S.; Wiersma, D.S.; Tesi, C. Development of light-responsive liquid crystalline elastomers to assist cardiac contraction. *Circ. Res.* **2019**, *124*, e44–e54. [CrossRef]
47. Chen, B.; Liu, C.; Xu, Z.; Wang, Z.; Xiao, R. Modeling the thermo-responsive behaviors of polydomain and monodomain nematic liquid crystal elastomers. *Mech. Mater.* **2024**, *188*, 104838. [CrossRef]
48. Lan, R.; Shen, W.; Yao, W.; Chen, J.; Chen, X.; Yang, H. Bioinspired humidity-responsive liquid crystalline materials: From adaptive soft actuators to visualized sensors and detectors. *Mater. Horiz.* **2023**, *10*, 2824–2844. [CrossRef] [PubMed]
49. Agrawal, A.; Chen, H.; Kim, H.; Zhu, B.; Adetiba, O.; Miranda, A.; Cristian Chipara, A.; Ajayan, P.M.; Jacot, J.G.; Verduzco, R. Electromechanically responsive liquid crystal elastomer nanocomposites for active cell culture. *ACS Macro Lett.* **2016**, *5*, 1386–1390. [CrossRef]
50. Liu, Y.; Wu, Y.; Liang, H.; Xu, H.; Wei, Y.; Ji, Y. Rewritable Electrically Controllable Liquid Crystal Actuators. *Adv. Funct. Mater.* **2023**, *33*, 2302110. [CrossRef]
51. Sun, J.; Wang, Y.; Liao, W.; Yang, Z. Ultrafast, high-contractile electrothermal-driven liquid crystal elastomer fibers towards artificial muscles. *Small* **2021**, *17*, 2103700. [CrossRef]
52. Liao, W.; Yang, Z. The integration of sensing and actuating based on a simple design fiber actuator towards intelligent soft robots. *Adv. Mater. Technol.* **2022**, *7*, 2101260. [CrossRef]
53. Wang, Y.; Liu, J.; Yang, S. Multi-functional liquid crystal elastomer composites. *Appl. Phys. Rev.* **2022**, *9*, 011301. [CrossRef]
54. Xu, T.; Pei, D.; Yu, S.; Zhang, X.; Yi, M.; Li, C. Design of MXene composites with biomimetic rapid and self-oscillating actuation under ambient circumstances. *ACS Appl. Mater. Interfaces* **2021**, *13*, 31978–31985. [CrossRef] [PubMed]
55. Manna, R.K.; Shklyaev, O.E.; Balazs, A.C. Chemical pumps and flexible sheets spontaneously form self-regulating oscillators in solution. *Proc. Natl. Acad. Sci.* **2021**, *118*, e2022987118. [CrossRef] [PubMed]
56. Bazir, A.; Baumann, A.; Ziebert, F.; Kulić, I.M. Dynamics of fiberboids. *Soft Matter* **2020**, *16*, 5210–5223. [CrossRef] [PubMed]
57. Vantomme, G.; Elands, L.C.; Gelebart, A.H.; Meijer, E.; Pogromsky, A.Y.; Nijmeijer, H.; Broer, D.J. Coupled liquid crystalline oscillators in Huygens' synchrony. *Nat. Mater.* **2021**, *20*, 1702–1706. [CrossRef] [PubMed]
58. Gelebart, A.H.; Jan Mulder, D.; Varga, M.; Konya, A.; Vantomme, G.; Meijer, E.; Selinger, R.L.; Broer, D.J. Making waves in a photoactive polymer film. *Nature* **2017**, *546*, 632–636. [CrossRef] [PubMed]
59. Shen, Q.; Trabia, S.; Stalbaum, T.; Palmre, V.; Kim, K.; Oh, I.-K. A multiple-shape memory polymer-metal composite actuator capable of programmable control, creating complex 3D motion of bending, twisting, and oscillation. *Sci. Rep.* **2016**, *6*, 24462. [CrossRef] [PubMed]
60. He, Q.; Wang, Z.; Wang, Y.; Wang, Z.; Li, C.; Annapooranan, R.; Zeng, J.; Chen, R.; Cai, S. Electrospun liquid crystal elastomer microfiber actuator. *Sci. Robot.* **2021**, *6*, eabi9704. [CrossRef] [PubMed]
61. Xu, P.; Chen, Y.; Sun, X.; Dai, Y.; Li, K. Light-powered self-sustained chaotic motion of a liquid crystal elastomer-based pendulum. *Chaos Solitons Fractals* **2024**, *184*, 115027. [CrossRef]
62. Wu, H.; Dai, Y.; Li, K.; Xu, P. Theoretical study of chaotic jumping of liquid crystal elastomer ball under periodic illumination. *Nonlinear Dyn.* **2024**, *112*, 7799–7815. [CrossRef]
63. Baumann, A.; Sánchez-Ferrer, A.; Jacomine, L.; Martinoty, P.; Le Houerou, V.; Ziebert, F.; Kulić, I.M. Motorizing fibres with geometric zero-energy modes. *Nat. Mater.* **2018**, *17*, 523–527. [CrossRef]
64. Zhao, J.; Dai, C.; Dai, Y.; Wu, J.; Li, K. Self-oscillation of cantilevered silicone oil paper sheet system driven by steam. *Thin-Walled Struct.* **2024**, *203*, 112270. [CrossRef]
65. Yu, Y.; Li, L.; Liu, E.; Han, X.; Wang, J.; Xie, Y.-X.; Lu, C. Light-driven core-shell fiber actuator based on carbon nanotubes/liquid crystal elastomer for artificial muscle and phototropic locomotion. *Carbon* **2022**, *187*, 97–107. [CrossRef]
66. Bartlett, N.W.; Tolley, M.T.; Overvelde, J.T.; Weaver, J.C.; Mosadegh, B.; Bertoldi, K.; Whitesides, G.M.; Wood, R.J. A 3D-printed, functionally graded soft robot powered by combustion. *Science* **2015**, *349*, 161–165. [CrossRef]
67. Wehner, M.; Truby, R.L.; Fitzgerald, D.J.; Mosadegh, B.; Whitesides, G.M.; Lewis, J.A.; Wood, R.J. An integrated design and fabrication strategy for entirely soft, autonomous robots. *Nature* **2016**, *536*, 451–455. [CrossRef]
68. Chen, Y.; Zhao, H.; Mao, J.; Chirarattananon, P.; Helbling, E.F.; Hyun, N.-s.P.; Clarke, D.R.; Wood, R.J. Controlled flight of a microrobot powered by soft artificial muscles. *Nature* **2019**, *575*, 324–329. [CrossRef]

69. Vantomme, G.; Gelebart, A.; Broer, D.; Meijer, E. A four-blade light-driven plastic mill based on hydrazone liquid-crystal networks. *Tetrahedron* **2017**, *73*, 4963–4967. [CrossRef]
70. Finkelmann, H.; Nishikawa, E.; Pereira, G.; Warner, M. A new opto-mechanical effect in solids. *Phys. Rev. Lett.* **2001**, *87*, 015501. [CrossRef]
71. Yu, Y.; Hu, H.; Dai, Y.; Li, K. Modeling the light-powered self-rotation of a liquid crystal elastomer fiber-based engine. *Phys. Rev. E* **2024**, *109*, 034701. [CrossRef] [PubMed]
72. Liu, J.; Shi, F.; Song, W.; Dai, Y.; Li, K. Modeling of self-oscillating flexible circuits based on liquid crystal elastomers. *Int. J. Mech. Sci.* **2024**, *270*, 109099. [CrossRef]
73. Liu, J.; Qian, G.; Dai, Y.; Yuan, Z.; Song, W.; Li, K. Nonlinear dynamics modeling of a light-powered liquid crystal elastomer-based perpetual motion machine. *Chaos Solitons Fractals* **2024**, *184*, 114957. [CrossRef]
74. Wu, H.; Zhao, C.; Dai, Y.; Li, K. Modeling of a light-fueled self-paddling boat with a liquid crystal elastomer-based motor. *Phys. Rev. E* **2024**, *109*, 044705. [CrossRef] [PubMed]
75. He, Q.; Yin, R.; Hua, Y.; Jiao, W.; Mo, C.; Shu, H.; Raney, J.R. A modular strategy for distributed, embodied control of electronics-free soft robots. *Sci. Adv.* **2023**, *9*, eade9247. [CrossRef] [PubMed]
76. Qiu, Y.; Chen, J.; Dai, Y.; Zhou, L.; Yu, Y.; Li, K. Mathematical Modeling of the Displacement of a Light-Fuel Self-Moving Automobile with an On-Board Liquid Crystal Elastomer Propulsion Device. *Mathematics* **2024**, *12*, 1322. [CrossRef]
77. Zhao, D.; Liu, Y.; Liu, C. Transverse vibration of nematic elastomer Timoshenko beams. *Phys. Rev. E* **2017**, *95*, 012703. [CrossRef] [PubMed]
78. Zhao, D.; Liu, Y. Effects of director rotation relaxation on viscoelastic wave dispersion in nematic elastomer beams. *Math. Mech. Solids* **2019**, *24*, 1103–1115. [CrossRef]
79. Jin, L.; Lin, Y.; Huo, Y. A large deflection light-induced bending model for liquid crystal elastomers under uniform or non-uniform illumination. *Int. J. Solids Struct.* **2011**, *48*, 3232–3242. [CrossRef]
80. Zhao, D.; Liu, Y. Photomechanical vibration energy harvesting based on liquid crystal elastomer cantilever. *Smart Mater. Struct.* **2019**, *28*, 075017. [CrossRef]
81. Wei, L.; Hu, J.; Wang, J.; Wu, H.; Li, K. Theoretical Analysis of Light-Actuated Self-Sliding Mass on a Circular Track Facilitated by a Liquid Crystal Elastomer Fiber. *Polymers* **2024**, *16*, 1696. [CrossRef]
82. Chartoff, R.P.; Menczel, J.D.; Dillman, S.H. Dynamic mechanical analysis (DMA). In *Thermal Analysis of Polymers: Fundamentals Applications*; John Wiley & Sons, Inc.: Hoboken, NJ, USA, 2009; pp. 387–495.
83. Menard, K.P.; Menard, N. *Dynamic Mechanical Analysis*; CRC Press: Boca Raton, FL, USA, 2020.
84. Lüdde, S.C.; Dreizler, M.R. *Theoretical Mechanics*; Springer: Berlin/Heidelberg, Germany, 2010.
85. Truesdell, C. *Essays in the History of Mechanics*; Springer Science & Business Media: Berlin/Heidelberg, Germany, 2012.
86. Dugas, R. *A History of Mechanics*; Courier Corporation: North Chelmsford, MA, USA, 1988.
87. Yu, Y.; Nakano, M.; Ikeda, T. Directed bending of a polymer film by light. *Nature* **2003**, *425*, 145. [CrossRef] [PubMed]
88. Herbert, K.M.; Fowler, H.E.; McCracken, J.M.; Schlafmann, K.R.; Koch, J.A.; White, T.J. Synthesis and alignment of liquid crystalline elastomers. *Nat. Rev. Mater.* **2022**, *7*, 23–38. [CrossRef]
89. Nägele, T.; Hoche, R.; Zinth, W.; Wachtveitl, J. Femtosecond photoisomerization of cis-azobenzene. *Chem. Phys. Lett.* **1997**, *272*, 489–495. [CrossRef]
90. Torras, N.; Zinoviev, K.; Marshall, J.; Terentjev, E.; Esteve, J. Bending kinetics of a photo-actuating nematic elastomer cantilever. *Appl. Phys. Lett.* **2011**, *99*, 254102. [CrossRef]
91. Zhao, T.; Zhang, Y.; Fan, Y.; Wang, J.; Jiang, H.; Lv, J.-A. Light-modulated liquid crystal elastomer actuator with multimodal shape morphing and multifunction. *J. Mater. Chem. C* **2022**, *10*, 3796–3803. [CrossRef]

Disclaimer/Publisher's Note: The statements, opinions and data contained in all publications are solely those of the individual author(s) and contributor(s) and not of MDPI and/or the editor(s). MDPI and/or the editor(s) disclaim responsibility for any injury to people or property resulting from any ideas, methods, instructions or products referred to in the content.

Article

Self-Vibration of Liquid Crystal Elastomer Strings under Steady Illumination

Haiyang Wu, Yuntong Dai and Kai Li *

School of Civil Engineering, Anhui Jianzhu University, Hefei 230601, China; hywu@stu.ahjzu.edu.cn (H.W.); daiytmechanics@ahjzu.edu.cn (Y.D.)
* Correspondence: kli@ahjzu.edu.cn

Abstract: Self-vibrating systems based on active materials have been widely developed, but most of the existing self-oscillating systems are complex and difficult to control. To fulfill the requirements of different functions and applications, it is necessary to construct more self-vibrating systems that are easy to control, simple in material preparation and fast in response. This paper proposes a liquid crystal elastomer (LCE) string–mass structure capable of continuous vibration under steady illumination. Based on the linear elastic model and the dynamic LCE model, the dynamic governing equations of the LCE string–mass system are established. Through numerical calculation, two regimes of the LCE string–mass system, namely the static regime and the self-vibration regime, are obtained. In addition, the light intensity, contraction coefficient and elastic coefficient of the LCE can increase the amplitude and frequency of the self-vibration, while the damping coefficient suppresses the self-oscillation. The LCE string—mass system proposed in this paper has the advantages of simple structure, easy control and customizable size, which has a wide application prospect in the fields of energy harvesting, autonomous robots, bionic instruments and medical equipment.

Keywords: self-vibration; liquid crystal elastomer; light-driven; string

Citation: Wu, H.; Dai, Y.; Li, K. Self-Vibration of Liquid Crystal Elastomer Strings under Steady Illumination. *Polymers* **2023**, *15*, 3483. https://doi.org/10.3390/polym15163483

Academic Editor: Tibor Toth-Katona

Received: 24 July 2023
Revised: 10 August 2023
Accepted: 18 August 2023
Published: 20 August 2023

Copyright: © 2023 by the authors. Licensee MDPI, Basel, Switzerland. This article is an open access article distributed under the terms and conditions of the Creative Commons Attribution (CC BY) license (https:// creativecommons.org/licenses/by/ 4.0/).

1. Introduction

Self-vibration exists widely in nature and engineering [1–7]. It is a non-attenuating vibration in which the process of vibration is accompanied by some periodically varying force by which the vibrating system can be replenished with energy to maintain the vibration. A self-vibration system usually includes vibration elements, steady energy sources and feedback mechanisms. Unlike forced vibration [8], self-vibration can independently obtain energy from the external steady environment to maintain its continuous vibration without additional periodic excitation. As a representative of nonlinear systems, self-vibration deepens the understanding of nonequilibrium dynamical processes [9,10], and also has guiding significance for constructing synchronous systems [11–13] and chaotic systems [14–16]. Self-vibration has autonomy, which is helpful to the design of autonomous components such as autonomous robots [17] and actuators [18,19]. Furthermore, self-vibration has significant application value in energy harvesting [20,21], soft robots [22,23], sensors [24], medical equipment [25,26] and other fields.

In recent years, many efforts have been made to construct self-vibration systems, among which a self-vibration system based on active materials has attracted extensive research interest. Active materials are kinds of material that can change their shapes or motion states when they are stimulated by external stimuli such as light [27,28], heat [29,30], electricity [31], magnetism [32,33] and so on. Common active materials include hydrogels [34], ionic gels [35,36], photoresponsive or thermal responsive polymers [37–43], dielectric elastomers [44], shape memory polymers [45] etc. Based on the response of active materials to external steady stimuli, people have built a variety of self-vibration modes, such as bending [46,47], swinging [48], rolling [49,50], twisting [51], vibrating [52], floating [53], buckling [54,55], jumping [56], stretching [57], shuttling [58], spinning [59] and curling [60]. In addition, some ingenious feedback mechanisms have been carefully designed, such as the self-shading mechanism [61], coupling mechanism of large deformation and chemical reaction [36], coupling mechanism of liquid volatilization and deformation [62] and photothermal surface tension gradient [63,64] to break the balance of the system, which leads to a stable and sustained response of the active material and further generates self-vibration.

Among these active materials that can be used to construct self-vibration systems, liquid crystal elastomers (LCEs) are widely considered because of their unique advantages. LCE is a unique material that is a liquid crystal polymer with a network structure formed after moderate cross-linking mesogens [65]. Its unique properties combine the characteristics of liquid crystals and elastomers, and it is capable of demonstrating an amazing ability to change shape. LCE presents a host of advantages, including substantial and reversible deformability [66,67], rapid deformation and straightforward controllability, which have garnered significant attention among researchers. Due to the rotation or phase transitions of liquid crystal monomer, liquid crystal elastomers can show reversible morphological changes when subjected to external stimuli, such as light [27,28], heat [29,30], electricity [31], magnetism [32,33] etc., displaying a variety of shapes and structures. Among these external stimuli, light stands out due to its fast response, environmental friendliness, easy accessibility, noiselessness [15,68,69] and precise control [70]. Considering the advantages of light, a rich variety of self-vibration systems based on light-driven LCE have been developed, such as bending [46,47], buckling [54,55], jumping [56], swimming [48] and other self-vibration systems. These self-vibration systems based on light-driven LCE have broad application prospects in bionic instruments [71], energy harvesting [20,21], actuators [19], soft robots [22,23] and other fields.

Although self-vibrating systems based on LCE have been widely developed, the design and construction of LCE self-vibration systems still have great limitations, such as complex structure, difficulty to control and difficulty to prepare. Therefore, it is necessary to construct more LCE self-vibration structures with simple structures and that are controllable and convenient. The tension string system has been widely studied as a classical self-vibration system. In this paper, we creatively propose a new self-vibrating system that is different from the previous self-vibration systems, which consists of two LCE strings and a mass block, and which can obtain sustained and stable vibration under steady illumination. Compared with existing self-oscillating systems [52,53], the system proposed by us has a simpler structure and is easier to implement. Our goal is to construct novel self-oscillating systems based on active materials that are simple in structure, easy to control, customizable in size and easy to prepare. Also, the effect of system parameters on self-oscillation is discussed to provide guidance for regulating this system. Depending on its excellent properties, the self-oscillating LCE string–mass system has significant application value in autonomous actuators, energy collectors, bionic instruments and other fields.

The paper is as follows. Firstly, in Section 2, based on the LCE dynamic model, a theoretical model of the LCE string system is established and the corresponding governing equations are derived. Then, in Section 3, two motion regimes of the LCE string system are described and the mechanism of self-vibration is explained in detail. In Section 4, the effects of system parameters on the amplitude and frequency of self-vibration are further discussed quantitatively. Finally, in Section 5, the results of this paper are summarized.

2. Theoretical Model and Formulation

In this section, firstly, a novelty light-driven self-vibration system consisting of two LCE strings and a mass block is described. Secondly, we derive the governing equations of the self-vibration system based on the dynamic LCE model, vibration theory and Newton's second law. Finally, the governing equation is dimensionless and the numerical calculation method is introduced.

2.1. Dynamics of Self-Vibration of LCE Strings

Figure 1 shows the physical model of the system of light-driven LCE self-vibration. The system, which can vibrate continuously and steadily under a given initial speed and designed illumination condition, consists of two LCE strings and a mass block, as shown in Figure 1a. One end of each LCE string is fixed to a horizontal rigid base, and the other end is attached to the mass block with a mass of m. The original length of each LCE string in the unstressed state is L_0. Considering that the gravity mg on the mass block is much less than the elastic force on it, the gravity is ignored. We take the initial position of the mass block as the origin of the coordinate system, with the horizontal direction as the x-axis and the vertical direction as the y-axis, as shown in Figure 1a. Since the two LCE strings are exactly the same, the tension of the two LCE strings is exactly the same, and the tension generated by the two strings in the vertical direction cancels each other, so the mass block only vibrates in the horizontal direction, and its displacement is x.

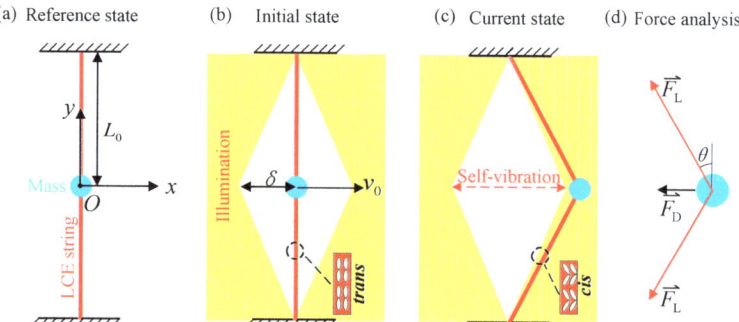

Figure 1. Diagram of self-vibration system of LCE strings: (**a**) Reference state; (**b**) Initial state; (**c**) Current state; (**d**) Force analysis. The mass block is subjected to the tension F_L of the LCE strings and the air damping force F_D. The LCE string–mass system can vibrate continuously and periodically under steady illumination.

As shown in Figure 1b,c, the yellow area represents the illumination zone and the gray rhomboid area represents the shading (non-illumination zone); the distance from the right end of the shading zone to the origin is δ. Due to the action of the initial velocity, the mass block continues to move to the right until it reaches the illuminated zone. UV light radiation can change the photochromic liquid crystal molecules in the material from straight trans configuration to bent cis configuration [65]. Thus, under continuous illumination, the chromophores (azobenzene) in the LCE fibers absorb light energy, followed by a continuous cycle of cis–trans isomerization. This process results in the transformation of the chromophores from the trans state to the cis state, thereby inducing the contraction of the LCE fibers. With the contraction and stretching of the LCE strings in the illuminated zone, the elastic potential energy of the system reaches its peak when the mass block reaches the maximum distance. The next moment, under the action of the tension of the LCE strings, the mass block moves in the opposite direction. When moving into the non-illumination zone, the light-driven contraction of the LCE strings resumes, at which point the tension of the LCE strings decreases until it reaches the illuminated zone on the other side, accumulating elastic potential energy, and then repeating the process. The LCE string–mass system can maintain continuous and stable self-vibration through the choice of proper system parameters and initial conditions.

The mass block is subjected to the tension of the LCE strings and the air damping force, as shown in Figure 1d. In the horizontal direction, the governing equation of mass block can be described as:

$$m\ddot{x}(t) = -2F_L \text{sgn}(x) \sin\theta - F_D, \tag{1}$$

where $\ddot{x}$ represents the acceleration of the mass block, F_L indicates the tension of the LCE strings, θ is the angle between the LCE strings and the y-axis, the sign function sgn(x) is a function that returns the sign of a real number x, and F_D denotes the air damping force.

It can be obtained by geometric relations that $\sin\theta = \frac{x}{\sqrt{L_0^2 + x^2}}$. where L_0 is the original length of each LCE string in the unstressed state and x indicates the horizontal displacement of mass block.

The tension of the LCE strings is proportional to its elongation, with the formula being:

$$F_L = K(\sqrt{L_0^2 + x(t)^2} - L_0 - L_0 \varepsilon_I(t)), \tag{2}$$

where K denotes the elastic coefficient of the LCE and ε_I refers to the light-driven contraction strain of the LCE strings.

The damping force is assumed to be linearly proportional to the velocity of the mass block and can be expressed as:

$$F_D = \beta \dot{x}(t), \tag{3}$$

where β represents the air damping coefficient and $\dot{x}$ denotes the velocity of the mass block.

Substituting Equations (2) and (3) into Equation (1), we can obtain:

$$m\ddot{x}(t) = -2K(\sqrt{L_0^2 + x(t)^2} - L_0 - L_0 \varepsilon_I(t)) \cdot \sin\theta - \beta \dot{x}(t). \tag{4}$$

2.2. Dynamic LCE Model

This section mainly deduces the strain equation of the LCE strings under illumination and non-illumination conditions. A linear model is adopted to describe the relationship between the cis number fraction $\phi(t)$ in LCE and the light-driven contraction of the LCE, namely:

$$\varepsilon_I = -C\phi(t), \tag{5}$$

where C indicates the contraction coefficient of the LCE.

The light-driven contraction strain of the LCE strings depends on the cis number fraction $\phi(t)$ in the LCE. UV light radiation can change the photochromic liquid crystal molecules (azobenzene) in the material from straight trans configuration to bent cis configuration, which is often accompanied with the contraction of a monodomain LCE along the mesogen aligning direction [65]. For simplicity, the LC cis–trans switching is assumed to be strain-independent. Furthermore, according to Nagele et al. [72], the cis number fraction depends on the thermal excitation from trans to cis, the thermal drive relaxation from cis to trans and the light-driven trans to cis isomerization. Assuming that the thermal excitation from trans to cis can be ignored, the governing equation for the evolution of the number fraction can be expressed as:

$$\frac{\partial \phi}{\partial t} = \eta_0 I (1 - \phi) - \frac{\phi}{T_0}, \tag{6}$$

where η_0 represents the light absorption constant, T_0 represents the thermally driven relaxation time from the cis to trans and I indicates the light intensity.

By solving Equation (6), we can get:

$$\phi(t) = \frac{\eta_0 T_0 I}{\eta_0 T_0 I + 1} + \left(\phi_0 - \frac{\eta_0 T_0 I}{\eta_0 T_0 I + 1}\right) \exp\left[-\frac{t}{T_0}(\eta_0 T_0 I + 1)\right], \tag{7}$$

where ϕ_0 denotes the initial cis number fraction in non-illumination zone.

In the illumination zone, the initial number fraction $\phi_0 = 0$, so Equation (7) can be simplified as:

$$\phi(t) = \frac{\eta_0 T_0 I}{\eta_0 T_0 I + 1}\left\{1 - \exp\left[-\frac{t}{T_0}(1 + \eta_0 T_0 I)\right]\right\}. \tag{8}$$

In the non-illumination zone, by setting the light intensity $I = 0$, we can obtain:

$$\phi(t) = \phi_0 \exp\left(-\frac{t}{T_0}\right). \tag{9}$$

In this case, ϕ_0 can be chosen as the maximum value of ϕ_0 in Equation (8) under continuous illumination. Then we can obtain:

$$\phi(t) = \frac{\eta_0 T_0 I}{\eta_0 T_0 I + 1} \exp\left(-\frac{t}{T_0}\right). \tag{10}$$

2.3. Nondimensionalization

For ease of calculation, we define the following dimensionless quantities: $\bar{x} = x/L_0$, $\bar{\dot{x}} = \dot{x}T_0/L_0$, $\bar{\ddot{x}} = \ddot{x}T_0^2/L_0$, $\bar{t} = t/T_0$, $\bar{\delta} = \delta/L_0$, $\bar{I} = IT_0\eta_0$, $\bar{\beta} = \beta T_0/m$ and $\bar{K} = KT_0^2/m$. So, the dimensionless form of governing equation can be written as:

$$\bar{\ddot{x}}(t) = -2\bar{K}(\sqrt{1+\bar{x}(t)^2} - 1 - \varepsilon_I(t))\frac{x}{\sqrt{1+\bar{x}(t)^2}} - \bar{\beta}\bar{\dot{x}}(t). \tag{11}$$

In the illuminated state, Equation (8) can be rewritten as:

$$\bar{\phi}(t) = 1 - \exp\left[-\bar{t}(\bar{I}+1)\right]. \tag{12}$$

In the non-illumination zone, we can obtain:

$$\bar{\phi}(t) = \exp(-\bar{t}). \tag{13}$$

Obviously, Equation (11) is a second-order nonlinear differential equation, and it is difficult to find the analytical solution of this kind of equation. Therefore, we use Matlab software (version R2018b) and the four-order Runge–Kutta method for numerical calculation. By adjusting the parameters within the program, for example $\bar{I}$, C, $\bar{K}$, $\bar{v}_0$, $\bar{\beta}$ and $\bar{\delta}$, we can obtain the displacement, velocity, elastic force, damping force and light-driven contraction strain of the self-vibration of the LCE string–mass system at each moment.

3. Two Motion Regimes and Mechanism of Self-Vibration

In this section, firstly, two typical motion regimes of the LCE string–mass system are described, namely the static regime and the self-vibration regime. Secondly, the corresponding mechanism of the self-vibration is elaborated in detail.

3.1. Two Motion Regimes

To study the self-vibration of the LCE string–mass system, it is necessary to calculate the typical values of the dimensionless system parameters. According to the existing experimental [72–75] and research results, the actual values of each system parameter are summarized in Table 1, and the corresponding dimensionless system parameters are listed in Table 2.

Table 1. Material properties and geometric parameters.

Parameter	Definition	Value	Unit
I	light intensity	0~100	kW/m²
C	contraction coefficient	0~0.5	/
K	elastic coefficient	1~50	N/m
T_0	thermally driven relaxation time	0.01~0.5	s
η_0	light absorption constant	0.00022	m²/s·W

Table 1. *Cont.*

Parameter	Definition	Value	Unit
m	mass	0~2	kg
β	damping coefficient	0~0.5	kg/s
v_0	initial velocity	0~0.4	m/s
δ	width of shade	0~0.1	m
L_0	original length of LCE	0.01~0.2	m

Table 2. Dimensionless parameters.

Parameter	$\bar{I}$	C	$\bar{K}$	$\bar{v}_0$	$\bar{\beta}$	$\bar{\delta}$
Value	0~1	0~0.5	0~10	0~1	0~0.2	0~0.5

Through the numerical solution of Equation (11), the time history curve vibration and phase trajectory diagram of the LCE string–mass system can be obtained, as shown in Figure 2. In this case, the other system parameters in the numerical calculation are set as $C = 0.25$, $\bar{K} = 2$, $\bar{v}_0 = 0.2$, $\bar{\beta} = 0.1$ and $\bar{\delta} = 0.2$. As can be seen from Figure 2, the system of the LCE strings has two different regimes, namely the static regime and the self-vibration regime. Figure 2a,b depict the static regime, where the vibration of the system finally stops and the corresponding phase trajectory diagram terminates at a point. In contrast, Figure 2c,d plot the self-vibration regime, in which the vibration of the system tends to stabilize after a period of time and maintains a fixed amplitude and period, and a limit cycle representing a single periodic motion appears in the corresponding phase trajectory diagram. The reason for the self-vibration phenomenon is that the system obtains enough light energy to compensate for the damping dissipation, so as to maintain its self-sustained vibration. The emergence of the phenomenon of self-vibration proves the rationality and feasibility of our constructed system. In the next section, we will elaborate on the mechanism of the self-vibration phenomenon.

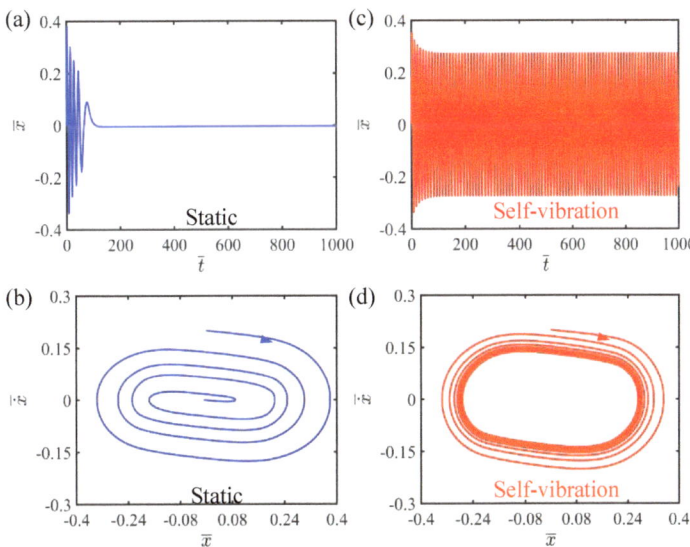

Figure 2. Two typical regimes: static regime and self-vibration regime. (**a**) Time history curve of the displacement with $\bar{I} = 0.2$; (**b**) Phase trajectory diagram with $\bar{I} = 0.2$; (**c**) Time history curve of the displacement with $\bar{I} = 0.5$; (**d**) The limit cycle in phase diagram with $\bar{I} = 0.5$. Two kinds of system regimes can be obtained with different light intensities, namely static regime and self-vibration regime.

3.2. Mechanism of Self-Vibration

This section aims to explain the mechanism of self-vibration, that is, the energy compensation mechanism of the LCE string–mass system. To better understand the energy compensation mechanism, it is necessary to plot the change curves of some key physical quantities in the process of self-vibration, as shown in Figure 3. In this case, the dimensionless parameters of the system are selected as $\bar{I} = 0.5$, $C = 0.25$, $\overline{K} = 2$, $\bar{v}_0 = 0.2$, $\bar{\beta} = 0.1$ and $\bar{\delta} = 0.2$. Figure 3a shows the curve of the mass block horizontal displacement over time, where the yellow area indicates the LCE strings are illuminated. It can be easily found that the LCE string–mass system at this time maintains a stable amplitude and period, and the mass block shuttles in the illumination zone on both sides. Figure 3b plots that when the displacement of the mass block is greater than the width of shade $\bar{\delta}$, the LCE strings are in the illumination zone, and the number fraction in the LCE gradually increases and tends to a limit value. When the displacement of the mass block is less than the width of shade $\bar{\delta}$, the LCE strings are in the non-illumination zone, and the number fraction in the LCE rapidly decreases to zero. As the mass block regularly enters and exits the illumination zone, the number fraction in the LCE strings also changes periodically. In addition, Figure 4 illustrates several characteristic snapshots for the self-vibration of the LCE string–mass system during one cycle under steady illumination.

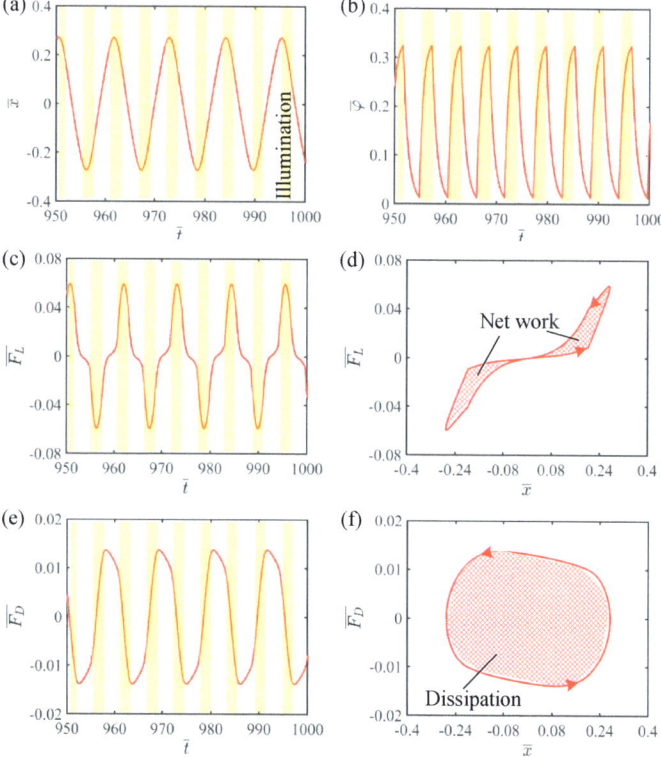

Figure 3. The mechanism of self-vibration of the LCE string–mass system. (**a**) Time history curve of the displacement. (**b**) Time history curve of the number fraction. (**c**) Variation of the tension of LCE strings with time. (**d**) Dependence of the tension of LCE strings on the displacement. (**e**) Variation of the damping force with time. (**f**) Dependence of the damping force on the displacement. The work done by the elastic force can compensate the damping dissipation, so the system can maintain stable vibration.

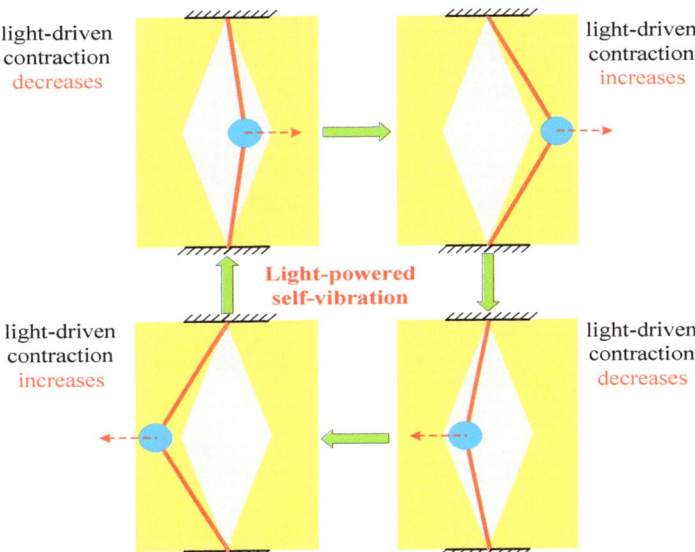

Figure 4. Snapshots of the LCE string–mass system in one cycle during the self-vibration. Under steady illumination, the system exhibits a continuous periodic self-vibration.

In order to understand the energy source and consumption of the LCE string–mass system, we plot the elastic force and damping force of the LCE with time and with displacement, as shown in Figure 3c–f. Figure 3c shows the variation of tension of the LCE strings over time. With the periodic vibration of the LCE string–mass system, the variation of tension of the LCE strings is also periodic. When the LCE strings enter the illumination zone, the tension of the LCE strings increases due to the light-driven contraction of the LCE strings. When the LCE strings leave the illumination zone, the light-driven contraction of the LCE strings recovers and the tension of the LCE strings decreases, as shown in Figure 3c. Figure 3d shows the hysteresis loop of the tension of the LCE string, the area of which represents the net work done by the tension of the LCE strings in one cycle of vibration, which is numerically calculated to be 0.018. Similarly to the tension of the LCE strings, Figure 3e plots the periodic change of damping force with time. Figure 3e shows the relationship between the damping force and displacement, and the hysteresis loop enclosed represents the work done by the damping force in one cycle of vibration, that is, the system damping dissipation. Through calculation, the area of the hysteresis loop in Figure 3f is also 0.018, which means that the energy lost by air damping during the self-vibration is compensated for by the work done by the tension of the LCE strings. Therefore, the self-vibration of the LCE string–mass system can be sustained.

4. Parametric Study

In this section, we quantitatively investigate the effects of system parameters such as light intensity, contraction coefficient, elastic coefficient, initial velocity, damping coefficient and width of shade on the amplitude A and frequency F of the self-vibration of the LCE string–mass system.

4.1. Effect of the Light Intensity

The light intensity influencing the self-vibration of the LCE string–mass system is investigated in this section. In this case, the values of the other parameters are $C = 0.25$, $\overline{K} = 2$, $\overline{v}_0 = 0.2$, $\overline{\beta} = 0.1$ and $\overline{\delta} = 0.2$. Figure 5a plots the limit cycles of self-vibration for $\overline{I} = 0.4$, $\overline{I} = 0.5$ and $\overline{I} = 0.6$. The horizontal width of the limit cycle represents the amplitude of the self-vibration, and the vertical height of the limit cycle indicates the velocity of the self-vibration. It can be seen from Figure 5a that the limit cycle is the largest for $\bar{I} = 0.6$, which indicates that the amplitude and kinetic energy of the self-vibration are largest in this case. Figure 5b shows the effect of the light intensity on amplitude and frequency. When the light intensity is below 0.396, the LCE strings cannot absorb enough light energy to offset the damping dissipation, and therefore cannot maintain continuous motion, thus entering a static state. When the light intensity is higher than 0.396, the LCE strings are able to absorb enough light energy to offset the damping dissipation and thus maintain a continuous stable vibration, i.e., the self-vibration regime. In the regime of self-vibration, the amplitude and frequency increase with the increase in light intensity. This is because the higher the light intensity, the greater the contraction and the greater the tension of the string. The greater tension is able to do more work on the system, producing more kinetic energy and thus a greater amplitude. The above results show that increasing the light intensity can make the light-driven system absorb more energy to achieve a larger amplitude, which is consistent with the current findings [3].

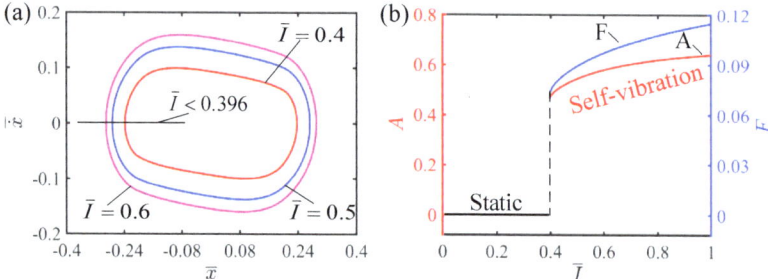

Figure 5. The effect of light intensity on the self-vibration. (**a**) Limit cycles with $\bar{I} = 0.4$, $\bar{I} = 0.5$ and $\bar{I} = 0.6$. (**b**) Variations of amplitude and frequency with different light intensities. The larger the light intensity, the larger the amplitude and frequency of self-vibration.

4.2. Effect of the Contraction Coefficient of LCE

This section presents a discussion on the effect of the contraction coefficient on the self-vibration of the LCE strings. Here, the values of the other parameters are $\bar{I} = 0.5$, $\bar{K} = 2$, $\bar{v}_0 = 0.2$, $\bar{\beta} = 0.1$ and $\bar{\delta} = 0.2$. Figure 6a plots the limit cycles of the self-vibration of the LCE string–mass system with different contraction coefficients. It can be observed from Figure 6a that the limit cycle with a larger contraction coefficient completely wraps the limit cycle with a smaller contraction coefficient, indicating that the larger the contraction coefficient, the larger the energy of the LCE string–mass system, and thus the larger the amplitude and kinetic energy. It can be seen from Figure 6b that the amplitude and frequency of the self-vibration change with the change of the contraction coefficient. When the contraction coefficient is less than 0.212, the system is in the static regime. On the contrary, when the contraction coefficient is greater than 0.212, the system is in the self-vibration regime. With the increase in the contraction coefficient, the amplitude and frequency also increase. The reason for this phenomenon is similar to the reason for the effect of light intensity on self-vibration: when the contraction coefficient is small, the LCE strings absorb insufficient light energy when they are in the illumination zone, and cannot obtain enough energy to compensate for the damping dissipation, so that the system eventually moves to the static regime. When the contraction coefficient is large, the LCE strings absorb enough energy in the illumination zone and have enough energy to compensate for the damping dissipation of the system, so as to maintain the self-vibration. As the contraction coefficient continues to increase, the light energy absorbed by the LCE strings further increases, and so does the amplitude.

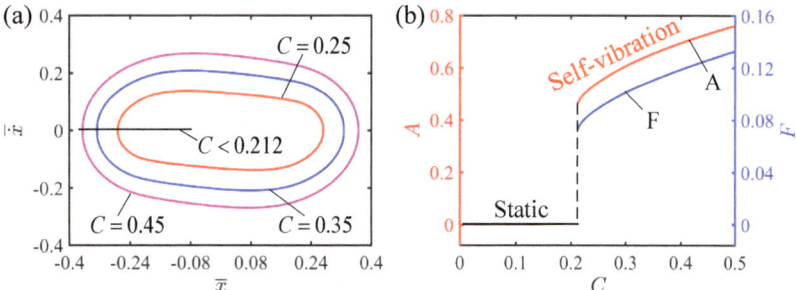

Figure 6. The effect of contraction coefficient on the self-vibration. (**a**) Limit cycles with $C = 0.25$, $C = 0.35$ and $C = 0.45$. (**b**) Variations of amplitude and frequency with different contraction coefficients. The larger contraction coefficient, the larger the amplitude and frequency of self-vibration.

4.3. Effect of the Elastic Coefficient of LCE

This section provides the influence of the elastic coefficient of the LCE strings on the self-vibration for $\bar{I} = 0.5$, $C = 0.25$, $\bar{v}_0 = 0.2$, $\bar{\beta} = 0.1$ and $\bar{\delta} = 0.2$. Figure 7a shows the limit cycle for different elastic coefficients of the LCE strings. When the elastic coefficient is less than 1.627, the phase trajectory diagram of the self-vibration is a fixed point, which indicates that the system is in the static regime. This is because when the elastic coefficient is small, the tension generated by the LCE strings in the illumination zone is small, which cannot provide enough elastic potential energy to compensate for the damping dissipation of the system, so the system finally reaches a static state. It can be seen from Figure 7b that the elastic coefficient has a significant influence on the amplitude and frequency of the self-vibration. With the increase in the elastic coefficient, the amplitude and frequency of the self-vibration increase. This is because as the elastic coefficient increases, the elastic force generated by the LCE strings increases, the elastic potential energy that the system is able to convert into kinetic energy increases, and therefore the amplitude of the self-vibration increases. Therefore, in the design of a tension system based on LCEs, it is key to select the appropriate elastic coefficient to obtain better performance.

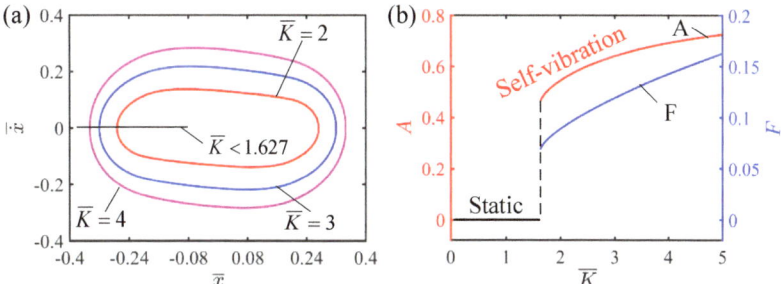

Figure 7. The effect of elastic coefficient on the self-vibration. (**a**) Limit cycles with $\bar{K} = 2$, $\bar{K} = 3$ and $\bar{K} = 4$ (**b**) Variations of amplitude and frequency with different elastic coefficients. The larger the elastic coefficient, the larger the amplitude and frequency of self-vibration.

4.4. Effect of the Initial Velocity

This section mainly focuses on the effect of initial velocity on the self-vibration of the LCE string–mass system, with parameters $\bar{I} = 0.5$, $C = 0.25$, $\bar{K} = 2$, $\bar{\beta} = 0.1$ and $\bar{\delta} = 0.2$. It can be observed from Figure 8a that the self-vibration can be successfully triggered at $\bar{v}_0 = 0.1$, $\bar{v}_0 = 0.2$ and $\bar{v}_0 = 0.3$. It is worth mentioning that the limit cycles at different velocities coincide completely. Figure 8b plots the relationship between the initial velocity and the amplitude and frequency of the self-vibration. It can be seen from Figure 8b that when the initial velocity is less than 0.066, the system is in the static regime, because the low initial velocity cannot allow the LCE strings to reach the illumination zone to absorb enough light energy, and it finally reaches the static regime. When the initial velocity is greater than 0.066, the system is in the self-vibration regime and the final amplitude and frequency are not affected. This is because the amplitude of the self-vibration depends on the energy conversion between the work done by the LCE strings and the damping dissipation, which belongs to the internal characteristics of the system, and the initial velocity does not affect the energy conversion of the system, so the amplitude does not change. Compared with other parameters, the initial velocity is more like a switch that triggers the self-vibration, which is responsible only for activating the system and does not affect the inherent characteristics of the system such as amplitude and frequency, which is in agreement with the results of existing studies [59].

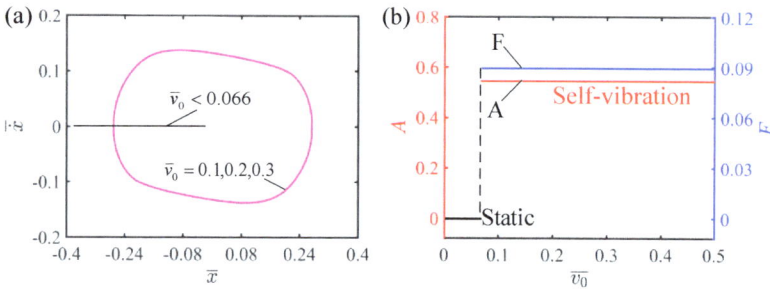

Figure 8. The effect of initial velocity on the self-vibration. (**a**) Limit cycles with $\bar{v}_0 = 0.1$, $\bar{v}_0 = 0.2$ and $\bar{v}_0 = 0.3$. (**b**) Variations of amplitude and frequency with different initial velocities. The initial velocity has no effect on the amplitude and frequency of self-vibration.

4.5. Effect of the Damping Coefficient

This section mainly studies the damping coefficient on the self-vibration of the LCE string–mass system. In the calculation, we set $\bar{I} = 0.5$, $C = 0.25$, $\bar{K} = 2$, $\bar{v}_0 = 0.2$ and $\bar{\delta} = 0.2$. The damping coefficient has a significant effect on the regime and amplitude of the system, as shown in Figure 9. Figure 9a draws the limit cycles for $\bar{\beta} = 0.06$, $\bar{\beta} = 0.08$ and $\bar{\beta} = 0.1$. It can be seen from Figure 9a that the smaller the damping coefficient is, the larger the limit cycle is. It can be seen from Figure 9b that there is a critical value between the static regime and the self-vibration regime. The system is in the static regime for $\bar{\beta} > 0.113$, while the system is in the self-vibration regime for $\bar{\beta} < 0.113$. In addition, it can be seen from Figure 9b that the smaller the damping coefficient is, the larger the amplitude and frequency of the self-vibration are. This can be explained in terms of energy compensation. The greater the damping coefficient, the greater the damping force of the system, which hinders the movement of the system and makes the LCE strings unable to reach the illuminated zone to absorb light energy. The energy of the system decreases continuously due to damping dissipation, so the system eventually reaches a static state. On the contrary, the smaller the damping coefficient, the smaller the damping dissipation of the system and the larger the converted kinetic energy, and thus the amplitude increases. Therefore, how to reduce the damping dissipation of the system through reasonable structural design is an important challenge.

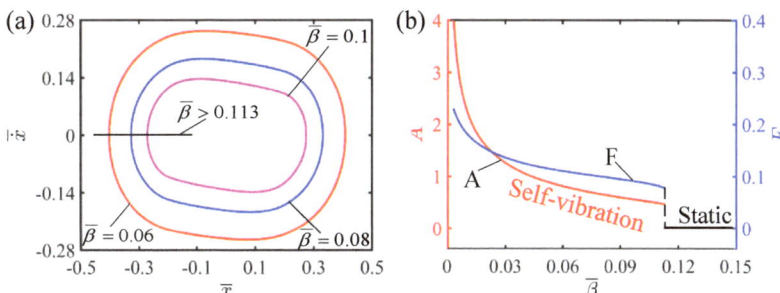

Figure 9. The effect of the damping coefficient on the self-vibration. (**a**) Limit cycles with $\bar{\beta} = 0.06$, $\bar{\beta} = 0.08$ and $\bar{\beta} = 0.1$. (**b**) Variations of amplitude and frequency with different damping coefficients. The larger the damping coefficient, the larger the amplitude and frequency of self-vibration.

4.6. Effect of the Width of Shade

The effect of the width of shade on the self-vibration is discussed in the current section. In this case, the other dimensionless parameters are selected as $\bar{l} = 0.5$, $C = 0.25$, $\bar{K} = 2$, $\bar{v}_0 = 0.2$ and $\bar{\beta} = 0.1$. It is not difficult to find that the width of shade affects the motion regime of the system. Figure 10a plots the limit cycle with different widths of shade. It can be seen from Figure 10b that there is a critical value between the static regime and the self-vibration regime. When the width of shade is greater than 0.272, the system cannot reach the illumination zone to absorb light energy, the initial kinetic energy is constantly consumed and finally the system reaches the static state. When the shadow width is less than 0.272, the system can reach the illumination to absorb light energy to compensate for the damping dissipation, so it can continue stable vibration, namely, the self-vibration regime. Figure 10b shows that, in the self-vibration regime, the amplitude of the system increases with the increase in the width of shade. This is because when the width of shade is small, the LCE strings soon enter the illumination zone, the elastic force of the LCE strings rapidly increases and inhibits the further displacement of the mass block, and thus the amplitude of the self-vibration is small. On the other hand, when the width of shade is large, there is larger displacement before the LCE strings enter the illumination zone, so that the whole amplitude of the self-vibration is larger.

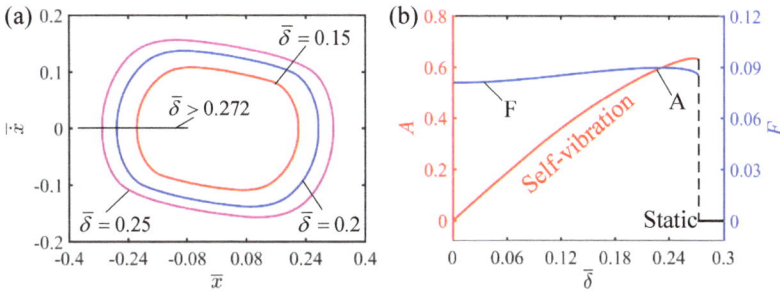

Figure 10. The effect of the width of shade on the self-vibration. (**a**) Limit cycles with $\bar{\delta} = 0.15$, $\bar{\delta} = 0.2$ and $\bar{\delta} = 0.25$. (**b**) Variations of amplitude and frequency with different widths of shade. The larger the width of shade, the larger the amplitude of self-vibration.

5. Conclusions

The self-vibration system can directly absorb energy from the steady external environment to maintain its continuous motion without external periodic stimuli, which has great application prospects in the fields of autonomous robotics, energy harvesting and bionic devices. Traditional self-vibration systems have the defects of complex structure, difficult manufacturing and poor controllability, so there is a great necessity to construct new self-vibration systems. In this paper, we construct a new self-vibration system, which consists of two LCE strings and a mass block, and it can achieve continuous and stable vibration under steady illumination. Based on the dynamic LCE model and linear elastic model, the theoretical model of self-vibration of the LCE string–mass system is established and the corresponding governing equations are derived. Based on the results of numerical simulations, two motion regimes of the LCE string–mass system, namely the static regime and self-vibration regime, are described, and the energy compensation mechanism of the self-vibration is revealed. In addition, the effects of the system parameters on the amplitude and frequency of the self-vibration are quantitatively discussed. The results show that the amplitude and frequency of the system increase with the increase in light intensity, contraction coefficient and elastic coefficient. By adjusting these coefficients, it is expected that faster, more powerful active machines can be realized. The damping coefficient inhibits the amplitude and frequency of the self-vibration, while the initial velocity does not affect the amplitude and frequency of the self-vibration regimes. Meanwhile, the values of the parameters can also determine the motion modes of the system, and there are critical values between the self-vibrating and static modes. In addition, future goals are to increase the credibility of our findings through experiments, as well as to build more active machines based on active materials and to realize their applications in fields such as energy harvesting, artificial muscles and autonomous robotics. The research in this paper deepens the understanding of self-vibration systems and helps to design new self-vibration systems. Meanwhile, the LCE string–mass system proposed in this paper has the advantages of simple structure, easy control and customizable size, and has the prospect of application in the field of autonomous robots and bionic instruments.

Author Contributions: Conceptualization, K.L.; methodology, H.W.; software, H.W.; validation, Y.D.; investigation, H.W.; data curation, H.W.; writing—original draft preparation, H.W.; writing—review and editing, Y.D. and K.L.; supervision, K.L. All authors have read and agreed to the published version of the manuscript.

Funding: This research was funded by University Natural Science Research Project of Anhui Province (Grant No. 2022AH020029), National Natural Science Foundation of China (Grant No. 12172001), and Anhui Provincial Natural Science Foundation (Grant Nos. 2008085QA23 and 2208085Y01).

Institutional Review Board Statement: Not applicable.

Data Availability Statement: The data that support the findings of this study are available upon reasonable request from the authors.

Conflicts of Interest: The authors declare no conflict of interest.

References

1. Wang, X.; Tan, C.F.; Chan, K.H.; Lu, X.; Zhu, L.; Kim, S.; Ho, G.W. In-built thermo-mechanical cooperative feedback mecha-nism for self-propelled multimodal locomotion and electricity generation. *Nat. Commun.* **2018**, *9*, 3438. [CrossRef]
2. Ge, F.; Yang, R.; Tong, X.; Camerel, F.; Zhao, Y. A multifunctional dyedoped liquid crystal polymer actuator: Light-guided transportation, turning in locomotion, and autonomous motion. *Angew. Chem. Int. Ed.* **2018**, *57*, 11758–11763. [CrossRef] [PubMed]
3. Zeng, H.; Lahikainen, M.; Liu, L.; Ahmed, Z.; Wani, O.M.; Wang, M.; Yang, H.; Priimagi, A. Light-fuelled freestyle self-oscillators. *Nat. Commun.* **2019**, *10*, 1–9. [CrossRef]
4. Li, M.-H.; Keller, P.; Li, B.; Wang, X.; Brunet, M. Light-Driven Side-On Nematic Elastomer Actuators. *Adv. Mater.* **2003**, *15*, 569–572. [CrossRef]
5. Nocentini, S.; Parmeggiani, C.; Martella, D.; Wiersma, D.S. Optically Driven Soft Micro Robotics. *Adv. Opt. Mater.* **2018**, *6*, 1800207. [CrossRef]
6. Preston, D.J.; Jiang, H.J.; Sanchez, V.; Rothemund, P.; Rawson, J.; Nemitz, M.P.; Lee, W.; Suo, Z.; Walsh, C.J.; Whitesides, G.M. A soft ring oscillator. *Sci. Robot.* **2019**, *4*, 5496. [CrossRef]
7. Ding, W. *Self-Excited Vibration*; Springer: Berlin/Heidelberg, Germany, 2010.
8. Zhao, J.; Xu, P.; Yu, Y.; Li, K. Controllable vibration of liquid crystal elastomer beams under periodic illumination. *Int. J. Mech. Sci.* **2020**, *170*, 105366. [CrossRef]

9. Fang, P.; Dai, L.; Hou, Y.; Du, M.; Luyou, W. The Study of Identification Method for Dynamic Behavior of High-Dimensional Nonlinear System. *Shock. Vib.* **2019**, *2019*, 1–9. [CrossRef]
10. Cross, M.; Greenside, H. *Pattern Formation and Dynamics in Nonequilibrium Systems*; Cambridge University Press: Cambridge, UK, 2009. [CrossRef]
11. Pivnenko, M.; Fedoryako, A.; Kutulya, L.; Seminozhenko, V. Resonance Phenomena in a Ferroelectric Liquid Crystal Near the Phase Transition SmA–SmC*. *Mol. Cryst. Liq. Cryst.* **1999**, *328*, 111–118. [CrossRef]
12. Li, K.; Zhang, B.; Cheng, Q.; Dai, Y.; Yu, Y. Light-Fueled Synchronization of Two Coupled Liquid Crystal Elastomer Self-Oscillators. *Polymers* **2023**, *15*, 2886. [CrossRef]
13. Thomson, W.T. *Theory of Vibration with Applications*, 5th ed.; Prentice-Hall: Hoboken, NJ, USA, 2005.
14. Xu, P.; Wu, H.; Dai, Y.; Li, K. Self-sustained chaotic floating of a liquid crystal elastomer balloon under steady illumination. *Heliyon* **2023**, *9*, e14447. [CrossRef] [PubMed]
15. Kumar, K.; Knie, C.; Bléger, D.; Peletier, M.A.; Friedrich, H.; Hecht, S.; Broer, D.; Debije, M.G.; Schening, A. A chaotic self-oscillating sunlight-driven polymer actuator. *Nat. Communicat.* **2016**, *7*, 11975. [CrossRef] [PubMed]
16. Grosso, M.; Keunings, R.; Crescitelli, S.; Maffettone, P.L. Prediction of Chaotic Dynamics in Sheared Liquid Crystalline Polymers. *Phys. Rev. Lett.* **2001**, *86*, 3184–3187. [CrossRef] [PubMed]
17. Kageyama, Y.; Ikegami, T.; Satonaga, S.; Obara, K.; Sato, H.; Takeda, S. Light-driven flipping of azobenzene assemblies-sparse crystal structures and responsive behavior to polarized light. *Chem. Eur. J.* **2020**, *26*, 10759–10768. [CrossRef] [PubMed]
18. Liu, Z.; Qi, M.; Zhu, Y.; Huang, D.; Zhang, X.; Lin, L.; Yan, X. Mechanical response of the isolated cantilever with a floating potential in steady electrostatic field. *Int. J. Mech. Sci.* **2019**, *161*, 105066. [CrossRef]
19. Lu, X.; Zhang, H.; Fei, G.; Yu, B.; Tong, X.; Xia, H.; Zhao, Y. Liquid-Crystalline Dynamic Networks Doped with Gold Nanorods Showing Enhanced Photocontrol of Actuation. *Adv. Mater.* **2018**, *30*, e1706597. [CrossRef] [PubMed]
20. Tang, R.; Liu, Z.; Xu, D.; Liu, J.; Yu, L.; Yu, H. Optical pendulum generator based on photomechanical liquid-crystalline actuators. *ACS Appl. Mater. Interfaces* **2015**, *7*, 8393–8397. [CrossRef]
21. Zhao, D.; Liu, Y. A prototype for light-electric harvester based on light sensitive liquid crystal elastomer cantilever. *Energy* **2020**, *198*, 117351. [CrossRef]
22. Yang, L.; Chang, L.; Hu, Y.; Huang, M.; Ji, Q.; Lu, P.; Liu, J.; Chen, W.; Wu, Y. An Autonomous Soft Actuator with Light-Driven Self-Sustained Wavelike Oscillation for Phototactic Self-Locomotion and Power Generation. *Adv. Funct. Mater.* **2020**, *30*, 1908842. [CrossRef]
23. Chun, S.; Pang, C.; Cho, S.B. A Micropillar-Assisted Versatile Strategy for Highly Sensitive and Efficient Triboelectric Energy Generation under In-Plane Stimuli. *Adv. Mater.* **2019**, *32*, e1905539. [CrossRef]
24. White, T.J.; Broer, D.J. Programmable and adaptive mechanics with liquid crystal polymer networks and elastomers. *Nat. Mater.* **2015**, *14*, 1087–1098. [CrossRef] [PubMed]
25. Charroyer, L.; Chiello, O.; Sinou, J.-J. Self-excited vibrations of a non-smooth contact dynamical system with planar friction based on the shooting method. *Int. J. Mech. Sci.* **2018**, *144*, 90–101. [CrossRef]
26. Zhang, Z.; Duan, N.; Lin, C.; Hua, H. Coupled dynamic analysis of a heavily-loaded propulsion shafting system with continuous bearing-shaft friction. *Int. J. Mech Sci.* **2020**, *172*, 105431. [CrossRef]
27. Zheng, R.; Ma, L.; Feng, W.; Pan, J.; Wang, Z.; Chen, Z.; Zhang, Y.; Li, C.; Chen, P.; Bisoyi, H.K.; et al. Autonomous Self-Sustained Liquid Crystal Actuators Enabling Active Photonic Applications. *Adv. Funct. Mater.* **2023**, 2301142, (online version). [CrossRef]
28. Ma, L.L.; Liu, C.; Wu, S.; Chen, P.; Chen, Q.; Qian, J.; Ge, S.; Wu, Y.; Hu, W.; Lu, Y. Programmable self-propelling actuators enabled by a dynamic helical medium. *Sci. Adv.* **2021**, *7*, 32. [CrossRef]
29. Guo, Y.; Liu, N.; Cao, Q.; Cheng, X.; Zhang, P.; Guan, Q.; Zheng, W.; He, G.; Chen, J. Photothermal Diol for NIR-Responsive Liquid Crystal Elastomers. *ACS Appl. Polym. Mater.* **2022**, *4*, 6202–6210. [CrossRef]
30. Cui, Y.; Yin, Y.; Wang, C.; Sim, K.; Li, Y.; Yu, C.; Song, J. Transient thermo-mechanical analysis for bimorph soft robot based on thermally responsive liquid crystal elastomers. *Appl. Math. Mech.* **2019**, *40*, 943–952. [CrossRef]
31. Na, Y.H.; Aburaya, Y.; Orihara, H.; Hiraoka, K. Measurement of electrically induced shear strain in a chiral smectic liq-uid-crystal elastomer. *Phys. Rev. E* **2011**, *83*, 061709. [CrossRef]
32. Haberl, J.M.; Sánchez-Ferrer, A.; Mihut, A.M.; Dietsch, H.; Hirt, A.M.; Mezzenga, R. Liquid-Crystalline Elastomer-Nanoparticle Hybrids with Reversible Switch of Magnetic Memory. *Adv. Mater.* **2013**, *25*, 1787–1791. [CrossRef]
33. Noorjahan; Pathak, S.; Jain, K.; Pant, R. Improved magneto-viscoelasticity of cross-linked PVA hydrogels using magnetic nanoparticles. *Colloids Surfaces A Physicochem. Eng. Asp.* **2018**, *539*, 273–279. [CrossRef]
34. Yoshida, R. Self-Oscillating Gels Driven by the Belousov-Zhabotinsky Reaction as Novel Smart Materials. *Adv. Mater.* **2010**, *22*, 3463–3483. [CrossRef] [PubMed]
35. Hua, M.; Kim, C.; Du, Y.; Wu, D.; Bai, R.; He, X. Swaying gel: Chemo-mechanical self-oscillation based on dynamic buckling. *Matter* **2021**, *4*, 1029–1041. [CrossRef]
36. Boissonade, J.; Kepper, P.D. Multiple types of spatio-temporal oscillations induced by differential diffusion in the Landolt reaction. *Phys. Chem. Chem. Phys.* **2011**, *13*, 4132–4137. [CrossRef] [PubMed]
37. Koibuchi, H. Bending of Thin Liquid Crystal Elastomer under Irradiation of Visible Light: Finsler Geometry Modeling. *Polymers* **2018**, *10*, 757. [CrossRef]

38. Camacho-Lopez, M.; Finkelmann, H.; Palffy-Muhoray, P.; Shelley, M. Fast liquid-crystal elastomer swims into the dark. *Nat. Mater.* **2004**, *3*, 307–310. [CrossRef]
39. Wang, Y.; Liu, J.; Yang, S. Multi-functional liquid crystal elastomer composites. *Appl. Phys. Rev.* **2022**, *9*, 011301. [CrossRef]
40. Bubnov, A.; Domenici, V.; Hamplová, V.; Kašpar, M.; Zalar, B. First liquid single crystal elastomer containing lactic acid derivative as chiral co-monomer: Synthesis and properties. *Polymer* **2011**, *52*, 4490–4497. [CrossRef]
41. Milavec, J.; Domenici, V.; Zupancic, B.; Rešetič, A.; Bubnov, A.; Zalar, B. Deuteron nmr resolved mesogen vs. crosslinker mo-lecular order and reorientational exchange in liquid single crystal elastomers. *Phys. Chem. Chem. Phys.* **2016**, *18*, 4071–4077. [CrossRef] [PubMed]
42. Rešetič, A.; Milavec, J.; Domenici, V.; Zupancic, B.; Bubnov, A.; Zalar, B. Stress-strain and thermomechanical characterization of nematic to smectic a transition in a strongly-crosslinked bimesogenic liquid crystal elastomer. *Polymer* **2018**, *158*, 96–102. [CrossRef]
43. Hu, Y.; Ji, Q.; Huang, M.; Chang, L.; Zhang, C.; Wu, G.; Zi, B.; Bao, N.; Chen, W.; Wu, Y. Light-Driven Self-Oscillating Actuators with Phototactic Locomotion Based on Black Phosphorus Heterostructure. *Angew. Chem. Int. Ed.* **2021**, *60*, 20511–20517. [CrossRef]
44. Wu, J.; Yao, S.; Zhang, H.; Man, W.; Bai, Z.; Zhang, F.; Wang, X.; Fang, D.; Zhang, Y. Liquid crystal elastomer metamaterials with giant biaxial thermal shrinkage for enhancing skin regeneration. *Adv. Mat.* **2021**, *33*, 2170356. [CrossRef]
45. Shen, Q.; Trabia, S.; Stalbaum, T.; Palmre, V.; Kim, K.; Oh, I.-K. A multiple-shape memory polymer-metal composite actuator capable of programmable control, creating complex 3D motion of bending, twisting, and oscillation. *Sci. Rep.* **2016**, *6*, 24462. [CrossRef] [PubMed]
46. Manna, R.K.; Shklyaev, O.E.; Balazs, A.C. Chemical pumps and flflexible sheets spontaneously form self-regulating oscillators in solution. *Proc. Nat. Acad. Sci. USA* **2021**, *118*, e2022987118. [CrossRef] [PubMed]
47. Li, Z.; Myung, N.V.; Yin, Y. Light-powered soft steam engines for self-adaptive oscillation and biomimetic swimming. *Sci. Robot.* **2021**, *6*, eabi4523. [CrossRef]
48. Serak, S.; Tabiryan, N.; Vergara, R.; White, T.J.; Vaia, R.A.; Bunning, T.J. Liquid crystalline polymer cantilever oscillators fueled by light. *Soft Matter* **2010**, *6*, 779–783. [CrossRef]
49. Ahn, C.; Li, K.; Cai, S. Light or Thermally Powered Autonomous Rolling of an Elastomer Rod. *ACS Appl. Mater. Interfaces* **2018**, *10*, 25689–25696. [CrossRef] [PubMed]
50. Bazir, A.; Baumann, A.; Ziebert, F.; Kulić, I.M. Dynamics of fifiberboids. *Soft Matter* **2020**, *16*, 5210–5223. [CrossRef]
51. Gelebart, A.H.; Mulder, D.; Varga, M.; Konya, A.; Vantomme, G.; Meijer, E.; Selinger, R. Making waves in a photoactive polymer film. *Nature* **2017**, *546*, 632–636. [CrossRef]
52. Zhou, L.; Dai, Y.; Fang, J.; Li, K. Light-powered self-oscillation in liquid crystal elastomer auxetic metamaterials with large volume change. *Int. J. Mech. Sci.* **2023**, *254*, 108423. [CrossRef]
53. Cheng, Q.; Cheng, W.; Dai, Y.; Li, K. Self-oscillating floating of a spherical liquid crystal elastomer balloon under steady illumination. *Int. J. Mech. Sci.* **2023**, *241*, 107985. [CrossRef]
54. Kuenstler, A.S.; Chen, Y.; Bui, P.; Kim, H.; DeSimone, A.; Jin, L.; Hayward, R.C. Blueprinting Photothermal Shape-Morphing of Liquid Crystal Elastomers. *Adv. Mater.* **2020**, *32*, e2000609. [CrossRef]
55. Ge, D.; Dai, Y.; Li, K. Self-Sustained Euler Buckling of an Optically Responsive Rod with Different Boundary Constraints. *Polymers* **2023**, *15*, 316. [CrossRef] [PubMed]
56. Kim, Y.; Berg, J.v.D.; Crosby, A.J. Autonomous snapping and jumping polymer gels. *Nat. Mater.* **2021**, *20*, 1695–1701. [CrossRef]
57. He, Q.; Wang, Z.; Wang, Y.; Wang, Z.; Li, C.; Annapooranan, R.; Zeng, J.; Chen, R.; Cai, S. Electrospun liquid crystal elastomer microfifiber actuator. *Sci. Robot.* **2021**, *6*, eabi9704. [CrossRef] [PubMed]
58. Yu, Y.; Du, C.; Li, K.; Cai, S. Controllable and versatile self-motivated motion of a fiber on a hot surface. *Extreme Mech. Lett.* **2022**, *57*, 101918. [CrossRef]
59. Ge, D.L.; Dai, Y.T.; Li, K. Light-powered self-spinning of a button spinner. *Int. J. Mechan. Sci.* **2023**, *238*, 107824. [CrossRef]
60. Liu, J.; Zhao, J.; Wu, H.; Dai, Y.; Li, K. Self-Oscillating Curling of a Liquid Crystal Elastomer Beam under Steady Light. *Polymers* **2023**, *15*, 344. [CrossRef] [PubMed]
61. Cheng, Y.; Lu, H.; Lee, X.; Zeng, H.; Priimagi, A. Kirigami-based light-induced shapemorphing and locomotion. *Adv. Mater.* **2019**, *32*, 1906233. [CrossRef]
62. Chakrabarti; Choi, G.P.; Mahadevan, L. Self-excited motions of volatile drops on swellable sheets. *Phys. Rev. Lett.* **2020**, *124*, 258002. [CrossRef]
63. Hauser, A.W.; Sundaram, S.; Hayward, R.C. Photothermocapillary Oscillators. *Phys. Rev. Lett.* **2018**, *121*, 158001. [CrossRef] [PubMed]
64. Kim, H.; Sundaram, S.; Kang, J.-H.; Tanjeem, N.; Emrick, T.; Hayward, R.C. Coupled oscillation and spinning of photothermal particles in Marangoni optical traps. *Proc. Natl. Acad. Sci. USA* **2021**, *118*, e2024581118. [CrossRef] [PubMed]
65. Warner, M.; Terentjev, E.M. *Liquid Crystal Elastomers*; Oxford University Press: Oxford, UK, 2007.
66. Bai, R.; Bhattacharya, K. Photomechanical coupling in photoactive nematic elastomers. *J. Mech. Phys. Solids* **2020**, *144*, 104115. [CrossRef]
67. Parrany, M. Nonlinear light-induced vibration behavior of liquid crystal elastomer beam. *Int. J. Mech. Sci.* **2018**, *136*, 179–187. [CrossRef]

68. Gelebart, A.H.; Vantomme, G.; Meijer, E.; Broer, D. Mastering the photothermal effect in liquid crystal networks: A general approach for self-sustained mechanical oscillators. *Adv. Mater.* **2017**, *29*, 1606712. [CrossRef]
69. Ghislaine, V.; Gelebart, A.; Broer, D.; Meijer, E. A four-blade light-driven plastic mill based on hydrazone liquid-crystal net-works. *Tetrahedron* **2017**, *73*, 4963–4967.
70. Finkelmann, H.; Nishikawa, E.; Pereira, G.G.; Warner, M. A New Opto-Mechanical Effect in Solids. *Phys. Rev. Lett.* **2001**, *87*, 015501. [CrossRef]
71. Dunn, M.L. Photomechanics of mono- and polydomain liquid crystal elastomer films. *J. Appl. Phys.* **2007**, *102*, 013506. [CrossRef]
72. Nägele, T.; Hoche, R.; Zinth, W.; Wachtveitl, J. Femtosecond photoisomerization of cis-azobenzene. *Chem. Phys. Lett.* **1997**, *272*, 489–495. [CrossRef]
73. Jampani, V.S.R.; Volpe, R.H.; de Sousa, K.R.; Machado, J.F.; Yakacki, C.M.; Lagerwall, J.P.F. Liquid crystal elastomer shell actuators with negative order parameter. *Sci. Adv.* **2019**, *5*, eaaw2476. [CrossRef]
74. Lee, V.; Bhattacharya, K. Actuation of cylindrical nematic elastomer balloons. *J. Appl. Phys.* **2021**, *129*, 114701. [CrossRef]
75. Torras, N.; Zinoviev, K.E.; Marshall, J.E.; Terentjev, E.M.; Esteve, J. Bending kinetics of a photo-actuating nematic elastomer cantilever. *Appl. Phys. Lett.* **2011**, *99*, 254102. [CrossRef]

Disclaimer/Publisher's Note: The statements, opinions and data contained in all publications are solely those of the individual author(s) and contributor(s) and not of MDPI and/or the editor(s). MDPI and/or the editor(s) disclaim responsibility for any injury to people or property resulting from any ideas, methods, instructions or products referred to in the content.

Article

Self-Vibration of a Liquid Crystal Elastomer Fiber-Cantilever System under Steady Illumination

Kai Li, Yufeng Liu, Yuntong Dai and Yong Yu *

School of Civil Engineering, Anhui Jianzhu University, Hefei 230601, China
* Correspondence: yyu@ahjzu.edu.cn

Abstract: A new type of self-oscillating system has been developed with the potential to expand its applications in fields such as biomedical engineering, advanced robotics, rescue operations, and military industries. This system is capable of sustaining its own motion by absorbing energy from the stable external environment without the need for an additional controller. The existing self-sustained oscillatory systems are relatively complex in structure and difficult to fabricate and control, thus limited in their implementation in practical and complex scenarios. In this paper, we creatively propose a novel light-powered liquid crystal elastomer (LCE) fiber-cantilever system that can perform self-sustained oscillation under steady illumination. Considering the well-established LCE dynamic model, beam theory, and deflection formula, the control equations for the self-oscillating system are derived to theoretically study the dynamics of self-vibration. The LCE fiber-cantilever system under steady illumination is found to exhibit two motion regimes, namely, the static and self-vibration regimes. The positive work done by the tension of the light-powered LCE fiber provides some compensation against the structural resistance from cantilever and the air damping. In addition, the influences of system parameters on self-vibration amplitude and frequency are also studied. The newly constructed light-powered LCE fiber-cantilever system in this paper has a simple structure, easy assembly/disassembly, easy preparation, and strong expandability as a one-dimensional fiber-based system. It is expected to meet the application requirements of practical complex scenarios and has important application value in fields such as autonomous robots, energy harvesters, autonomous separators, sensors, mechanical logic devices, and biomimetic design.

Keywords: self-vibration; liquid crystal elastomer; light-powered; fiber-cantilever

Citation: Li, K.; Liu, Y.; Dai, Y.; Yu, Y. Self-Vibration of a Liquid Crystal Elastomer Fiber-Cantilever System under Steady Illumination. *Polymers* **2023**, *15*, 3397. https://doi.org/10.3390/polym15163397

Academic Editors: Valeriy V. Ginzburg and Alexey V. Lyulin

Received: 19 July 2023
Revised: 8 August 2023
Accepted: 11 August 2023
Published: 13 August 2023

Copyright: © 2023 by the authors. Licensee MDPI, Basel, Switzerland. This article is an open access article distributed under the terms and conditions of the Creative Commons Attribution (CC BY) license (https://creativecommons.org/licenses/by/4.0/).

1. Introduction

Self-excited oscillation refers to a recurring oscillatory phenomenon that arises from external steady excitations. Conventional mechanical oscillation is usually subjected to periodic external stimulus that generates periodic forced motion in time and space. In contrast to forced oscillation, self-oscillation can actively adjust its own motion, provide feedback in response to steady external stimulus, and obtain regular energy to maintain its periodic motion [1–4]. Self-oscillation can not only obtain energy directly and independently from the external environment to maintain its own motion mode, but also its vibration frequency and amplitude depend only on the inherent parameters of the structure. It does not require other complex controllers to achieve periodic oscillation [5,6], so from the perspective of dynamics theory, self-oscillation is of great significance for understanding new behaviors such as bifurcation, chaos, synchronization, and other non-equilibrium dynamics in nonlinear systems. It is a typical non-equilibrium dynamical process in nonlinear systems [7]. Self-oscillating systems have broad application prospects and revolutionary impact on autonomous robots [8–12], energy harvesters [13,14], independent separators, sensors [15], mechanical logic devices [16], and biomimetic design.

In recent years, active materials such as hydrogels [17,18], dielectric elastomers [19], ion gels [20], and thermally responsive polymer materials [21] have exhibited different responses under different stimulus conditions. These responses generally change the morphology and motion state of the active materials themselves. People have established various self-oscillating systems and multiple self-sustained motion modes using the properties of active materials, including bending [22–24], swimming [25], swinging [26], rolling [2,9,10,27], rotating [28,29], twisting [30,31], vibration [6], and even synchronized motion of several coupled self-oscillators [32,33]. In general, in all dynamic systems, there is energy dissipation [34], and in practical environments, the vibrations tend to approach an equilibrium state. Therefore, designing different types of self-oscillating systems is a challenging process. In a constant environment, how to enable the system to absorb energy autonomously, compensate for the damping dissipation, and maintain periodic motion is the key to realize self-oscillation. A large number of self-excited oscillatory systems have been established based on various feedback mechanisms. These different feedback mechanisms typically lead to different self-sustained motion modes, such as self-shadowing [35–37], coupling of liquid evaporation and membrane deformation [38], coupling mechanism of air expansion and liquid column motion [39], and coupling of plate bending and chemical reaction [40], all of which can cause self-excited oscillations.

The advantages of light in various stimuli are its sustainability, accuracy, controllability [41,42], and non-contact. Optically-responsive materials that can convert near-infrared and visible light into thermal energy, such as carbon nanotubes, graphene, and liquid crystal elastomers (LCEs) [43–48] have good photomechanical effects [49–54]. Among them, LCEs are important optically responsive materials, synthesized from anisotropic rod-shaped liquid crystal molecules and stretchable long-chain polymers. When liquid crystal monomers are subjected to external stimuli such as light, heat, electricity, and magnetism, they will rotate or undergo phase transitions, thereby changing their configuration and generating macroscopic deformation [55,56]. LCEs typically offer advantages of large deformation, fast deformation response, recoverable deformation, low noise, easy remote control, and easy manipulation. Based on LCEs, photomechanical effects have been utilized to build various self-sustained oscillatory systems, including but not limited to shuttling [57], bending [58], rotation [29,30,55], spinning [59], curling [60], oscillating [61,62], buckling [63–65], rolling [28], floating [66], twisting [67], vibration [68], swimming [25], chaos [69], and even several synchronous motions coupled with self-excited oscillations [2,27,34]. These LCE-based self-sustained oscillatory systems have attracted much attention in both fundamental and applied research [55,70–72].

Although a large number of self-sustained oscillatory systems have been constructed, these systems generally have complex structures, are difficult to manufacture and control, and may not meet the requirements of complex practical applications. In this article, we propose a novel and simple LCE fiber-cantilever system that exhibits self-sustained oscillation under steady illumination and essentially functions as a "self-shadowing" system. Compared to previous self-oscillating systems such as balls [66] and tubes [42], the structure of one-dimensional fiber is relatively simple, making it easy to assemble and disassemble. It should also be noted that the proposed LCE fiber-cantilever system may exhibit a dependence on the angle of illumination in practice. Furthermore, the system is highly extensible, holding potential for constructing more complex LCE fiber-based systems to achieve advanced self-sustained motions. The objective of this research is to build the LCE fiber-cantilever system and investigate its self-oscillation characteristics under stable illumination. Meanwhile, we discuss the underlying mechanisms of self-oscillation and systematically explore the impacts of various physical and geometric parameters on the system's amplitude and frequency.

The organization of this paper is as follows. First, in Section 2, considering the dynamic LCE model and beam theory, the theoretical model and control equations for the LCE fiber-cantilever system are established. Then, in Section 3, two motion regimes of the LCE fiber-cantilever system are obtained by numerical calculations, and the mechanism of its self-vibration is explained in detail. Next, in Section 4, the influences of various system parameters on the amplitude and frequency of self-vibration are discussed in detail. Finally, the results are summarized.

2. Theoretical Model and Formulation

In this section, we first propose a light-powered self-oscillation system containing an LCE fiber, an oblique bending cantilever, and a mass block. Then, we present a theoretical model for the self-oscillation system based on the dynamic LCE model [8] and beam theory [73]. The dynamic control equations of the system, the evolution law of the *cis* number fraction in LCE, and the nondimensionalization of the system parameters are then given in turn.

2.1. Dynamics of System

Figure 1 schematically describes the proposed LCE fiber-cantilever system, in which an LCE fiber, a lightweight cantilever beam and a mass block are included. The lightweight cantilever of length L_B at an angle θ from the horizontal, is fixed on a vertical rigid base. The mass block with mass m at the cantilever end is connected by the LCE fiber fixed on another vertical rigid base to form a tension string system. The bending effect of gravity on the cantilever can be ignored as it is much smaller than other forces. Both the torsion and displacement of the cantilever along the length are small, so the mass block is assumed to move in a plane. We take the initial position of the mass block as the origin of the coordinate system and establish the coordinate axis along the direction of cantilever deflection. The initial length of LCE fiber is L_0. In addition, the masses of the LCE fiber and the cantilever are much less than the mass m, so they are neglected.

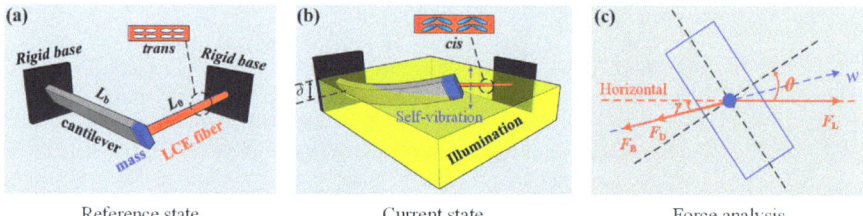

Figure 1. Schematic of an LCE fiber-cantilever system containing an LCE fiber, a lightweight cantilever beam, and a mass block: (**a**) Reference state; (**b**) Current state; (**c**) Force analysis. F_L denotes the tension of the LCE fiber, F_B represents the force exerted by the beam on the mass block, F_D represents the air damping force, γ is the angle between the cantilever deflection and the horizontal direction, and θ is the inclined angle of cantilever.

The system is placed under steady illumination as shown in Figure 1b, with the yellow region representing the illumination zone with a height of δ. Generally, chromophores in the LCE fiber upon illumination undergo series of *trans-cis-trans* isomerization cycles ending up in the change of the orientation of the *trans-isomer* long axis [74]. In case of non-polarized light illumination, the long axes orient towards the illumination direction, while in case of illumination with polarized light, the long axes orient perpendicular to the light polarization, because of the direction-dependent absorption of the chromophore. These changes can change the order parameter of the LCE and lead in some geometries to contraction of the fiber. As the LCE fiber contracts, the cantilever bends further into the dark zone. When the LCE fiber is in the dark, the azobenzene molecules in it switch from *cis* to *trans*, causing the light-driven contraction of the LCE fiber to recover. Subsequently, the tension of the LCE fiber decreases and the cantilever returns to the illumination zone due to the structural resistance. Through the proper adjustment of the system parameters and initial conditions, the LCE fiber-cantilever system can maintain continuous self-oscillation.

The mass block is subjected to the tension of LCE fiber, the structural resistance form cantilever, and the air damping force, as depicted in Figure 1c. In the deflection direction, the control equation for the nonlinear dynamics model of mass block can be expressed as follows:

$$m\ddot{w} = F_L \cdot \cos\gamma - F_B - F_D \tag{1}$$

where $\ddot{w}$ refers to the acceleration of the mass block, F_L denotes the tension of the LCE fiber, F_B represents the force exerted by the beam on the mass block, F_D represents the air damping force, and γ is the angle between the cantilever deflection and the horizontal direction.

Through the beam deflection theory, the moment of inertia formula, and the trigonometric function, it can be calculated $\gamma = \arctan[r^2 \tan\theta] - \theta$, where r refers to the ratio of cantilever height to width.

The tension of LCE fiber is related to its elongation and cross-sectional area, which can be described as

$$F_L = \frac{E_L A_L \cdot \Delta L}{L} = \frac{E_L A_L \{[L_0 + 2w(t) \cdot \cos\gamma] - L_0[1 + \varepsilon_L(t)]\}}{L_0[1 + \varepsilon_L(t)]} \tag{2}$$

where F_L refers to the elastic modulus of the LCE fiber, A_L refers to the cross-sectional area of the LCE fiber, L_0 is the original length of LCE fiber, $w(t)$ represents the cantilever-end deflection, i.e., the displacement of the mass block, and $\varepsilon_L(t)$ represents the light-driven contraction strain of LCE fiber.

It is assumed that the cantilever beam is always in a state of small deformation, while the theory of linear elasticity is applied, thus the structural resistance from cantilever is proportional to the displacement, that is

$$F_B = \frac{3E_B I_B}{L_B^3} \cdot w(t) \tag{3}$$

where L_B is the cantilever length, $E_B I_B$ is the bending stiffness of the cantilever.

The damping force is assumed to be linearly proportional to the velocity of the mass block, with the formula being

$$F_D = \beta \cdot \dot{w}(t) \tag{4}$$

where β denotes the air damping coefficient and $\dot{w}$ is the velocity of the mass block.

Thus far, substituting Equations (2)–(4) into Equation (1), we have

$$m\frac{d^2 w(t)}{dt^2} = E_L A_L \cdot \cos\gamma \cdot \frac{\{[L_0 + 2w(t) \cdot \cos\gamma] - L_0[1 + \varepsilon_L(t)]\}}{L_0[1 + \varepsilon_L(t)]}. \tag{5}$$

2.2. Dynamic LCE Model

This section mainly describes the dynamic model of the light-driven contraction in LCE fiber. The fiber radius is assumed to be much smaller than the penetration depth of light, and no absorption gradient within the fiber is considered. The LCE fiber-cantilever system uses a linear model, which is adopted to describe the relationship between the *cis* number fraction $\varphi(t)$ in LCE and the light-driven contraction of LCE, that is

$$\varepsilon_L = -C_0 \cdot \varphi(t) \tag{6}$$

where C_0 is the contraction coefficient.

The light-driven contraction in LCE depend on the cis number fraction $\varphi(t)$ [75,76]. The study by Yu et al. found that the *trans-to-cis* isomerization of LCE could be induced by UV or laser with wavelength less than 400 nm [77]. In this study, a 'push-pull' mechanism is considered to calculate the *cis* number fraction [76]. The number fraction $\varphi(t)$ of the *cis*-isomer depends on the thermal excitation from *trans* to *cis*, the thermally driven relaxation from *cis* to *trans*, and the light driven relaxation from *trans* to *cis*. Supposing that the thermal excitation from *trans* to *cis* can be ignored, the governing equation for the evolution of the *cis* number fraction can be formulated as

$$\frac{\partial \varphi}{\partial t} = \eta_0 I_0 (1 - \varphi) - \frac{\varphi}{T_0} \tag{7}$$

where T_0 refers to the thermally driven relaxation time from the *cis* to *trans*, I_0 denotes the light intensity, and η_0 is the light absorption constant. By solving Equation (7), the *cis* number fraction can be described as

$$\varphi(t) = \frac{\eta_0 T_0 I_0}{\eta_0 T_0 I_0 + 1} + (\varphi_0 - \frac{\eta_0 T_0 I_0}{\eta_0 T_0 I_0 + 1}) \exp[-\frac{t}{T_0}(\eta_0 T_0 I_0 + 1)] \tag{8}$$

where φ_0 represents the initial *cis* number fraction at $t = 0$.

In illuminated state, for initially zero-number fraction, i.e., $\varphi_0 = 0$, Equation (8) can be simplified as

$$\varphi(t) = \frac{\eta_0 T_0 I_0}{\eta_0 T_0 I_0 + 1} \{1 - \exp[-\frac{t}{T_0}(\eta_0 T_0 I_0 + 1)]\} \tag{9}$$

In non-illuminated state, namely $I_0 = 0$, Equation (8) can be simplified as

$$\varphi(t) = \varphi_0 \exp(-\frac{t}{T_0}) \tag{10}$$

where the undetermined φ_0 can be set to be the maximum value of $\varphi(t)$ in Equation (9), namely, $\varphi_0 = \frac{\eta_0 T_0 I_0}{\eta_0 T_0 I_0 + 1}$. Then Equation (10) can be rewritten as

$$\varphi(t) = \frac{\eta_0 T_0 I_0}{\eta_0 T_0 I_0 + 1} \exp(-\frac{t}{T_0}) \tag{11}$$

2.3. Nondimensionalization

We introduce the following dimensionless quantities by defining: $\overline{w} = \frac{w}{L_0}$, initial velocity $\overline{\dot{w}}_0 = \frac{T_0 \dot{w}_0}{L_0}$, $\overline{t} = \frac{t}{T_0}$, spring constant $\overline{K}_L = \frac{E_L A T_0^2}{m L_0}$, flexural stiffness $\overline{K}_B = \frac{3 E_B I_B T_0^2}{m L_B^3}$, $\overline{\beta} = \frac{\beta T_0}{m}$, $\overline{I}_0 = \eta_0 T_0 I_0$, $\overline{\delta} = \frac{\delta}{L_0}$, and $\overline{\varphi} = \frac{\varphi(\eta_0 T_0 I_0 + 1)}{\eta_0 T_0 I_0}$, to simplify the governing equations Equations (5) and (9)–(11).

The dimensionless form of Equation (5) can be expressed as

$$\overline{\ddot{w}}(\overline{t}) = \overline{K}_L \cdot \cos \gamma \cdot [\frac{1}{1 - C_0 \cdot \overline{\varphi}(\overline{t})} + \frac{\overline{w}(\overline{t}) \cdot \cos \gamma}{1 - C_0 \cdot \overline{\varphi}(\overline{t})} - 1] - \overline{K}_B \cdot \overline{w}(\overline{t}) - \overline{\beta} \cdot \overline{\dot{w}}(\overline{t}) \tag{12}$$

In illuminated state, Equation (9) can be rewritten as

$$\overline{\varphi} = 1 - \exp[-\overline{t}(\overline{I}_0 + 1)] \tag{13}$$

and in non-illuminated state, Equation (11) becomes

$$\overline{\varphi} = \exp(-\overline{t}) \tag{14}$$

Equations (12)–(14) are utilized to regulate the self-vibration of the LCE fiber-cantilever system in the presence of steady illumination. These equations involve a time-varying fractional quantity associated with the cis isomer and closely linked to the light intensity. To solve these intricate linear equations, the fourth-order Runge–Kutta method is employed for numerical computations using the Matlab software. Moreover, Equations (13) and (14) are employed to determine the cis number fraction φ and time length $\overline{t}$, enabling the calculation of tension F_L, air damping force F_D, and structural resistance F_B of the LCE fiber. By iterating calculation with given parameters $\overline{\dot{w}}_0$, $\overline{K}_L$, $\overline{K}_B$, $\overline{\beta}$, $\overline{I}_0$, C_0, θ, r, and $\overline{\delta}$, the dynamics of the LCE fiber-cantilever system can be obtained.

3. Two Motion Regimes and Mechanism of Self-Vibration

In this section, through solving the control equation Equation (12), we first propose two typical motion regimes of the LCE fiber-cantilever system, which are distinguished as static regime and self-vibration regime. Next, the corresponding mechanism of self-vibration is elaborated in detail.

3.1. Two Motion Regimes

In order to further study the self-vibration behavior of the LCE fiber-cantilever system, we first need to determine the typical values for the dimensionless system parameters. Based on the existing experiments and information [78–80], Table 1 gathers the typical values of the system parameters required in current paper. The corresponding dimensionless parameters are listed in Table 2. In the following section, these values of parameters are used to study the self-vibration of the LCE fiber-cantilever system under steady illumination. It is worth noting that the small deformation hypothesis can be verified under these given parameters.

Table 1. Material properties and geometric parameters.

Parameter	Definition	Value	Unit
I_0	Light intensity	0~10	kW/m^2
C_0	Contraction coefficient	0~0.5	/
K_L	Spring constant	0.1~1	N/m
K_B	Flexural stiffness	0.3~3	N/m
β	Damping coefficient	0~0.001	kg/s
w_0	Initial velocity	0~0.5	mm/s
δ	Height of illumination zone	0~0.1	m
r	Ratio of cantilever height to width	1~20	/
θ	Inclined angle of cantilever	0~1.2	rad
T_0	Cis- to trans- thermal relaxation time	1~100	ms
η_0	Light-absorption constant	0.001	m^2/(s·W)

Table 2. Dimensionless parameters.

Parameter	$\bar{I}_0$	C_0	$\bar{K}_L$	$\bar{K}_B$	$\bar{\beta}$	$\bar{w}_0$	$\bar{\delta}$	r	θ
Value	0~1	0~0.5	0~1.2	0~1	0~0.2	0~5	0~0.1	1~20	0~$\frac{\pi}{2}$

By solving Equations (12)–(14), the time histories and phase trajectories for the LCE fiber-cantilever system can be obtained, with examples for $\bar{I}_0 = 0.25$ and $\bar{I}_0 = 0.5$ shown in Figure 2. The other parameters used in the calculation are set as $C_0 = 0.25$, $\bar{K}_L = 0.2$, $\bar{K}_B = 0.7$, $\bar{\beta} = 0.02$, $\bar{w}_0 = 0$, $\bar{\delta} = 0.03$, $r = 2$ and $\theta = \frac{\pi}{4}$. In Figure 2a,b, the amplitude of the cantilever-end deflection gradually decreases with time due to the damping dissipation, and the system eventually reaches a stationary position at equilibrium, which is referred to as the static regime. In contrast, Figure 2c,d show that the system initially vibrates from a static equilibrium position and then progressively increases in vibration amplitude over time until it remains constant. On exposure to steady illumination, the LCE fiber-cantilever system eventually presents a continuous periodic vibration, which we refer to as the self-vibration regime.

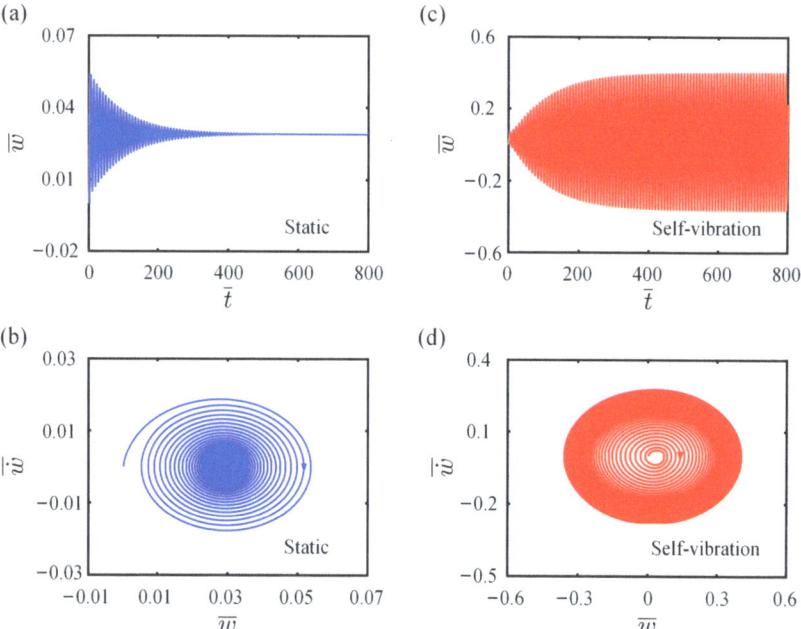

Figure 2. Two typical motion regimes of the LCE fiber-cantilever system under steady illumination: static regime and self-vibration regime. (**a**) Time-history curve of the displacement with $\bar{I}_0 = 0.25$; (**b**) Phase trajectory diagram with $\bar{I}_0 = 0.25$; (**c**) Time-history curve of the displacement with $\bar{I}_0 = 0.5$ and (**d**) Phase trajectory diagram with $\bar{I}_0 = 0.5$.

3.2. Mechanism of the Self-Vibration

In this section, the mechanism of self-vibration will be explained in detail. To better understand the energy compensation mechanism of the LCE fiber-cantilever system, we plot the relationship curves for some key physical quantities in the self-vibration process. In this case, the system parameters are selected as $\bar{I}_0 = 0.5$, $C_0 = 0.25$, $\overline{K}_L = 0.25$, $\overline{K}_B = 0.7$, $\bar{\beta} = 0.02$, $\bar{w}_0 = 0$, $\bar{\delta} = 0.03$, $r = 2$, and $\theta = \frac{\pi}{4}$. Figure 3a illustrates the cantilever-end deflection over time, with the yellow area indicating that the LCE fiber is in the illumination zone. As the system vibrates continuously, the LCE fiber also oscillates back and forth between the illumination and dark zones, and the change in the *cis* number fraction $\bar{\varphi}$ over time is drawn in Figure 3b. It is clearly observed that as the illumination condition changes, the *cis* number fraction changes rapidly at first and then slowly approaches a critical value determined by the contraction coefficient C_0. In addition, Figure 4 illustrates several characteristic snapshots for the self-vibration of the LCE fiber-cantilever system during one cycle under steady illumination.

Figure 3c presents the periodic time variation of the tension of the LCE fiber. The tension decreases first and then increases in the illumination zone, while the opposite is true in the dark zone. The hysteresis loop shown in Figure 3d indicates that the LCE fiber-cantilever system maintains its oscillation as the LCE fiber absorbs light energy and does work. The area enclosed by the loop represents the net work done by the tension of the LCE fiber in one cycle, with a value of approximately 0.0029. Like the tension of the LCE fiber, it is clear from Figure 3e that the damping force also presents a periodic time variation. Figure 3f plots the dependence of the damping force on the cantilever-end deflection, which also forms a closed loop representing the damping dissipation, with a value being calculated to be about 0.0029. The net work done by the tension of LCE fiber is exactly equal to the damping dissipation, implying that the energy consumed by the system motion is compensated by the light energy absorbed by the LCE fiber, thus maintaining the self-vibration.

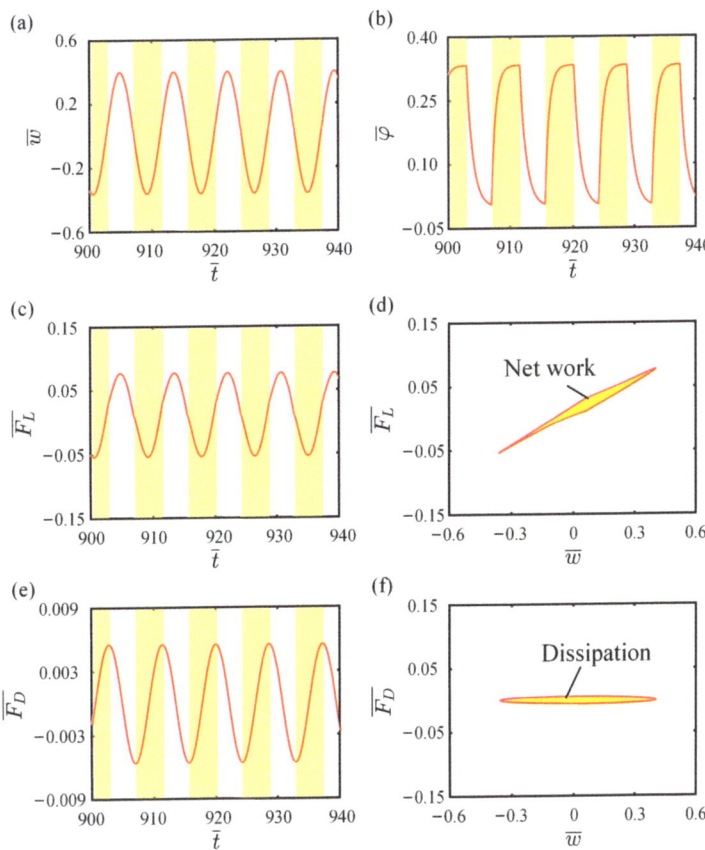

Figure 3. Self-vibration mechanism of the LCE fiber-cantilever system (**a**) Time-history curve of the cantilever-end deflection. (**b**) Time variation of the light-driven contraction of LCE fiber. (**c**) Time variation of the tension of LCE fiber. (**d**) Dependence of the tension of LCE fiber on the cantilever-end deflection. (**e**) Time variation of the damping force. (**f**) Dependence of the damping force on the cantilever-end deflection.

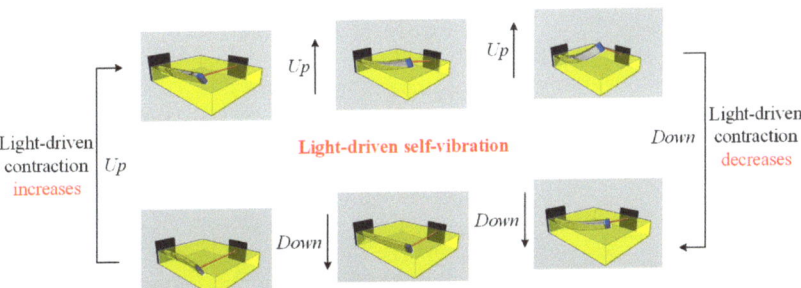

Figure 4. Snapshots of the LCE fiber-cantilever system in one cycle during the self-vibration. Under steady illumination, the system exhibits a continuous periodic self-vibration due to the periodic variation of light-driven contraction.

4. Parametric Study

In the mechanical model of the self-vibration for the LCE fiber-cantilever system described above, there are nine dimensionless system parameters: $\bar{I}_0, C_0, \bar{K}_L, \bar{K}_B, \bar{\beta}, \bar{w}_0, \bar{\delta}, r$, and θ. In this section, we investigate in detail the effects of these system parameters on the self-vibration of the LCE fiber-cantilever system, including its frequency and amplitude. The dimensionless self-vibration frequency and amplitude are denoted by f and A, respectively.

4.1. Effect of Light Intensity

The effect of light intensity on the self-vibration is discussed in current subsection. In this case, the values of the other parameters are, $C_0 = 0.25, \bar{K}_L = 0.25, \bar{K}_B = 0.7, \bar{\beta} = 0.02, \bar{w}_0 = 0, \bar{\delta} = 0.03, r = 2$, and $\theta = \frac{\pi}{4}$. The limit cycles of the self-vibration are depicted in Figure 5a, where $\bar{I}_0 = 0.39$ is the critical value of light intensity between the static and self-vibration regimes. When the light intensity is below 0.39, the system is in static regime, while above 0.39, the system is in self-vibration regime. When the light intensity is relatively small, the LCE fiber does not absorb enough light energy to offset the damping dissipation, thus it cannot maintain its continuous motion and comes to rest. Conversely, when the light intensity is large enough, the light energy absorbed by the system can compensate for the damping dissipation, so as to maintain its own motion. Figure 5b describes the effect of light intensity on the self-vibration amplitude and frequency. With the increasing light intensity, the amplitude increases, while the frequency remains essentially constant. Larger light intensity allows the system to absorb more light energy, thereby maintaining oscillation with higher amplitude. These results suggest that increasing the light intensity is crucial for improving the energy utilization efficiency of the LCE fiber-cantilever system.

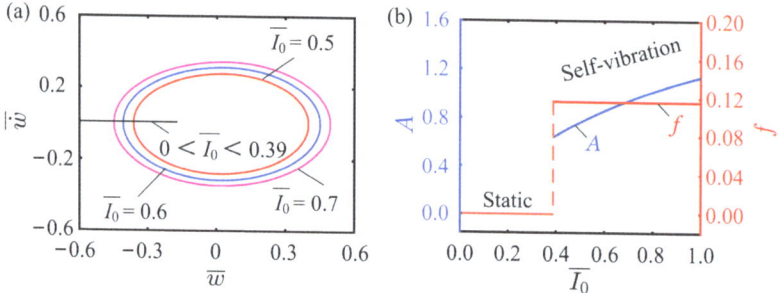

Figure 5. Effect of light intensity on the self-vibration. (**a**) Limit cycles with $\bar{I}_0 = 0.5, \bar{I}_0 = 0.6$ and $\bar{I}_0 = 0.7$. (**b**) Variations of amplitude and frequency with different light intensities.

4.2. Effect of Contraction Coefficient

This subsection mainly discusses the effect of contraction coefficient on the self-vibration. Here, the values of the other parameters are $\bar{I}_0 = 0.5, \bar{K}_L = 0.25, \bar{K}_B = 0.7, \bar{\beta} = 0.02, \bar{w}_0 = 0, \bar{\delta} = 0.03, r = 2$, and $\theta = \frac{\pi}{4}$. Figure 6a plots the limit cycles for different contraction coefficients. Obviously, there exists a critical value for contraction coefficient to trigger the self-vibration, which is numerically determined to be 0.207. A small contraction coefficient means a low light energy input, and there is not enough energy to compensate for the damping dissipation. For $C_0 = 0.25, C_0 = 0.35$, and $C_0 = 0.45$, the self-vibration can be triggered. Figure 6b presents the dependencies of the self-vibration amplitude and frequency on the contraction coefficient. The larger the contraction coefficient, the higher the amplitude. As the contraction coefficient increases, the LCE fiber makes more efficient use of the illumination, absorbs more light energy, and shifts the system from a static regime to a self-vibration regime, resulting in an increase in the amplitude. The result implies that increasing the contraction coefficient of LCE material can improve the efficient conversion of light energy to mechanical energy.

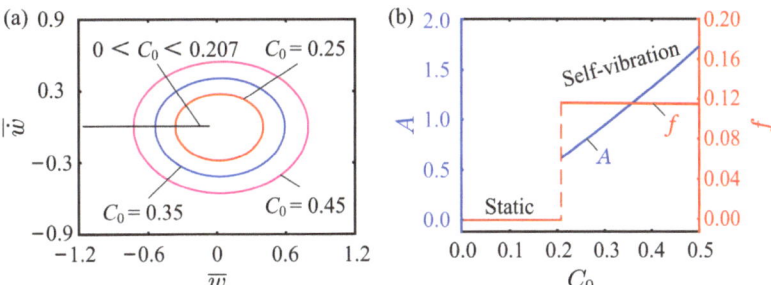

Figure 6. Effect of contraction coefficient on the self-vibration. (**a**) Limit cycles with $C_0 = 0.25$, $C_0 = 0.35$, and $C_0 = 0.45$. (**b**) Variations of amplitude and frequency with different contraction coefficients.

4.3. Effect of Spring Constant

This subsection mainly focuses on the effect of spring constant on the self-vibration. In this case, the values of the other parameters are $\bar{I}_0 = 0.5$, $C_0 = 0.25$, $\overline{K}_B = 0.7$, $\bar{\beta} = 0.02$, $\bar{w}_0 = 0$, $\bar{\delta} = 0.03$, $r = 2$, and $\theta = \frac{\pi}{4}$. Figure 7a displays the limit cycles for different spring constants, among which two critical spring constants exist for triggering the self-vibration. It is clear to see that the system is in the static regime when the spring constant is below 0.214 or above 0.951. This can be explained by the relationship between the spring constant and the tension of the LCE fiber. When the spring constant is small, the tension of the LCE fiber is small, which is not enough to force the system to remain in oscillation. When the spring constant is large, the tension of the LCE fiber can be equal to the structural resistance, thus allowing the whole system to equilibrate the forces and reach a static regime. Figure 7b illustrates that the spring constant has a significant effect on the amplitude and frequency of the self-vibration. As the spring constant increases, the amplitude increases, while the frequency decreases. This is because the spring constant determines the driving force of the system, which in turn affects the oscillatory behavior of the system. Therefore, when we design the LCE fiber-cantilever system, the adjustment of the spring constant can be used to control its amplitude and frequency to achieve better performance. For example, in some robotic applications, the LCE fiber-cantilever system is required to realize stable motion or grasp an object, we can select the appropriate spring constant according to the desired motion mode and the weight of the object, so as to keep the system stable and have good accuracy during operation. In addition, when designing suspended structures or other oscillatory systems, the amplitude and frequency can also be controlled according to the variation of the spring constant to achieve better performance.

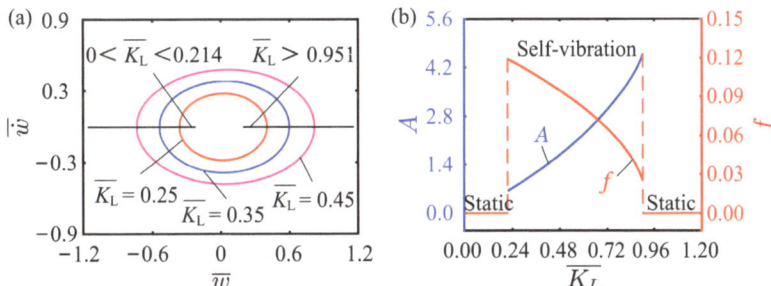

Figure 7. Effect of spring constant on the self-vibration. (**a**) Limit cycles with $\overline{K}_L = 0.25$, $\overline{K}_L = 0.35$, and $\overline{K}_L = 0.45$. (**b**) Variations of amplitude and frequency with different spring constants.

4.4. Effect of Flexural Stiffness

The influence of flexural stiffness on the self-vibration is provided for $\bar{I}_0 = 0.5$, $C_0 = 0.25$, $\bar{K}_L = 0.25$, $\bar{\beta} = 0.02$, $\bar{w}_0 = 0$, $\bar{\delta} = 0.03$, $r = 2$, and $\theta = \frac{\pi}{4}$. The limit cycles for different flexural stiffnesses are drawn in Figure 8a. The flexural stiffness has two critical values for the transition between the static and self-vibration regimes, which are numerically calculated to be around 0.19 and 0.81. When the flexural stiffness is small, the structural resistance of the cantilever is small, and the net work done by the tension of the LCE fiber is not sufficient to maintain the self-vibration. When the flexural stiffness is large, the structural resistance from the cantilever is so great that the tension of the LCE fiber cannot drive the system to oscillate. Figure 8b plots the variations of self-vibration amplitude and frequency with different flexural stiffnesses. As the flexural stiffness increases, the amplitude decreases, while the frequency increases. This can be explained by the beam theory, where the greater the flexural stiffness of the beam, the greater the recovery force on the beam, thus preventing further bending of the beam. As a result, the amplitude decreases. Therefore, to improve the system stability, it is a good way to choose the appropriate flexural stiffness of the beam.

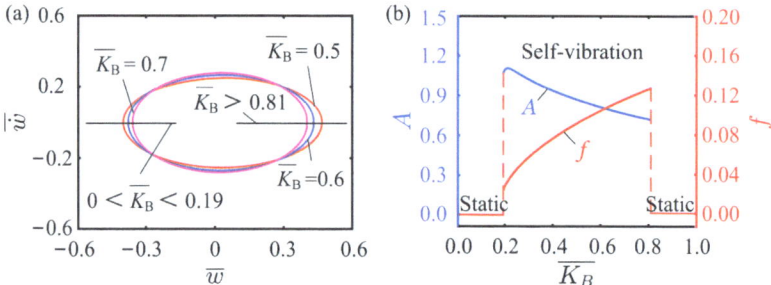

Figure 8. Effect of flexural stiffness on the self-vibration. (**a**) Limit cycles with $\bar{K}_B = 0.5$, $\bar{K}_B = 0.6$, and $\bar{K}_B = 0.7$. (**b**) Variations of amplitude and frequency with different bending stiffnesses.

4.5. Effect of Damping Coefficient

Figure 9 presents the influence of damping coefficient on the self-vibration, with parameters $\bar{I}_0 = 0.5$, $C_0 = 0.25$, $\bar{K}_L = 0.25$, $\bar{K}_B = 0.7$, $\bar{w}_0 = 0$, $\bar{\delta} = 0.03$, $r = 2$, and $\theta = \frac{\pi}{4}$. The limit cycles for different damping coefficients can be observed in Figure 9a. It is not difficult to find that the variation of damping coefficient does not affect the motion regime of the LCE fiber-cantilever system. For different damping coefficients, the system is always in a self-vibration regime. The dependencies of the self-vibration amplitude and frequency on the damping coefficient are depicted in Figure 9b. With the increase of damping coefficient, the amplitude decreases sharply and then slowly, presenting the characteristics of an exponential function. In contrast, changes in the damping coefficient have little effect on the frequency. This suggests that the damping coefficient plays an important role in influencing the amplitude and energy level of self-vibration systems. Proper adjustment of the damping coefficient can control the vibration amplitude and energy level of the system to ensure the system stability. Moreover, as the damping coefficient has little effect on the frequency, the damping coefficient and frequency need to be considered comprehensively during the system design process to obtain the optimal scheme. These research results not only provide important application value in the field of engineering design and manufacture, but also provide new ideas and methods for the in-depth understanding of complex systems.

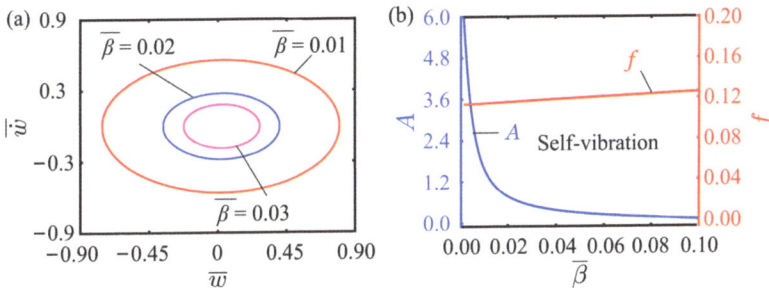

Figure 9. Effect of damping coefficient on the self-vibration. (**a**) Limit cycles with $\bar{\beta} = 0.01$, $\bar{\beta} = 0.02$, and $\bar{\beta} = 0.03$. (**b**) Variations of amplitude and frequency with different damping coefficients.

4.6. Effect of Initial Velocity

The effect of initial velocity $\bar{w}_0$ on the self-vibration is displayed in Figure 10, with other parameters being $\bar{I}_0 = 0.5$, $C_0 = 0.25$, $\bar{K}_L = 0.25$, $\bar{K}_B = 0.7$, $\bar{\beta} = 0.02$, $\bar{\delta} = 0.03$, $r = 2$, and $\theta = \frac{\pi}{4}$. $\bar{w}_0 = 0$, $\bar{w}_0 = 0.5$, and $\bar{w}_0 = 1$ are found to successfully trigger the self-vibration, and the limit cycles are plotted in Figure 10a. It is worth mentioning that the limit cycles for these three initial velocities overlap. As can be seen in Figure 10b, the variation of the initial velocity does not affect the amplitude and frequency of the system. Since the self-vibration results from the energy conversion between the damping dissipation and the network done by the tension of the LCE fiber, the self-vibration amplitude and frequency are determined by the internal properties of the system, which is consistent with other self-vibration systems. The initial velocity therefore has no effect on the final amplitude of the system.

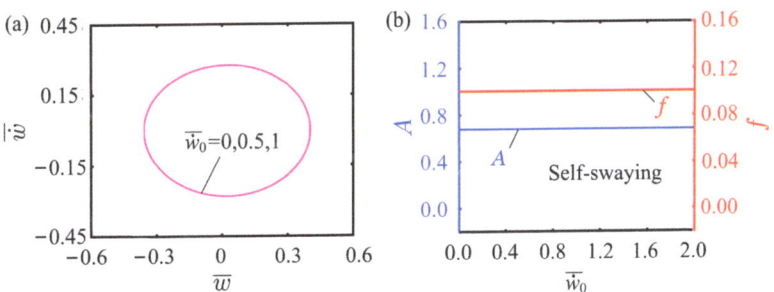

Figure 10. Effect of initial velocity on the self-vibration. (**a**) Limit cycles with $\bar{w}_0 = 0$, $\bar{w}_0 = 0.5$, and $\bar{w}_0 = 1$. (**b**) Variations of amplitude and frequency with different initial velocities.

4.7. Effect of Illumination Zone Height

This subsection presents a discussion on the effect of illumination zone height on the self-vibration. In the calculation, we set other parameters as $\bar{I}_0 = 0.5$, $C_0 = 0.25$, $\bar{K}_L = 0.25$, $\bar{K}_B = 0.7$, $\bar{\beta} = 0.02$, $\bar{w}_0 = 0$, $r = 2$, and $\theta = \frac{\pi}{4}$. As observed from Figure 11a, for the phase transition between the static and self-vibration regimes, two critical illumination zone heights exist with values of 0.001 and 0.037, respectively. When the illumination zone height is less than 0.001 or greater than 0.037, the system is in astatic regime. When the illumination zone height is within the interval of 0.001 and 0.037, the system is in a self-vibration regime. The effect of illumination zone height on the amplitude and frequency is shown in the Figure 11b. Obviously, the amplitude and frequency do not vary with increasing the illumination zone height. This is contributed to the fact that as the illumination zone expands, the tension of the LCE fiber increases, and the structural resistance from cantilever also increases accordingly. Consequently, the system encounters greater resistance during self-vibration, resulting in a drop in amplitude. In conclusion, adjusting the appropriate range of the illumination zone can be more effective in improving the efficiency of light utilization.

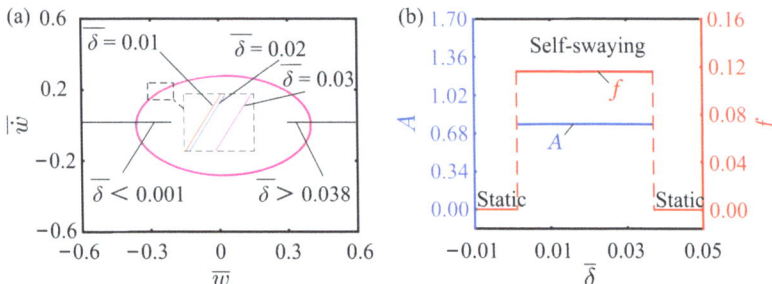

Figure 11. Effect of illumination zone height on the self-vibration. (**a**) Limit cycles with $\bar{\delta} = 0.01$, $\bar{\delta} = 0.02$, and $\bar{\delta} = 0.03$. (**b**) Variations of amplitude and frequency with different illumination zone heights.

4.8. Effect of Ratio of Cantilever Height to Width

This subsection mainly discusses how the ratio of cantilever height to width affects the self-vibration. In this case, the other dimensionless parameters are selected as $\bar{I}_0 = 0.5$, $C_0 = 0.25$, $\bar{K}_L = 0.25$, $\bar{K}_B = 0.7$, $\bar{\beta} = 0.02$, $\bar{w}_0 = 0$, $\bar{\delta} = 0.03$, and $\theta = \frac{\pi}{4}$. Figure 12a shows the three limit cycles for ratios of cantilever height to width of $r = 2$, $r = 4$, and $r = 6$. The system is in the static regime when the ratio is below 1.48, while it is in the self-vibration regime when the ratio exceeds 1.48. This is due to the small deflection angle of the cantilever end when the ratio of cantilever height to width is small. The longitudinal deflection of the cantilever end is too small for the system to leave the illumination zone, so the system becomes static. Figure 12b depicts how the ratio of cantilever height to width affects the self-vibration amplitude and frequency. As the ratio of cantilever height to width increases, the self-vibration amplitude will first decrease rapidly, and then a marginal effect occurs, slowing down the reduction rate. At the same time, the self-vibration frequency will first increase rapidly, and then a marginal effect appears, slowing down its increase. These findings underscore the significance of meticulous selection of the ratio of cantilever height to width and suggest that opting for an appropriate ratio can effectively enhance the efficiency of converting light energy into mechanical energy.

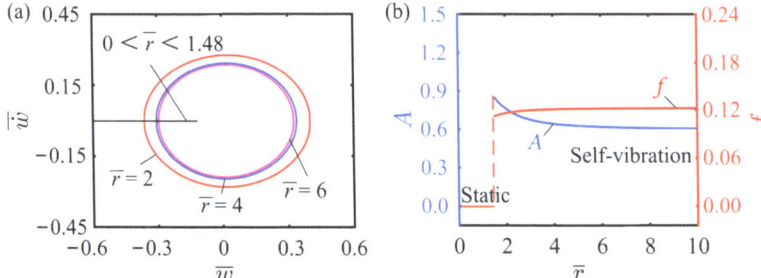

Figure 12. Effect of ratio of cantilever height to width on the self-vibration. (**a**) Limit cycles with $r = 2$, $r = 4$, and $r = 6$. (**b**) Variations of amplitude and frequency with different ratios of cantilever height to width.

4.9. Effect of Inclined Angle of Cantilever

The inclined angle of cantilever affecting the self-vibration is investigated in this subsection, where the other dimensionless parameters are chosen as $\bar{I}_0 = 0.5$, $C_0 = 0.25$,

$\overline{K}_L = 0.25$, $\overline{K}_B = 0.7$, $\overline{\beta} = 0.02$, $\overline{w}_0 = 0$, $\overline{\delta} = 0.03$, and $r = 2$. Figure 13a illustrates the limit cycles for different inclined angles, in which $\theta = \frac{2\pi}{45}$ and $\theta = \frac{123\pi}{360}$ are the two critical inclined angles for the phase transition between the static and the self-vibration regimes. The self-vibration can be triggered with inclined angles of $\theta = \frac{\pi}{6}$, $\theta = \frac{\pi}{4}$, and $\theta = \frac{\pi}{3}$, while the system is in the static regime with $\theta < \frac{2\pi}{45}$ and $\theta > \frac{123\pi}{360}$. Clearly observed from Figure 13b that as the inclined angle of cantilever increases, the self-vibration frequency first increases and then decreases, and conversely the amplitude first decreases and then increases, indicating that there is an optimal inclined angle for the self-excited oscillation. In summary, setting an appropriate inclined angle of cantilever can promote the self-vibration. Too large- or too small- inclined angle of cantilever is not conducive to the self-vibration of the system.

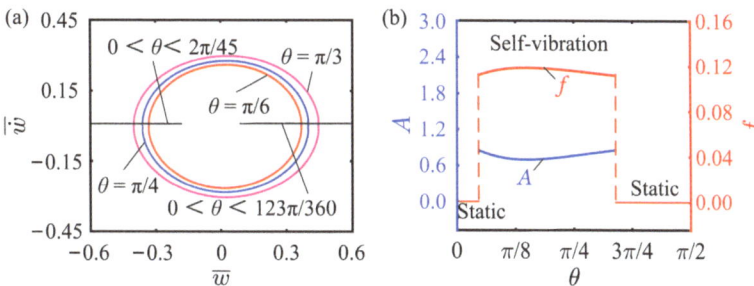

Figure 13. Effect of inclined angle of cantilever on the self-vibration. (**a**) Limit cycles with $\theta = \frac{\pi}{6}$, $\theta = \frac{\pi}{4}$, and $\theta = \frac{\pi}{3}$. (**b**) Variations of amplitude and frequency with different inclined angles of cantilever.

5. Conclusions

Self-excited oscillatory systems can maintain continuous motion by absorbing energy from the stable external environment, and possess potential applications in biomedicine, advanced robotics, rescue operations, military industry, and other fields. In order to overcome the disadvantages of existing self-sustained oscillatory systems that are relatively complex in structure and difficult to fabricate and control, we creatively propose a novel light-powered LCE fiber-cantilever system composed of an LCE fiber, a lightweight cantilever beam, and a mass block under steady illumination. The dynamic control equations for the LCE fiber-cantilever system are derived based on the established LCE dynamic model, beam theory, and deflection formula. The solutions of the nonlinear control equations are obtained using the Runge–Kutta numerical calculation method with MATLAB software. The results show that the LCE fiber-cantilever system evolves into two motion regimes, namely the static and self-vibration regimes. We have described these two motion regimes specifically and also revealed the energy compensation mechanism of the system. In a constant illumination, the positive work done by the tension of the LCE fiber is used to compensate for the structural resistance from the cantilever and the air damping, resulting in the contraction and relaxation.

Further numerical calculations show that the light intensity, contraction coefficient, spring constant, flexural stiffness, damping coefficient, ratio of cantilever height to width, and the inclined angle of the cantilever have a considerable effect on the self-vibration amplitude of the system. The spring constant of the LCE fiber and the flexural stiffness of the cantilever beam significantly affect the self-vibration frequency of the system. The illumination zone height has little effect on the amplitude and frequency, and the amplitude and frequency are not affected by the initial velocity. The LCE fiber-cantilever system constructed in this paper is a simple, easy-to-assemble and disassemble, easy-to-prepare, and highly expandable one-dimensional fiber-based system. It is expected to meet the application requirements of practical complex scenarios and has important application value in the fields of autonomous robotics, energy harvesters, autonomous separators, sensors, mechanical logic devices, and bionic design.

Author Contributions: The contribution of the authors are as follows: Data curation, Visualization, Validation, Methodology, Software, K.L.; Validation, Methodology, Software, Writing—Original draft preparation, Y.L.; Validation, Writing—Reviewing and Editing, Y.D.; Conceptualization, Investigation, Supervision, Writing—Reviewing and Editing, Y.Y. All authors have read and agreed to the published version of the manuscript.

Funding: This study was supported by University Natural Science Research Project of Anhui Province under Grant Nos. 2022AH020029 and KJ2020A0453, National Natural Science Foundation of China under Grant Nos. 12172001 and 12202002, and Anhui Provincial Natural Science Foundation under Grant Nos. 2208085Y01 and 2008085QA23.

Institutional Review Board Statement: Not applicable.

Data Availability Statement: Not applicable.

Acknowledgments: Not applicable.

Conflicts of Interest: The authors declare that they have no known competing financial interest or personal relationships that could have appeared to influence the work reported in this paper.

References

1. Ding, W. *Self-Excited Vibration*; Springer: Berlin/Heidelberg, Germany, 2010.
2. Sangwan, V.; Taneja, A.; Mukherjee, S. Design of a robust self-excited biped walking mechanism. *Mech. Mach. Theory* **2004**, *39*, 1385–1397. [CrossRef]
3. Wang, X.; Tan, C.F.; Chan, K.H.; Lu, X.; Zhu, L.; Kim, S.; Ho, G.W. In-built thermo-mechanical cooperative feedback mechanism for self-propelled multimodal locomotion and electricity generation. *Nat. Commun.* **2018**, *9*, 3438. [CrossRef]
4. Nocentini, S.; Parmeggiani, C.; Martella, D.; Wiersma, D.S. Optically driven soft micro robotics. *Adv. Opt. Mater.* **2018**, *6*, 1800207. [CrossRef]
5. Wang, X.; Ho, G.W. Design of untethered soft material micromachine for life-like locomotion. *Mat. Today* **2022**, *53*, 197–216. [CrossRef]
6. Hu, W.; Lum, G.Z.; Mastrangeli, M.; Sitti, M. Small-scale soft-bodied robot with multimodal locomotion. *Nature* **2018**, *554*, 81–85. [CrossRef] [PubMed]
7. Cheng, Y.; Lu, H.; Lee, X.; Zeng, H.; Priimagi, A. Kirigami-based light-induced shape-morphing and locomotion. *Adv. Mater.* **2019**, *32*, 1906233. [CrossRef] [PubMed]
8. Yang, L.; Chang, L.; Hu, Y.; Huang, M.; Ji, Q.; Lu, P.; Liu, J.; Chen, W.; Wu, Y. An autonomous soft actuator with light-driven self-sustained wavelike oscillation for phototactic self-locomotion and power generation. *Adv. Funct. Mater.* **2020**, *30*, 1908842. [CrossRef]
9. Shin, B.; Ha, J.; Lee, M.; Park, K.; Park, G.H.; Choi, T.H.; Cho, K.-J.; Kim, H.-Y. Hygrobot: A self-locomotive ratcheted actuator powered by environmental humidity. *Sci. Robot.* **2018**, *3*, eaar2629. [CrossRef]
10. Liao, B.; Zang, H.; Chen, M.; Wang, Y.; Lang, X.; Zhu, N.; Yang, Z.; Yi, Y. Soft Rod-Climbing Robot Inspired by Winding Locomotion of Snake. *Soft Robot.* **2020**, *7*, 500–511. [CrossRef] [PubMed]
11. He, Q.; Yin, R.; Hua, Y.; Jiao, W.; Mo, C.; Shu, H.; Raney, J. A modular strategy for distributed, embodied control of electronics-free soft robots. *Sci. Adv.* **2023**, *9*, eade9247. [CrossRef]
12. Chun, S.; Pang, C.; Cho, S.B. A micropillar-assisted versatile strategy for highly sensitive and effificient triboelectric energy generation under in-plane stimuli. *Adv. Mater.* **2020**, *32*, 1905539. [CrossRef]
13. Zhao, D.; Liu, Y. A prototype for light-electric harvester based on light sensitiveliquid crystal elastomer cantilever. *Energy* **2020**, *198*, 117351. [CrossRef]
14. Tang, R.; Liu, Z.; Xu, D.; Liu, J.; Yu, L.; Yu, H. Optical pendulum generator based on photomechanical liquid-crystalline actuators. *ACS Appl. Mater. Interf.* **2015**, *7*, 8393–8397. [CrossRef]
15. White, T.J.; Broer, D.J. Programmable and adaptive mechanics with liquid crystal polymer networks and elastomers. *Nat. Mater.* **2015**, *14*, 1087–1098. [CrossRef] [PubMed]
16. Rothemund, P.; Ainla, A.; Belding, L.; Preston, D.J.; Kurihara, S.; Suo, Z.; Whitesides, G.M. A soft, bistable valve for autonomous control of soft actuators. *Sci. Robot.* **2018**, *3*, eaar7986. [CrossRef] [PubMed]
17. Yoshida, R. Self-oscillating gels driven by the Belousov-Zhabotinsky reaction as novel smart materials. *Adv. Mater.* **2010**, *22*, 3463–3483. [CrossRef] [PubMed]
18. Hua, M.; Kim, C.; Du, Y.; Wu, D.; Bai, R.; He, X. Swaying gel: Chemo-mechanical self-oscillation based on dynamic buckling. *Matter* **2021**, *4*, 1029–1041. [CrossRef]

19. Wu, J.; Yao, S.; Zhang, H.; Man, W.; Bai, Z.; Zhang, F.; Wang, X.; Fang, D.; Zhang, Y. Liquid crystal elastomer metamaterials with giant biaxial thermal shrinkage for enhancing skin regeneration. *Adv. Mater.* **2021**, *33*, 2170356. [CrossRef]
20. Boissonade, J.; Kepper, P.D. Multiple types of spatio-temporal oscillations induced by differential diffusion in the Landolt reaction. *Phys. Chem.* **2011**, *13*, 4132–4137. [CrossRef]
21. Shen, Q.; Trabia, S.; Stalbaum, T.; Palmre, V.; Kim, K.; Oh, I. A multiple-shape memory polymer-metal composite actuator capable of programmable control, creating complex 3D motion of bending, twisting, and oscillation. *Sci. Rep.* **2016**, *6*, 24462. [CrossRef]
22. Hu, Y.; Ji, Q.; Huang, M.; Chang, L.; Zhang, C.; Wu, G.; Zi, B.; Bao, N.; Chen, W.; Wu, Y. Light-driven self-oscillating actuators with pototactic locomotion based on black phosphorus heterostructure. *Angew. Chem. Int. Ed.* **2021**, *60*, 20511–20517. [CrossRef] [PubMed]
23. Sun, J.; Hu, W.; Zhang, L.; Lan, R.; Yang, H.; Yang, D. Light-driven self-oscillating behavior of liquid-crystalline networks riggered by dynamic isomerization of molecular motors. *Adv. Funct. Mater.* **2021**, *31*, 2103311. [CrossRef]
24. Manna, R.K.; Shklyaev, O.E.; Balazs, A.C. Chemical pumps and flexible sheets spontaneously form self-regulating oscillators in solution Proc. *Natl. Acad. Sci. USA* **2021**, *118*, e2022987118. [CrossRef] [PubMed]
25. Li, Z.; Myung, N.V.; Yin, Y. Light-powered soft steam engines for self-adaptive oscillation and biomimetic swimming. *Sci. Robot.* **2021**, *6*, eabi4523. [CrossRef]
26. Zeng, H.; Lahikainen, M.; Liu, L.; Ahmed, Z.; Wani, O.M.; Wang, M.; Priimagi, A. Light-fuelled freestyle self-oscillators. *Nat. Commun.* **2019**, *10*, 5057. [CrossRef]
27. Chen, Y.; Zhao, H.; Mao, J.; Chirarattananon, P.; Helbling, E.F.; Hyun, N.P.; Clarke, D.R.; Wood, R.J. Controlled flight of a microrobot powered by soft artificial muscles. *Nature* **2019**, *575*, 324–329. [CrossRef] [PubMed]
28. Wang, Y.; Sun, J.; Liao, W.; Yang, Z. Liquid Crystal Elastomer Twist Fibers toward Rotating Microengines. *Adv. Mater.* **2022**, *34*, 2107840. [CrossRef] [PubMed]
29. Bazir, A.; Baumann, A.; Ziebert, F.; Kulić, I.M. Dynamics of fiberboids, Soft. *Matter* **2020**, *16*, 5210–5223.
30. Hu, Z.; Li, Y.; Lv, J. Phototunable self-oscillating system driven by a self-winding fiber actuator. *Nat. Commun.* **2021**, *12*, 3211. [CrossRef]
31. Zhao, Y.; Chi, Y.; Hong, Y.; Li, Y.; Yang, S.; Yin, J. Twisting for soft intelligent autonomous robot in unstructured environments. *Proc. Natl. Acad. Sci. USA* **2022**, *119*, e2200265119. [CrossRef]
32. Ghislaine, V.; Lars, C.M.E.; Anne, H.G.; Meijer, E.W.; Alexander, Y.P.; Henk, N.; Dirk, J.B. Coupled liquid crystalline oscillators in Huygens' synchrony. *Nat. Mater.* **2021**, *20*, 1702–1706.
33. O'Keeffe, K.P.; Hong, H.; Strogatz, S.H. Oscillators that sync and swarm. *Nat. Commun.* **2017**, *8*, 1504. [CrossRef] [PubMed]
34. Li, K.; Zhang, B.; Cheng, Q.; Dai, Y.; Yu, Y. Light-Fueled Synchronization of Two Coupled Liquid Crystal Elastomer Self-Oscillators. *Polymers* **2023**, *15*, 2886. [CrossRef]
35. Vick, D.; Friedrich, L.J.; Dew, S.K.; Brett, M.J.; Robbie, K.; Seto, M.; Smy, T. Self-shadowing and surface diffusion effects in obliquely deposited thin films. *Thin Solid Film.* **1999**, *339*, 88–94. [CrossRef]
36. Kuenstler, A.; Chen, Y.; Bui, P.; Kim, H.; DeSimone, A.; Jin, L.; Hayward, R. Blueprinting photothermal shape-morphing of liquid crystal elastomers. *Adv. Mater.* **2020**, *32*, 2000609. [CrossRef]
37. Liu, X.; Liu, Y. Spontaneous photo-buckling of a liquid crystal elastomer membrane. *Int. J. Mech. Sci.* **2021**, *201*, 106473. [CrossRef]
38. Chakrabarti, A.; Choi, G.P.T.; Mahadevan, L. Self-excited motions of volatile drops on swellable sheets. *Phys. Rev. Lett.* **2020**, *124*, 258002. [CrossRef] [PubMed]
39. Lv, X.; Yu, M.; Wang, W.; Yu, H. Photothermal pneumatic wheel with high loadbearing capacity. *Comp. Comm.* **2021**, *24*, 100651. [CrossRef]
40. Wang, Y.; Liu, J.; Yang, S. Multi-functional liquid crystal elastomer composites. *Appl. Phys. Rev.* **2022**, *9*, 011301. [CrossRef]
41. Lendlein, A.; Jiang, H.; Jünger, O.; Langer, R. Light-induced shape-memory polymers. *Nature* **2005**, *434*, 879–882. [CrossRef]
42. Yu, Y.; Li, L.; Liu, E.; Han, X.; Wang, J.; Xie, Y.; Lu, C. Light-driven core-shell fiber actuator based on carbon nanotubes/liquid crystal elastomer for artificial muscle and phototropic locomotion. *Carbon* **2022**, *187*, 97–107. [CrossRef]
43. Ge, F.; Yang, R.; Tong, X.; Camerel, F.; Zhao, Y. A multifunctional dye-doped liquid crystal polymer actuator: Light-guided transportation, turning in locomotion, and autonomous motion. *Angew. Chem. Int. Ed.* **2018**, *57*, 11758–11763. [CrossRef] [PubMed]
44. Bubnov, A.; Domenici, V.; Hamplová, V.; Kašpar, M.; Zalar, B. First liquid single crystal elastomer containing lactic acid derivative as chiral co-monomer: Synthesis and properties. *Polymers* **2011**, *52*, 4490–4497. [CrossRef]
45. Milavec, J.; Domenici, V.; Zupančič, B.; Rešetič, A.; Bubnov, A.; Zalar, B. Deuteron NMR resolved mesogen vs. crosslinker molecular order and reorientational exchange in liquid single crystal elastomers. *Phys. Chem. Chem. Phys.* **2016**, *18*, 4071–4077. [CrossRef]
46. Rešetič, A.; Milavec, J.; Domenici, V.; Zupančič, B.; Bubnov, A.; Zalar, B. Stress-strain and thermomechanical characterization of nematic to smectic A transition in a strongly-crosslinked bimesogenic liquid crystal elastomer. *Polymers* **2018**, *158*, 96–102. [CrossRef]
47. Wang, Y.; Yin, R.; Jin, L.; Liu, M.; Gao, Y.; Raney, J.; Yang, S. 3D-Printed Photoresponsive Liquid Crystal Elastomer Composites for Free-Form Actuation. *Adv. Funct. Mater.* **2023**, *33*, 2210614. [CrossRef]

48. Wang, Y.; Dang, A.; Zhang, Z.; Yin, R.; Gao, Y.; Feng, L.; Yang, S. Repeatable and Reprogrammable Shape Morphing from Photoresponsive Gold Nanorod/Liquid Crystal Elastomers. *Adv. Mater.* **2020**, *32*, 2004270. [CrossRef]
49. Ula, S.W.; Traugutt, N.A.; Volpe, R.H.; Patel, R.R.; Yu, K.; Yakacki, C.M. Liquid crystal elastomers, an introduction and review of emerging technologies. *Liq. Cryst. Rev.* **2018**, *6*, 78–107. [CrossRef]
50. Warner, M.; Terentjev, E.M. *Liquid Crystal Elastomers*; Oxford University Press: Oxford, UK, 2007.
51. Domenici, V.; Milavec, J.; Bubnov, A.; Pociecha, D.; Zupančič, B.; Rešetič, A.; Hamplová, V.; Gorecka, E.; Zalar, B. Effect of co-monomers' relative concentration on self-assembling behaviour of side-chain liquid crystalline elastomers. *RSC Adv.* **2014**, *4*, 44056–44064. [CrossRef]
52. Domenici, V.; Milavec, J.; Zupančič, B.; Bubnov, A.; Hamplová, V.; Zalar, B. Brief overview on 2H NMR studies of polysiloxane-based side-chain nematic elastomers. *Magn. Reson. Chem.* **2014**, *52*, 649–655. [CrossRef]
53. Milavec, J.; Rešetič, A.; Bubnov, A.; Zalar, B.; Domenici, V. Dynamic investigations of liquid crystalline elastomers and their constituents by 2H NMR spectroscopy. *Liq. Cryst.* **2018**, *45*, 2158–2173. [CrossRef]
54. Rešetič, A.; Milavec, J.; Domenici, V.; Zupančič, B.; Zalar, B. Deuteron NMR investigation on orientational order parameter in polymer dispersed liquid crystal elastomers. *Phys. Chem. Chem. Phys.* **2020**, *22*, 23064–23072. [CrossRef]
55. Parrany, M. Nonlinear light-induced vibration behavior of liquid crystal elastomer beam. *Int. J. Mech. Sci.* **2018**, *136*, 179–187. [CrossRef]
56. Bishop, R.E.D.; Daniel, C.J. *The Mechanics of Vibration*; Cambridge University Press: Cambridge, UK, 2011.
57. Yu, Y.; Du, C.; Li, K.; Cai, S. Controllable and versatile self-motivated motion of a fiber on a hot surface. *EML* **2022**, *57*, 101918. [CrossRef]
58. Xu, T.; Pei, D.; Yu, S.; Zhang, X.; Yi, M.; Li, C. Design of MXene composites with biomimetic rapid and self-oscillating actuation under ambient circumstances. *ACS Appl. Mater. Interf.* **2021**, *13*, 31978–31985. [CrossRef] [PubMed]
59. Ge, D.; Dai, Y.; Li, K. Light-powered self-spinning of a button spinner. *Int. J. Mech. Sci.* **2023**, *238*, 107824. [CrossRef]
60. Liu, J.; Zhao, J.; Wu, H.; Dai, Y.; Li, K. Self-Oscillating Curling of a Liquid Crystal Elastomer Beam under Steady Light. *Polymers* **2023**, *15*, 344. [CrossRef] [PubMed]
61. Shen, B.; Kang, S.H. Designing self-oscillating matter. *Matter* **2021**, *4*, 766–769. [CrossRef]
62. Zhou, L.; Dai, Y.; Fang, J.; Li, K. Light-powered self-oscillation in liquid crystal elastomer auxetic metamaterials with large volume change. *Int. J. Mech. Sci.* **2023**, *254*, 108423. [CrossRef]
63. Ge, D.; Dai, Y.; Li, K. Self-Sustained Euler Buckling of an Optically Responsive Rod with Different Boundary Constraints. *Polymers* **2023**, *15*, 316. [CrossRef]
64. Nayfeh, A.H.; Emam, S.A. Exact solution and stability of post buckling configurations of beams. *Nonlinear Dyn.* **2008**, *54*, 395–408. [CrossRef]
65. He, Q.; Wang, Z.; Wang, Y.; Wang, Z.; Li, C.; Annapooranan, R.; Zeng, J.; Chen, R.; Cai, S. Electrospun liquid crystal elastomer microfiber actuator. *Sci. Robot.* **2021**, *6*, eabi9704. [CrossRef]
66. Cheng, Q.; Cheng, W.; Dai, Y.; Li, K. Self-oscillating floating of a spherical liquid crystal elastomer balloon under steady illumination. *Int. J. Mech. Sci.* **2023**, *241*, 107985. [CrossRef]
67. Gelebart, A.H.; Mulder, D.J.; Varga, M.; Konya, A.; Vantomme, G.; Meijer, E.W.; Selinger, R.S.; Broer, D.J. Making waves in a photoactive polymer film. *Nature* **2017**, *546*, 632–636. [CrossRef]
68. Cunha, M.P.D.; Peeketi, A.R.; Ramgopal, A.; Annabattula, R.K.; Schenning, A.P.H.J. Light-driven continual oscillatory rocking of a polymer film. *Chem. Open.* **2020**, *9*, 1149–11525.
69. Xu, P.; Wu, H.; Dai, Y.; Li, K. Self-sustained chaotic floating of a liquid crystal elastomer balloon under steady illumination. *Heliyon* **2023**, *9*, e14447. [CrossRef]
70. Kim, Y.; Berg, J.; Crosby, A.J. Autonomous snapping and jumping polymer gels. *Nat. Mater.* **2021**, *20*, 1695–1701. [CrossRef] [PubMed]
71. Dawson, N.J.; Kuzyk, M.G.; Neal, J.; Luchette, P.; Palffy-Muhoray, P. Cascading of liquid crystal elastomer photomechanical optical devices. *Opt. Commun.* **2011**, *284*, 991–993. [CrossRef]
72. Zhao, D.; Liu, Y. Photomechanical vibration energy harvesting based on liquid crystal elastomer cantilever. *Smart Mater. Struct.* **2019**, *28*, 075017. [CrossRef]
73. Zhao, D.; Liu, Y. Effects of director rotation relaxation on viscoelastic wave dispersion in nematic elastomer beams. *Math. Mech. Solids* **2019**, *24*, 1103–1115. [CrossRef]
74. Ichimura, K.; Morino, S.; Akiyama, H. Three-dimensional orientational control of molecules by slantwise photoirradiation. *Appl. Phys. Lett.* **1998**, *73*, 921–923. [CrossRef]
75. Nagele, T.; Hoche, R.; Zinth, W.; Wachtveitl, J. Femtosecond photoisomerization of cisazobenzene. *Phys. Rev. Lett.* **1997**, *272*, 489–495.
76. Finkelmann, H.; Nishikawa, E.; Pereira, G.G.; Warner, M. A new opto-mechanical effect in solids. *Phys. Rev. Lett.* **2001**, *87*, 015501. [CrossRef]
77. Yu, Y.; Nakano, M.; Ikeda, T. Photomechanics: Directed bending of a polymer film by light-miniaturizing a simple photomechanical system could expand its range of applications. *Nature* **2003**, *425*, 145. [CrossRef]

78. Serak, S.V.; Tabiryan, N.V.; Vergara, R.; White, T.J.; Vaia, R.; Bunning, T. Liquid crystalline polymer cantilever oscillators fueled by light. *Soft Matter* **2010**, *6*, 779–783. [CrossRef]
79. Braun, L.B.; Hessberger, T.; Pütz, E.; Müller, C.; Giesselmann, F.; Serra, C.A.; Zentel, R. Actuating thermo- and photo-responsive tubes from liquid crystalline elastomers. *J. Mater. Chem. C* **2018**, *6*, 9093–9101. [CrossRef]
80. Camacho, L.M.; Finkelmann, H.; Palffy, M.P.; Shelley, M. Fast liquid-crystal elastomer swims into the dark. *Nat. Mater.* **2004**, *5*, 307–310. [CrossRef]

Disclaimer/Publisher's Note: The statements, opinions and data contained in all publications are solely those of the individual author(s) and contributor(s) and not of MDPI and/or the editor(s). MDPI and/or the editor(s) disclaim responsibility for any injury to people or property resulting from any ideas, methods, instructions or products referred to in the content.

MDPI AG
Grosspeteranlage 5
4052 Basel
Switzerland
Tel.: +41 61 683 77 34

Polymers Editorial Office
E-mail: polymers@mdpi.com
www.mdpi.com/journal/polymers

Disclaimer/Publisher's Note: The title and front matter of this reprint are at the discretion of the Guest Editors. The publisher is not responsible for their content or any associated concerns. The statements, opinions and data contained in all individual articles are solely those of the individual Editors and contributors and not of MDPI. MDPI disclaims responsibility for any injury to people or property resulting from any ideas, methods, instructions or products referred to in the content.

www.ingramcontent.com/pod-product-compliance
Lightning Source LLC
LaVergne TN
LVHW072335090526
838202LV00019B/2423